PROGRESS IN

Molecular Biology and Translational Science

Volume 83

PROGRESS IN
Molecular Biology and Translational Science

Molecular Biology of Protein Folding, Part A

edited by

P. Michael Conn
Oregon National Primate Research Center
Oregon Health and Science University
Beaverton, Oregon

Volume 83

AMSTERDAM • BOSTON • HEIDELBERG • LONDON
NEW YORK • OXFORD • PARIS • SAN DIEGO
SAN FRANCISCO • SINGAPORE • SYDNEY • TOKYO
Academic Press is an imprint of Elsevier

Academic Press is an imprint of Elsevier
32 Jamestown Road, London, NW1 7BY, UK
Radarweg 29, PO Box 211, 1000 AE Amsterdam, The Netherlands
30 Corporate Drive, Suite 400, Burlington, MA 01803, USA
525 B Street, Suite 1900, San Diego, CA 92101-4495, USA

This book is printed on acid-free paper. ∞

Copyright © 2008, Elsevier Inc. All rights reserved

No part of this publication may be reproduced, stored in a retrieval system or
transmitted in any form or by any means electronic, mechanical, photocopying,
recording or otherwise without the prior written permission of the publisher

Permissions may be sought directly from Elsevier's Science & Technology Rights
Department in Oxford, UK: phone (+44) (0) 1865 843830; fax (+44) (0) 1865 853333;
email: permissions@elsevier.com. Alternatively you can submit your request online by
visiting the Elsevier web site at http://elsevier.com/locate/permissions, and selecting
Obtaining permission to use Elsevier material

Notice
No responsibility is assumed by the publisher for any injury and/or damage to persons
or property as a matter of products liability, negligence or otherwise, or from any use
or operation of any methods, products, instructions or ideas contained in the material
herein. Because of rapid advances in the medical sciences, in particular, independent
verification of diagnoses and drug dosages should be made

Library of Congress Cataloging-in-Publication Data
A catalog record for this book is available from the Library of Congress

British Library Cataloguing in Publication Data
A catalogue record for this book is available from the British Library

ISBN: 978-0-12-374594-1
ISSN: 0079-6603

For information on all Academic Press publications
visit our website at elsevierdirect.com

Printed and bound in the USA
08 09 10 11 12   10 9 8 7 6 5 4 3 2 1

Working together to grow
libraries in developing countries

www.elsevier.com | www.bookaid.org | www.sabre.org

ELSEVIER   BOOK AID International   Sabre Foundation

Contents

Stability and Design of α-Helical Peptides 1
Andrew J. Doig

I. Introduction	2
II. Design of Peptide Helices	7
III. Helix–Coil Theory	15
IV. Forces Affecting α-Helix Stability	25
References	38

Folding and Wrapping Soluble Proteins: Exploring the Molecular Basis of Cooperativity and Aggregation 53
Ariel Fernández, Xi Zhang, and Jianping Chen

I. Folding, Cooperativity, and Wrapping of Soluble Proteins: An Overview	54
II. Wrapping the Folded Structure	57
III. Exploring the Molecular Basis of Folding Cooperativity	62
IV. Protein Under-Wrapping, Misfolding, and Aggregation	80
References	85

Rescuing Proteins of Low Kinetic Stability by Chaperones and Natural Ligands: Phenylketonuria, a Case Study 89
Aurora Martinez, Ana C. Calvo, Knut Teigen, and Angel L. Pey

I. Introduction	90
II. Protein Folding	93
III. Misfolding	100
IV. Ligand Binding	110

V. Strategies to Correct Misfolding 116
VI. Concluding Remarks 123
   References 124

The Endoplasmic Reticulum: Crossroads for Newly Synthesized Polypeptide Chains 135

Tito Calì, Omar Vanoni, and Maurizio Molinari

I. Protein Translocation and Maturation in the Mammalian ER.............. 136
II. Substrate Recognition and Dislocation into the
   Cytosol for ERAD 152
   References 163

All-Atom Protein Folding with Free-Energy Forcefields........................ 181

A. Verma, S. M. Gopal, A. Schug, T. Herges,
K. Klenin, and W. Wenzel

I. Introduction 182
II. Protein Structure and Folding.. 186
III. Kinetic Folding Methods.. 199
IV. Free-Energy Forcefields and Simulation Methods 205
V. Free-Energy Protein Folding... 216
VI. Summary.. 242
VII. Outlook... 246
   References 247

Folding Considerations for Therapeutic Protein Formulations.. 255

Lioubov G. Korotchkina, Karthik Ramani,
and Sathy V. Balu-Iyer

I. Introduction 256
II. Folding and Stability of Therapeutic Proteins 257
III. Case Study: Rational Formulation Development of Human
   Recombinant Factor VIII 266
   References 268

Index... 271

Stability and Design of α-Helical Peptides

ANDREW J. DOIG

Manchester Interdisciplinary Biocentre, The University of Manchester, 131 Princess Street, Manchester M1 7DN, United Kingdom

I. Introduction	2
A. Structure of the α-Helix	2
B. Capping Motifs	4
C. Metal Binding	4
D. The 3_{10}-Helix	6
E. The π-Helix	6
II. Design of Peptide Helices	7
A. History	7
B. Host–Guest Studies	8
C. Helix Lengths	8
D. The Helix Dipole	8
E. Acetylation and Amidation	10
F. Side-Chain Spacings	10
G. Solubility	10
H. Concentration Determination	11
I. Design of Peptides to Measure Helix Parameters	12
J. Helix Templates	13
K. Design of 3_{10}-Helices	13
L. Design of π-Helices	15
III. Helix–Coil Theory	15
A. Zimm–Bragg Model	16
B. Lifson–Roig Model	16
C. The Unfolded State and Polyproline II Helix	19
D. Single-Sequence Approximation	20
E. N- and C-Caps	20
F. Capping Boxes	21
G. Side-Chain Interactions	21
H. N1, N2, and N3 Preferences	21
I. Helix Dipole	22
J. 3_{10}- and π-Helices	22
K. AGADIR	23
L. Lomize–Mosberg Model	24
M. Extension of the Zimm–Bragg Model	25
N. Tertiary Interactions	25
IV. Forces Affecting α-Helix Stability	25
A. Helix Interior	25
B. Caps	29
C. Phosphorylation	32
D. Noncovalent Side-Chain Interactions	32

E. Covalent Side-Chain Interactions	33
F. Capping Motifs	33
G. Ionic Strength	35
H. Enthalpy Change for the Helix–Coil Transition	36
I. Trifluoroethanol	37
J. pK_a Values	37
K. Relevance to Proteins	38
References	38

The α-helix is the most abundant secondary structural element in proteins and has been intensively studied using peptides that form isolated helices in aqueous solution, usually composed mostly of Ala, and with helix content measured using circular dichroism. Structural features that affect helix stability include: residue preferences for helix interiors and N- and C-terminal positions, capping motifs unique to helix termini, side-chain interactions, metal binding, helix length, and covalent crosslinks. The quantitative interpretation of helical peptide data, usually in the form of a free energy, requires the use of helix–coil theory that includes energetic terms for all the structural features present. The application of helix–coil theory to extensive experimental work has yielded scales of residue preferences for different positions in the helix, phosphorylation effects and many side-chain interaction energies. Calorimetry and metal binding has subdivided interior preferences into enthalpic and entropic terms.

I. Introduction

Proteins assembled regular local folds of the polypeptide chain called secondary structure. The α-helix was first described by Pauling, Corey, and Branson in 1950 (1), and their model was supported by X-ray analysis of hemoglobin (2), and myoglobin, where most secondary structure is helical (3). α-Helices are found in nearly all globular proteins. It is the most abundant secondary structure, with ≈30% of residues found in α-helices (4). Here, we discuss structural features of the helix and their study in peptides. Some earlier reviews on this field are (5–11).

A. Structure of the α-Helix

A helix by definition combines a linear translation with an orthogonal circular rotation. In the α-helix, the linear translation is a rise of 5.4 Å per turn of the helix and a circular rotation is 3.6 residues per turn. Side chains spaced $(i, i+3)$, $(i, i+4)$, and $(i, i+7)$ are therefore close in space, since these are close to multiples of 3.6, giving the potential of interactions between them.

Spacings of $(i, i + 2)$, $(i, i + 5)$, and $(i, i + 6)$ place the side-chain pairs on opposite faces of the helix avoiding any interaction in a monomeric helix. The helix is primarily stabilized by $(i, i + 4)$ backbone CO–NH hydrogen bonds.

The conformation of a polypeptide can be described by the backbone dihedral angles ϕ and ψ. Most ϕ, ψ combinations are not possible for steric reasons, leaving only the α and β regions. In an α-helix a succession of residues in the α region position the backbone NH and CO groups toward each other for $(i, i + 4)$ hydrogen bond formation. The sterically most stable conformations thus also give strong, linear hydrogen bonds.

The residues at the N-terminus of the α-helix are called N'–N-cap–N1–N2–N3–N4, etc., where the N-cap is the residue with nonhelical ϕ, ψ angles immediately preceding the N-terminus of an α-helix and N1 is the first residue with helical ϕ, ψ angles (12). The C-terminal residues are similarly called C4–C3–C2–C1–C-cap–C', etc. Unfortunately, different definitions of the cap locations are sometimes used. For example, Fonseca et al. (13) call the N-cap residue N1, so their results are offset by one position. The N1, N2, N3, C1, C2, and C3 residues are distinct from interior positions because their amide groups participate in $(i, i + 4)$ backbone–backbone hydrogen bonds using either only their CO (at the N-terminus) or NH (at the C-terminus) groups. These free hydrogen bond donors and acceptors have strong effects on helix structure and stability (14). Bonds between sp^3 hybridized atoms show preferences for dihedral angles of $+60°$ (gauche$^-$), $-60°$ (gauche$^+$), or $180°$ (trans). The following rules are generally observed for N-capping in α-helices: Thr and Ser N-cap side chains adopt the gauche$^-$ rotamer, hydrogen bond to the N3 NH and have ψ restricted to $164 \pm 8°$. Asp and Asn N-cap side chains either adopt the gauche$^-$ rotamer and hydrogen bond to the N3 NH with $\psi = 172 \pm 10°$, or adopt the trans rotamer and hydrogen bond to both the N2 and N3 NH groups with $\psi = 107 \pm 19°$. With all other N-caps, the side chain is found in the gauche$^+$ rotamer so that the NH groups are free to interact with water. An $(i, i + 3)$ hydrogen bond from N3 NH to the N-cap backbone C=O is therefore more likely to form at the N-terminus when an unfavorable N-cap is present (15). ψ is unrestricted.

Side-chains rotamer preferences vary between secondary structures (16–19). Within the α-helix, rotamer preferences vary greatly between different positions of N-cap, N1, N2, N3, and interior (15, 20). At helix interior positions, the gauche$^+$ rotamer is usually most abundant (64%), followed by trans (33%) with gauche$^-$ very rare (3%), though exceptions are seen. Ser and Thr are most likely to be gauche$^-$ (19% and 14% of the time, respectively), as their hydroxyl groups can hydrogen bond to the helix backbone in this conformation. The β-branched side chains Val, Ile, and Thr are most restricted with trans very rare.

B. Capping Motifs

The amide NH groups at the helix N-terminus are satisfied predominantly by local side-chain hydrogen bond acceptors. In contrast, carbonyl CO groups at the C-terminus are satisfied primarily by backbone NH groups from the sequence following the helix (*14*). At the N-terminus, helix geometry favors side chain-to-backbone hydrogen bonding and selects for compatible polar residues (*15, 20*). Accordingly, the N-terminus promotes selectivity in all polar positions, especially N-cap and N3 in the capping box. In contrast, at the C-terminus, side-chain-to-backbone hydrogen bonding is disfavored. Backbone hydrogen bonds are satisfied instead by post-helical backbone groups. The C-terminus need only select for C' residues that can adopt positive values of the backbone dihedral angle ϕ, most notably Gly (*21*). Capping motifs are defined as specific patterns found at or near the ends of helices (*12, 22–26*). They involve hydrogen bonding and hydrophobic interactions, as shown in Table I. In addition, Fonseca *et al.* (*13*) performed an extensive survey of statistical side-chain pair preferences at helix N- and C-termini.

One example of capping and a strong side-chain preference at the helix N-terminus is the capping box, where the side chain of the N-cap forms a hydrogen bond with the backbone of N3 and the side chain of N3 forms a hydrogen bond with the backbone of the N-cap (*27*). An additional hydrophobic interaction between residues N' and N4 or N4 and N" may also be present and is also known as a "hydrophobic staple" (*28, 29*). The Probox motif involves three hydrophobic residues and a Pro residue at the N-cap (*30*). N-cap Pro residues are usually associated with Ile or Leu at position N", Val at position N3 and a hydrophobic residue at position N4.

The two most abundant capping motifs found at helix C-termini are the Schellman and the α_L motifs (*21, 31, 32*). The Schellman motif has two backbone–backbone hydrogen bonds between the amide NH at C" and the carbonyl CO at C3, and between the amide NH at C' and the carbonyl CO at C2. The associated hydrophobic interaction is between C3 and C". In a Schellman motif, polar residues are highly favored at the C1 position and the C' residue is typically glycine. If C" is polar, the alternative α_L motif is observed. This has hydrogen bond between the amide NH at C' and the carbonyl CO at C3. As in the Schellman motif, the C' residue is typically glycine, which adopts a positive value of ϕ dihedral angle. The hydrophobic interaction in α_L can form between C3 and any of several residues external to the helix ($C^{3\prime}$, $C^{4\prime}$, or $C^{5\prime}$) (*21*).

C. Metal Binding

A variety of peptide models have been studied that bind metals. In metal–ligand binding "hard metals" prefers "hard ligands." For example, Ca and Mg prefer ligands with oxygen as the coordinating atoms (Asp, Glu) (*33*). In contrast, soft metals, such as Cu and Zn, bind mostly to His, Cys, and Trp ligands and sometimes indirectly via water molecules (*34*).

TABLE I
THE MOST COMMON CAPPING MOTIFS AT α-HELIX TERMINI

Sequence	Related position	Designation	Interactions	References
N-capping				
p-XXp	N-cap → N3	Capping box	Reciprocal H-bonds between residues at N-cap and N3, where S/T at N-cap and E at N3 is the most frequently observed.	(26, 27)
hp-XXph	N' → N4	Expanded capping box/hydrophobic staple	Reciprocal H-bonds between residues at N-cap and N3 accompanied by hydrophobic interactions between residues at N' and N4.	(28, 272)
hP-XXVh	N' → N4	Probox motif	N-cap Pro residues are usually associated to Ile and Leu, at position N', Val at position N3 and a hydrophobic residue at position N4.	(30)
C-capping				
hxp-xGh	C3 → C"	Schellman motif	H-bonds between NH at C" and the carbonyl CO at C3 and between the amide NH at C' and the carbonyl CO at C2, respectively. Accompanied by hydrophobic interaction between C3 and C".	(31, 234)
hXp-XGp	C2 → C"	αL motifs	H-bond between NH at C' and the carbonyl CO at C3, C" polar.	(21)
X-Pro	C-cap → C'	Pro-capping motif	A stabilizing electrostatic interaction of the residues at positions C-cap and C' with the helix macrodipole.	X = Asn, Cys, His, Phe, Tyr, Trp, Ile, Val, and Leu

p, polar amino acids; h, hydrophobic amino acids; X, any amino acids; P, proline; G, glycine.

In studies of helical peptide models, soft ligands have been mostly used. In the presence of Cd ions, a synthetic peptide containing Cys–His ligands ($i, i + 4$) apart at the C-terminal region promoted helicity from 54% to 90%. The helicity of a similar peptide containing His–His ligands increased by up to 90% as a result of Cu and Zn binding (35). The addition of a *cis*-Ru(III) ion to a 6-mer peptide, Ac–AHAAAHA–NH$_2$, changed the peptide conformation from random coil to 37% helix (36). An 11-residue peptide formed 80% helix content by the addition of Cd ions, using aminodiacetic acids to bind the Cd (37). As (III) stabilizes helices when bound to Cys side chains spaced ($i, i + 4$) by -0.7 to -1.0 kcal mol$^-$ (38). Trimeric coiled coils can bind Cd(II) at two sites with different affinities (39).

D. The 3$_{10}$-Helix

3$_{10}$-Helices are stabilized by ($i, i + 3$) hydrogen bonds, instead of ($i, i + 4$) found in α-helices, making the cylinder of the 3$_{10}$-helix narrower than α and their hydrogen bonds nonlinear. Three to four percent of residues in crystal structures are in 3$_{10}$ helices (1, 40). Most 3$_{10}$-helices are only 3 or 4 residues long, compared to a mean of 10 residues in α-helices (4). 3$_{10}$-Helices are commonly found as N- or C-terminal extensions to an α-helix (4, 41, 42). Strong amino acid preferences have been observed for different locations within the interiors (40) and N- and C-caps (15) of 3$_{10}$-helices in crystal structures.

The 3$_{10}$-helix has been proposed to be an intermediate in α-helix formation (43, 44). 3$_{10}$-helix formation can be induced by the introduction of a C$_{\alpha-\alpha}$-disubstituted α-amino acid, of which α-aminoisobutyric acid (AIB) is the prototype. For most amino acids, the α-helical geometry ($\phi = -57°, \psi = -70°$) is of lower energy than the 3$_{10}$ geometry ($\phi = -49°, \psi = -26°$). There is no barrier between the α and 3$_{10}$ conformations in the Ramachandran plot, and a peptide can therefore be gradually transformed from one helix to the other (45).

E. The π-Helix

In contrast to the widely occurring α- and 3$_{10}$-helices, the π-helix is extremely rare. The π-helix is unfavorable for three reasons: its dihedral angles energetically unfavorable relative to the α-helix (46, 47), its three dimensional structure has a 1 Å hole down the center that is too narrow for access by a water molecule resulting in the loss of van der Waals interactions, and a higher number of residues (four) must be correctly oriented before the first ($i, i + 5$) hydrogen bond is formed, making helix initiation more entropically unfavorable than for α- or 3$_{10}$-helices (48). Despite these disadvantages, evidence for π-helices has been put forward (49–52). In particular, π-helices are more likely to be associated with function than other parts of a protein (53).

II. Design of Peptide Helices

A. History

The first protein crystal structures showed an abundance of α-helices, leading to speculation whether fragments of the helical sequences could be stable in isolation. This would require the amide–amide hydrogen bond to be strong enough to oppose the loss of conformational entropy arising from restricting the peptide into a helical structure (54). Early studies of helices from myoglobin (55) and staphylococcal nuclease (56) found no helix formation. Estimates of helix/coil parameters from a host–guest system (see below) suggested that polypeptides would need to be hundreds of residues long to form stable helices. The first work on peptide helices was on long homopolymers of Glu or Lys which show coil to helix transitions on changing the pH from charged to neutral. The neutral polypeptides are metastable and prone to aggregation, ultimately to β-sheet amyloid (57).

The first experimental indication of helix formation by a short peptide was from Brown and Klee in 1971 (58), who reported that the C-peptide of ribonuclease A, which contains the first 13 residues of the protein and which forms a helix in the protein, was helical at 0°C, as shown by circular dichroism (CD). This observation was not followed up for 10 years until extensive work on the sequence features responsible for helix formation in this peptide, and in the larger S-peptide, was performed by Baldwin and coworkers (59–69). NMR studies by Rico and coworkers precisely defined the helical structure (70–78). Some important features responsible for the helicity of the C-peptide that emerged from this work included: an $(i, i + 4)$ Phe–His interaction, an $(i, i + 3)$ Glu–His salt bridge, an $(i, i + 8)$ Glu–Arg salt bridge across two turns of the helix, and a helix termination signal at Met13. The stabilizing effects of salt bridges were inferred from pH titrations and CD, where a decrease in helicity was seen when a residue participating in a salt bridge was neutralized. Quantification of these features, in terms of a free energy, was not possible at that time, as helix/coil theory that included side-chain interactions and termination signals (caps) had not been developed.

Perhaps the most interesting result from the C- and S-peptide work, was the importance of Ala. The replacement of interior helical residues with Ala was stabilizing, indicating that a major reason why this helix was folded in isolation was the presence of three successive alanines from positions 4–6. This led to the successful design of isolated, monomeric helical peptides in aqueous solution, first containing several salt bridges and a high alanine content, based on $(EAAAK)_n$ (79, 80) and then a simple sequence with a high alanine content solubilized by several Lysines (81). These "AK peptides" are based on the sequence $(AAKAA)_n$, where n is typically 2–5. The Lys side chains are spaced

($i, i + 5$) so they are on opposite faces of the helix, giving no charge repulsion and may be substituted with Arg or Gln to give a neutral peptide. Hundreds of AK peptides have been studied, giving most of the results on helix stability in peptides.

An alternative strategy is to stabilize the helix with large numbers of salt bridges, with sequences based on (EEEEKKKK)$_n$ or EAK (8, 82–84). While these do form stable helices, substitutions within them usually disrupt these stabilizing bonds.

B. Host–Guest Studies

Extensive work from the Scheraga group has obtained helix/coil parameters using a host–guest method. Long random copolymers were synthesized of a water soluble, nonionic guest (poly [N^5-(3-hydroxypropyl)-L-glutamine] (PHPG) or poly [N^5-(4-hydroxybutyl)-L-glutamine] (PHBG)), together with a low (10–50%) content of the guest residue. Using the s and σ Zimm–Bragg helix/coil parameters (see below) for the host homopolymer, it was possible to calculate those for the guest using helix/coil theory as a function of temperature. The results from the host–guest work are in disagreement with most of those from short peptides of fixed sequence (see below).

C. Helix Lengths

Helix formation in peptides is cooperative, with a nucleation penalty. Helix stability therefore tends to increase with length, in homopolymers at least. As the length of a homopolymer increases, the mean fraction helix will level off below 100%, as long helices are likely to break in two. In heteropolymers, helices are readily terminated by the introduction of a strong capping residue or a residue with a low intrinsic helical preference.

The length distribution of helices in proteins is very different to proteins (4, 85). Most helices are short, with 5–14 residues most abundant. There is a general trend for a decrease in frequency as the length increases beyond 13 residues. Helix lengths longer than 25 are rare. This is a consequence of the organization of proteins into domains of similar size, rather than showing different rules for stability; helices do not extend beyond the boundaries of the domain, so terminate. There is also a preference to have close to an integral number of turns so that their N- and C-caps are on the same side of the helix (85).

D. The Helix Dipole

The secondary amide group in a protein backbone is polarized with the oxygen negatively charged and hydrogen positively charged. In a helix the amides are all oriented in the same direction with the positive hydrogens pointing to the N-terminus and negative oxygens pointing to the C-terminus. This can be regarded as giving a positive charge at the helix N-terminus and a negative charge at the helix C-terminus (86–88). In general, therefore, negatively charged groups are stabilizing at the N-terminus and positive at the C-terminus, as shown by titrations that measure helix content as a function of pH and amino acid preferences for helix terminal positions. An alternative interpretation of these results is that favored side chains are those that can make hydrogen bonds to the free amide NH groups at N1, N2, and N3 or free CO groups at C1, C2, and C3 (89). Charged groups can form stronger hydrogen bonds than neutral groups, thus providing an alternative rationalization of the pH titration results. These hypotheses are not mutually exclusive, as a charged side chain can also function as a hydrogen bond acceptor or donor. A free energy simulation of Tidor (90) suggested that helix stabilizing interactions, as a result of a Tyr → Asp substitution at an N-cap site, arose from hydrogen bonding interactions from its direct hydrogen bonding partners, and from more distant electrostatic interactions with groups within the first two turns of the helix.

Measurements of the amino acid preferences for the N-cap, N1, N2, and N3 positions in the helix allow a comparison to be made of the relative importance of helix dipole and hydrogen bonding interactions (91–94). The helix dipole model implies that the side chains most favored at the helix N-terminus are those with a negative charge while positive charges are disfavored. If hydrogen bonding is the only important feature, then favored side chains are those that can make hydrogen bonds to the free amide NH groups at N1, N2, and N3. In general, the N-cap results suggest that hydrogen bonding is more important than helix dipole interactions; the best N-caps are Asn, Asp, Ser, and Thr (91) which can accept hydrogen bonds from the N2 and N3 NH groups (15). Glu has only a moderate N-cap preference despite its negative charge. In contrast, N1, N2, and N3 results suggest that helix dipole interactions are more important. The contrasting results between the different helix N-terminal positions can be rationalized by considering the geometry of the hydrogen bonds. N-cap hydrogen bonds are close to linear (15) and so are strong, while N1 and N2 hydrogen bonds are close to 90° (20), making them much weaker. Helix dipole effects are likely to be present at all sites, as also shown by every pH titration where a more negative side chain is favored over a more positive side chain, but this can be overwhelmed by strong hydrogen bonds, as at the N-cap. In the absence of strong hydrogen bonds, helix dipole effects dominate. Hydrogen bonds can therefore make a substantial contribution to protein stability, but only if their geometry is close to linear (as it is for backbone to backbone hydrogen bonds in α-helices and β-sheets).

E. Acetylation and Amidation

A simple, yet effective, way to increase the helicity of a peptide is to acetylate its N-terminus (24, 95). This is readily done with acetic anhydride/pyridine, after completion of a peptide synthesis, but before cleavage of the peptide from the resin and deprotection of the side chains. Acetylation removes the positive charge that is present at the helix terminus at low or neutral pH; this would interact unfavorably with the positive helix dipole and free N-terminal NH groups. The extra CO group from the acetyl group can form an additional hydrogen bond to the NH group, putting the acetyl at the N-cap position. This has a strong stabilizing effect by approximately 1.0 kcal mol^{-1} compared to Ala (91, 96, 97). The acetyl group is one of the best N-caps. We found that only Asn and Asp to be more stabilizing (91).

Amidation of the peptide C-terminus is achieved by using different types of resin in solid-phase peptide synthesis, resulting in the replacement of COO^- with $CONH_2$. This is similar structurally to acetylation: the helix is extended by one hydrogen bond and an unfavorable charge–charge repulsion with the helix dipole removed. The energetic benefit of amidation is rather smaller, however, with the amide group being no better than Ala and in the middle if the C-cap residues are ranked in order of stabilization effect (91). As most helical peptides studied to date are both acetylated and amidated, and acetylation is more stabilizing than amidation, the helicity of peptides is generally skewed so that residues near the N-terminus are more helical than those near the C-terminus. This is because conformations that are helical all the way to the N-terminus are favored, as they will have an acetyl N-cap. Acetylation and amidation are often also beneficial in removing a pH titration that can obscure other effects.

F. Side-Chain Spacings

Side chains in the helix are spaced 3.6 residues per turn of the helix. Side chains spaced $(i, i+3)$, $(i, i+4)$, and $(i, i+7)$ are therefore close in space and interactions between them can affect helix stability. Spacings of $(i, i+2)$, $(i, i+5)$, and $(i, i+6)$ place the side-chain pairs on opposite faces of the helix avoiding any interaction.

G. Solubility

Helical peptides may have low solubility in water, particularly when uncharged. Peptides designed to be helices can even become highly insoluble amyloid with a high β-sheet content. We have occasionally found that Ala-based peptides designed to be helical instead form amyloid, though this may only take place after several years of storage in solution. The measurement of interactions in a helix will be compromised by peptide oligomerization so it is generally essential to check that the peptides are monomeric. This can be done rigorously by sedimentation equilibrium which determines the oligomeric state of a molecule in solution. This is difficult, however, with the short peptides often used as their molecular weights are at the lower limit for this technique. A simpler method is to check a spectroscopic technique that depends on peptide structure, most obviously CD, as a function of concentration. If the signal depends linearly on peptide concentration across a large range, including that used to study the peptide structure, it is safe to assume that the peptide is monomeric. For example, if helicity measurements are made at 10 µM, CD spectra can be acquired from 5 to 100 µM. An oligomer that does not change state, such as a coiled coil, across the concentration range cannot be excluded, however. Light scattering can detect also aggregation. A monomeric peptide should have a flat baseline in a UV spectrum outside the range of any chromophores in the peptide. In stock solutions of a peptide with a single tyrosine isolated from the helix region by Gly should have $A_{300}/A_{275} < 0.02$ and $A_{250} < A_{275} < 0.2$ (98).

Consideration of solubility is essential when designing helical peptides. While Ala has the highest helix propensity and would provide an ideal theoretical background for substitutions, poly(Ala) is insoluble. Solubility can be achieved most easily by including polar side chains spaced (i, $i + 5$) in the sequence where they cannot interact. Lys, Arg, and Gln are used most often for this purpose. Gln may be preferred if unwanted interactions with charged Lys or Arg may be a problem, but some AQ peptides lack sufficient solubility and AQ peptides are less helical. It is not always easy to predict peptide solubility, as leaving solution may depend on hydrogen bond formation and packing, rather than simply hydrophobicity. The spacing of side chains in the helix are best visualized with a helical wheel, to ensure that the designed helix does not have a nonpolar face that may lead to dimerization. The following web page provides a useful resource for this: http://www.site.uottawa.ca/~turcotte/resources/HelixWheel/.

H. Concentration Determination

An accurate measurement of helix content depends on an accurate spectroscopic measurement and, equally importantly, peptide concentration. This is usually achieved by including a Tyr side chain at one end of the peptide. The extinction coefficient of Tyr at 275 nm is 1450 M^{-1} cm^{-1} (99). If Trp is present measurements at 281 nm can be used where the extinction coefficient of Trp is 5690 M^{-1} cm^{-1} and Tyr is 1250 M^{-1} cm^{-1} (100). Phe absorbance is negligible at this wavelength. These UV absorbances are ideally made in 6 M

GuHCl, pH 7.0, 25°C, though we have found very little variation with solvent so measurements in water are identical within error. The main source of error is in pipetting small volumes; this is typically around 2%. Pipetting larger volumes with well maintained fixed volume pipettors can help minimize this error.

Though the inclusion of aromatic residues is required for concentration determination, this can have the unwanted side effect of perturbing a CD spectrum, leading to an inaccurate measure of helix content. A simple solution to this is to separate the terminal Tyr from the rest of the sequence by one or more Gly residues (*101*). If the aromatic residues must be included within the helical region, the CD spectrum should be corrected to remove this perturbation (e.g., (*102*)).

I. Design of Peptides to Measure Helix Parameters

Numerous peptides have been studied to measure the forces responsible for helix formation (see below). In general, one or more peptides are synthesized that contain the interaction of interest, while all other terms that can contribute to helix stability are known. The helix content of the peptide is measured and the statistical weight of the interaction is varied until predictions from helix–coil theory match experiment. The weight of the interaction (and hence its free energy, as $-RT \ln(\text{weight})$) is then known. In practice it is wise to also synthesize a control peptide that lacks the interaction, but is otherwise very similar. The helix content of this peptide should be predictable from helix–coil theory using known parameters.

It is important to minimize the error in determining helix–coil parameters. Errors can be calculated by assuming an error in measurement of percent helix ($\pm 3\%$ is reasonable) and refitting the results across this experimental error range. The inclusion of multiple identical interactions in the same peptide can increase accuracy. The helix contents of the control and interaction peptides should be close to 50%. This may be difficult to achieve if the residues being used have low helix preferences. The best way to maximize sensitivity is to calculate it in advance using possible sequences, helix–coil theory, an error of $\pm 3\%$ and guessing a sensible value for the interaction energy. The interaction of interest can be placed at various positions in the peptide, its length can be changed by adding further AAKAA sequences, terminal residues, such as a Tyr can be moved, and solubilizing side chains can be changed. Peptide design can be lengthy, considering sensitivity and solubility, but is a valuable process. The complex nature of the helix–coil equilibrium with frayed conformations highly populated means that considerable variations can be seen for apparently small changes, such as moving an interaction toward one terminus of a sequence.

J. Helix Templates

A major penalty to helix formation is the loss of entropy arising from the requirement to fix three consecutive residues to form the first hydrogen bond of the helix. Following this nucleation, propagation is much more favored as only a single residue need be restricted to form each additional hydrogen bond. A way to avoid this barrier is to synthesize a template molecule that facilitates helix initiation, by fixing hydrogen bond acceptors, or donors in the correct orientation for a peptide to bond in a helical geometry. The ideal template nucleates a helix with an identical geometry to a real helix. Kemp's group applied this strategy and synthesized a proline-like template that nucleated helices when a peptide chain was covalently attached to a carboxyl group (Fig. 1A) (*103–107*). The template is in an equilibrium between *cis* and *trans* isomers of the proline-like part of the molecule. Only the *trans* isomer can bond to the helix so helix content is determined by measuring the *cis/trans* ratio by NMR. The templates moderately destabilize the helix (*108*). Bartlett *et al.* (*109*) reported on a hexahydroindol-4-one template (Fig. 1B) that induce 49–77% helicity at 0°C, depending on the method of determination, in an appended hexameric peptide.

Seven different N-capping templates were tested based on ease of synthesis and an abundance of hydrogen bond acceptors. Different scaffolds based on sugars, cyclic hydrocarbons, and amino acids were used with a variety of hydrogen bond acceptors including esters, carboxyls, amides, and a sulfonic acid (*110*). A cyclic peptide appeared to be most useful, judged by its effect on helix formation and potential for introducing future sequence variants (Fig. 1C).

Several other templates were less successful and could only induce helicity in organic solvents (*111–113*). Their syntheses are often lengthy and difficult, partly due to the challenging requirement of orienting several dipoles to act as hydrogen bond acceptors or donors.

K. Design of 3_{10}-Helices

Peptides can be induced to form 3_{10}-helix by the incorporation of disubstituted C_α-amino acids, for which the simplest is AIB (*114–117*). The presence of steric interactions from the two methyl groups on the α-carbon in AIB results in the 3_{10} geometry being energetically favored over the α. Peptides rich in $C_{\alpha,\alpha}$-disubstituted α-amino acids are readily crystallized and many of their structures have been solved (reviewed by (*45, 118*)). Aib is conformationally restricted so shorter Aib-based helices are more stable than Ala-based α-helices (*119*). Many examples of helix stabilization or 3_{10}-helix formation by Aib have been reported. The $\alpha/3_{10}$ equilibrium has been studied in peptides. Yokum *et al.* (*120*) synthesized a peptide composed of $C_{\alpha,\alpha}$-disubstituted α-amino acids that

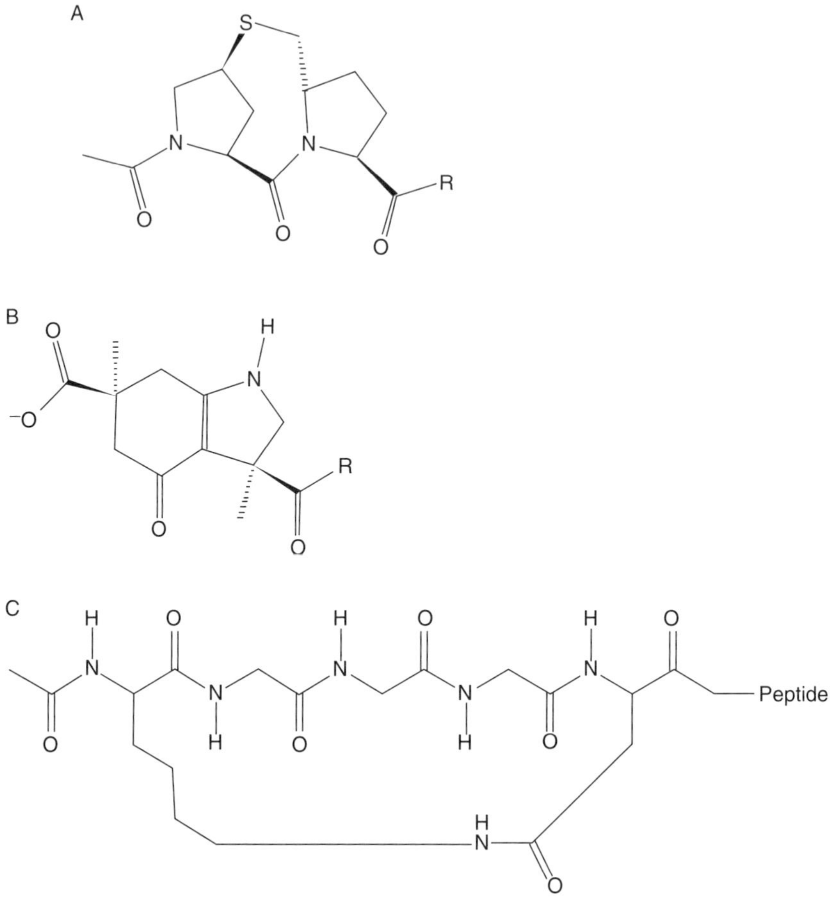

FIG. 1. Helix templates: (A) Kemp, (B) Bartlett, and (C) Mutter.

forms mixed 3_{10}-/α-helices in mixed aqueous/organic solvents. Millhauser and coworkers (*121–124*) have studied the 3_{10}-/α equilibrium in peptides by ESR and NMR and argued that mixed 3_{10}-/α-helices are common in polyalanine-based helices. Yoder *et al.* (*125*) studied a series of L-(αMe)-Val homopolymers and showed that their peptides can form coil, 3_{10}-helix, or α-helix depending on concentration, peptide length and solvent. Kennedy *et al.* (*126*) used FTIR spectroscopy to monitor 3_{10}- and α-helix formation in poly(AIB)-based peptides. Hungerford *et al.* (*127*) synthesized peptides that showed an α to 3_{10} transition upon heating. 3_{10}-Helices have also been proposed as thermodynamic intermediates where helical peptides of moderate stability exist as a mixture of α- and 3_{10}-structures (*43, 128*).

Guidelines for the inclusion of the 20 natural amino acids in 3_{10}-helices can be taken from their propensities in proteins, both at interior positions (*40*) and at N-caps (*15*). These are similar, but not identical to α preferences. Side-chain interactions in the 3_{10}-helix are spaced $(i, i + 3)$, with $(i, i + 4)$ on opposite sides of the helix. This offers scope to preferentially stabilize 3_{10} over α, by including stabilizing $(i, i + 3)$ interactions and $(i, i + 3)$ repulsions. As the 3_{10}- and α-helix structures are so similar, with only a small change in backbone dihedral angle, peptides designed to form 3_{10}-helix are likely to form a mixture of 3_{10} and α, with a central α segment and 3_{10} at the helix termini common. This complex equilibrium, and the spectroscopic similarity of α and 3_{10} makes the analysis of these peptides difficult.

L. Design of π-Helices

To our knowledge, no peptide has yet been made that forms π-helix. There are 4.4 side chains per turn of the π-helix, so $(i, i + 5)$ interactions may weakly stabilize π-helix over α. Given the other strongly destabilizing features of the π-helix, however, we doubt whether this effect will ever be strong enough.

III. Helix–Coil Theory

Peptides that form helices in solution do not show a simple two state equilibrium between a fully folded and fully unfolded structure. Instead, they form a complex mixture of all helix, all coil or, most frequently, central helices with frayed coil ends. In order to interpret experiments on helical peptides and make theoretical predictions on helices it is therefore essential to use a helix–coil theory that considers every possible location of a helix within a sequence. The first wave of work on helix–coil theory was in the late 1950s and early 1960s (reviewed in detail by Poland and Scheraga (*129*)). In 1992 Qian and Schellman (*130*) reviewed current understanding of helix–coil theories. Our review covered the development of helix–coil theory since this date (*131*).

The simplest way to analyze the helix–coil equilibrium is the two state model where the equilibrium is assumed to be between a 100% helix conformation and 100% coil. This is incorrect and its use gives serious errors. This is because helical peptides are generally most often found in partly helical conformations, often with a central helix and frayed, disordered ends, rather than in the fully folded or fully unfolded states.

A. Zimm–Bragg Model

The two major types of helix–coil model are (i) those which count hydrogen bonds, principally ZB (*132*) and (ii) those that consider residue conformations, principally Lifson–Roig (*133*). In the ZB theory the units being considered are peptide groups and they are classified on the basis of whether their NH groups participate in hydrogen bonds within the helix. The ZB coding is shown in Fig. 2. A unit is given a code of 1 (e.g., peptide unit 5 in Fig. 2) if its NH group forms a hydrogen bond and 0 otherwise. The first hydrogen-bonded unit proceeding from the N-terminus has a statistical weight of σs, successive hydrogen-bonded units have weights of s and nonhydrogen-bonded units have weights of 1. The s-value is a propagation parameter and σ is an initiation parameter. The most fundamental feature of the thermodynamics of the helix–coil transition is that the initiation of a new helix is much more difficult than the propagation of an existing helix. This is because three residues need to be fixed in a helical geometry to form the first hydrogen bond while adding an additional hydrogen bond to an existing helix require that only one residue is fixed. These properties are thus captured in the ZB model by having σ smaller than s. The statistical weight of a homopolymeric helix of an N hydrogen bonds is σs^{N-1}. The cost of initiation, σ, is thus paid only once for each helix while extending the helix simply multiplies its weight by one additional s-value for each extra hydrogen bond.

B. Lifson–Roig Model

In the LR model each residue is assigned a conformation of helix (h) or coil (c), depending on whether it has helical ϕ, ψ angles. Every conformation of a peptide of N residues can therefore be written as a string of N cs or hs, giving 2^N conformations in total. Residues are assigned statistical weights depending on their conformations and the conformations of surrounding residues. A residue in an h conformation with an h on either side has a weight of w. This can be thought of as an equilibrium constant between the helix interior and the coil. Coil residues are used as a reference and have a weight of 1. In order to form an (i, i + 4) hydrogen bond in a helix three successive residues need to be fixed in a helical conformation. M consecutive helical residues will therefore

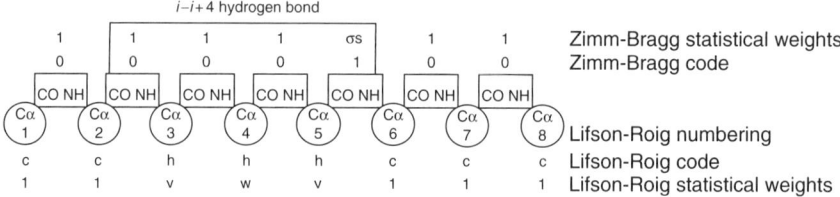

FIG. 2. Zimm–Bragg and Lifson–Roig codes and weights for the α-helix.

have $M - 2$ hydrogen bonds. The two residues at the helix termini (i.e., those in the center of *chh* or *hhc* conformations) are therefore assigned weights of v (Fig. 2). The ratio of w to v gives the approximately the effect of hydrogen bonding (1.7:0.036 for Ala (97) or $-RT \ln(1.7/0.036) = -2.1$ kcal mol^{-1}). A helical homopolymer segment of M residues has a weight of $v^2 w^{M-2}$ and a population in the equilibrium of $v^2 w^{M-2}$ divided by the sum of the weights of every conformation (i.e., the partition function). In this way the population of every conformation is calculated and all properties of the helix–coil equilibrium evaluated. The LR model is easier to handle conceptually for heteropolymers since the w and v parameters are assigned to individual residues. The substitution of one amino acid at a certain position thus changes the w- and v-values at that position. In the ZB model the initiation parameter σ is associated with several residues and s with a peptide group, rather than a residue. It is therefore easier to use the LR model when making substitutions. Indeed, most recent work has been based on this model. A further difference is that the ZB model assigns weights of zero to all conformations that contain a *chc* or *chhc* sequence. This excludes a very large number of conformations that contain a residue with helical ϕ, ψ angles but with no hydrogen bond. In LR theory, these are all considered. The ZB and LR weights are related by the following formulae (130): $s = w/(1 + v)$; $\sigma = v^2/(1 + v)^4$.

The complete helix–coil equilibrium is handled by determining the statistical weight for every possible conformation that contains a helix plus a reference weight of 1 for the coil conformation. Each conformation considered in the helix–coil equilibrium is given a statistical weight. This indicates the stability of that conformation, with the higher the weight, the more probable the conformation. Weights are defined relative to the all coil conformation which is given a weight of 1. The statistical weight of a conformation can thus be regarded as an equilibrium constant relative to the coil; a weight >1 indicates the conformation is more stable than coil, <1 means less stable and =1 means equally stable. The population of each conformation is given by the statistical weight of that conformation divided by the sum of the statistical weights for every conformation (i.e., the partition function). Thus the greater the statistical weight, the more stable the conformation. The key to using helix–coil theory is the partition function. All the properties of a system at equilibrium are contained within the partition function which makes it very valuable. Partition functions are extremely powerful concepts in statistical thermodynamics since they include all properties of an equilibrium. Any property of the equilibrium can be extracted from the partition function by applying the appropriate mathematical function. In this case the properties could be the mean number of hydrogen bonds, the mean helix length, the probability that each residue is within a helix, etc. In particular, the mean number of residues with a weight x is given by $\partial \ln Z/\partial \ln x$. CD is commonly used to give the mean helix content of a helical peptide, namely the fraction of residues that have a weight of w. LR-based models can thus be related to experimental data by equating the measured mean helix content to $(\partial \ln Z / \partial \ln w)/N$, where N is the number of residues in the peptide. Statistical weights can be regarded as equilibrium constants for the equilibrium between coil and the structure (as the reference coil weight is defined as 1). They can therefore be converted to free energies as $-RT \ln$ (weight).

Equations for partition functions are determined as follows: Write a table of the conformations and assigned statistical weights for residues. For the Lifson–Roig model, the weights of a residue depend on its conformation and the weights of its two neighboring residues as follows:

Conformation	Weight of central residue in triplet
ccc	1
cch	1
chc	v
chh	v
hcc	1
hch	1
hhc	v
hhh	w

The same information is rewritten in the form of a matrix (M):

$$M = \begin{matrix} \\ h\bar{h} \\ h\bar{c} \\ c\bar{h} \\ c\bar{c} \end{matrix} \begin{matrix} \bar{h}h & \bar{h}c & \bar{c}h & \bar{c}c \\ \begin{pmatrix} w & v & 0 & 0 \\ 0 & 0 & 1 & 1 \\ v & v & 0 & 0 \\ 0 & 0 & 1 & 1 \end{pmatrix} \end{matrix}.$$

For each triplet, the state of the leftmost residue is shown at the start of each row in the matrix. The state of the rightmost residue in each triplet is shown at the end of the top of each column. The state of the center residue of each triplet is shown as the barred residue in both the rows and columns. When the states of the center residues differ (i.e., one is c, while the other is h), the entry in the matrix is zero. Otherwise, the matrix gives the weight, taken from the table. This apparently simple change in presentation now (amazingly) allows the generation of the partition function (Z) for a polypeptide of N residues as

$$Z = (0 \ \ 0 \ \ 1 \ \ 1) M^N \begin{pmatrix} 0 \\ 1 \\ 1 \\ 1 \end{pmatrix}.$$

The end vectors are present to ensure that the first and last residues in the peptide cannot have a w-weighting. A w-weighting indicates that the residue is between two residues that hydrogen bond to each other and this is impossible for terminal residues. Further extensions to helix–coil theory are dealt with in the same way, by defining residue weight assignments, rewriting as a matrix and determining end vectors. The work may be made easier by combining any identical columns. For example, the Lifson–Roig matrix above has identical third and fourth columns so can be simplified to a 3 × 3 matrix:

$$M = \begin{array}{c} \bar{h}h \\ h\bar{h} \\ h\bar{c} \\ \overline{c(h \cup c)} \end{array} \begin{array}{ccc} \bar{h}h & \bar{h}c & \bar{c}(h \cup c) \\ \begin{pmatrix} w & v & 0 \\ 0 & 0 & 1 \\ v & v & 1 \end{pmatrix} \end{array}.$$

Here, $h \cup c$ shows the residues being combined. The matrix entries for the combined positions are the entries for the noncombined entries, with zero being discounted if combined with a nonzero entry. New end vectors are now required, as their lengths must be the same as the order of the square matrix.

C. The Unfolded State and Polyproline II Helix

The treatment of peptide conformations is based on Flory's isolated-pair hypothesis (*134*). This states that while ϕ and ψ for a residue are strongly interdependent, giving preferred areas in a Ramachandran plot, each ϕ, ψ pair is independent of the ϕ, ψ angles of its neighbors. Pappu *et al.* (*135*) examined the isolated-pair hypothesis in detail by exhaustively enumerating the conformations of poly(Ala) chains. Each residue was considered to populate 14 mesostates, defined by ranges of ϕ, ψ values. By considering all 14^N mesostate strings, all conformations were considered for up to seven alanines. The number of allowed conformations was found to be considerably fewer than the maximum, thus showing that the isolated-pair hypothesis is invalid. The chains mostly populated extended or helical conformations as many partly helical conformations are sterically disallowed. Such effects are not included in helix–coil theories, thus presenting a considerable challenge for the future. Helix–coil theories assign the same weight (1) to every coil residue; steric exclusion means that these should vary and be lower than 1 in many cases.

The polyproline II helix may well be an important conformation for unfolded proteins (*136–145*). In particular, denatured alanine-rich peptides may form polyproline II helix (*140, 145, 146*). Firestine *et al.* (*147*) used Ac–Lys$_2$–Ala$_7$–Lys$_2$–Gly–Tyr–NH$_2$ as a model for an unfolded peptide with a sequence typical of those used to study isolated helices. CD showed that the α-helix and PII helix conformations were the most prevalent for each residue in peptide dilute phosphate buffer, though with large variations when guest residues are substituted into the peptide. It may therefore be valid to consider residues in helical peptides to be in three possible states (helix, coil, or polyproline II), rather than two (helix or coil). No current helix–coil model takes this into account. A scale of amino acid preferences for the polyproline II helix has been published (148).

D. Single-Sequence Approximation

Since helix nucleation is difficult, conformations with multiple helical segments are expected to be rare in short peptides. In the one-, or single-, helical-sequence approximation, peptide conformations containing more than one helical segment are assumed not to be populated and are excluded from the partition function (i.e., assigned statistical weights of zero). As peptide length increases, the approximation is no longer valid since multiple helical segments can be long enough to overcome the initiation penalty. The single-sequence approximation will also break down when a sequence with a high preference for a helix terminus, such as a cap, is within the middle of the chain. The error from using the single-sequence approximation will therefore show a wide variation with sequence and could be potentially serious if a sequence has a high preference to populate more than one helix simultaneously. Conformations with two or more helices may also often include helix–helix tertiary interactions that are ignored in all helix–coil models.

E. N- and C-Caps

In the original LR model weights are assigned to residues in the center of hhh triplets (a weight of w for propagation) or in the center of chh or hhc triplets (a weight of v for initiation). Residues in all other triplets have weights of 1. LR-based models have been extended by assigning weights to additional conformations. N-capping can therefore be included in LR theory be assigning a weight of n to the central residue in a cch triplet (96). Similarly, the C-cap is the first residue in a nonhelical conformation (c) at the C-terminus of a helix. C-cap weights (c-values) are assigned to central residues in hcc triplets. Application of the model to experimental data where the C-terminal amino acid of a helical peptide was varied, allowed the determination of the c-values and hence free energies of C-capping (as $-RT \ln c$) of all the amino acids (91).

A problem with the original definitions of the capping weights above is that they apply to isolated h or hh conformations that are best regarded as part of the random coil. A helical hydrogen bond can only form when a minimum of three consecutive h residues are present. Andersen and Tong (149) and Rohl et al. (9) therefore changed the definition of the N-cap to apply only to the c

residue in a *chhh* quartet. The N-cap residues in *chc* or *chhc* conformations have weights of zero in the Andersen–Tong model and 1 in the Rohl et al. model.

F. Capping Boxes

The N-terminal capping box (27) includes a side chain–backbone hydrogen bond from N3 to the N-cap ($i, i - 3$). This is included in the LR model by assigning a weight of wr to the *chhh* conformation, where r is the weight for the Ser backbone to Glu side-chain bond (9).

G. Side-Chain Interactions

As helices have 3.6 residues per turn, side chains spaced ($i, i + 3$) or ($i, i + 4$) are close in space. Side-chain interactions are thus possible when four or five consecutive residues are in a helix. They are included in the LR-based model by giving a weight of wq to *hhhh* quartets and wp to *hhhhh* quintets. The side-chain interaction is between the first and last side chains in these groups; the w-weight is maintained to maintain the equivalence between the number of residues with a w-weighting and the number of backbone helix hydrogen bonds (150).

Scholtz et al. (151) used a model based on the one-helical-sequence approximation of the LR model to quantitatively analyze salt-bridge interactions in alanine-based peptides. Only a single interaction between residues of any spacing was considered, though this was appropriate for the sequences they studied. Shalongo and Stellwagen (152) also proposed incorporating side-chain interaction energies into the LR model, using a clever recursive algorithm.

H. N1, N2, and N3 Preferences

The helix N-terminus shows significantly different residue frequencies for the N-cap, N1, N2, N3, and helix interior positions (12, 20, 153, 154). A complete theory for the helix should therefore include distinct preferences for the N1, N2, and N3 positions. In the original LR model, the N1 and C1 residues are both assigned the same weight, v. Shalongo and Stellwagen (152) separated these as v_N and v_C. Andersen and Tong (149) did the same and derived complete scales for these parameters from fitting experimental data, though some values were tentative. The helix initiation penalty is $v_N v_C$ and so v_N- and v_C-values are all small (≈ 0.04).

We added weights for the N1, N2, and N3 (n_1, n_2, and n_3) positions as follows (155): The n_1-value is assigned to a helical residue immediately following a coil residue. The penalty for helix initiation is now $n_1 v$, instead of v^2, as v

remains the C1-weight. An N2 helical residue is assigned a weight of n_2w, instead of w. The weight w is maintained in order to keep the useful definition of the number of residues with a w-weighting being equal to the number of residues with an $(i, i + 4)$ main chain–main chain hydrogen bond. The n_2-value is an adjustment to the weight of an N2 residue that takes into account the structures that can be adopted by side chains uniquely at this position. Similarly, an N3 residue is now assigned the weight n_3w, instead of w.

I. Helix Dipole

Helix dipole effects were added to the LR model by Scholtz et al. (151), though they used the one-sequence approximation so that only one or no dipoles in total are present. In LR models helix dipole effects are subsumed within other energies. For example, N-cap, N1, N2, and N3 energies will include a contribution from the helix dipole interaction so the energy of interaction of charged groups at this position with the dipole should not be counted in addition.

J. 3_{10}- and π-Helices

The Lifson–Roig formalism can easily be adapted to describe helices of other cooperative lengths (48). The fundamental difference between a 3_{10}-helix and an α-helix is that the 3_{10}-helix has an $(i, i + 3)$ hydrogen bonding pattern rather than the $(i, i + 4)$ pattern characteristic of the α-helix. For a given number of units in helical conformations, a 3_{10}-helix will consequently have one more hydrogen bond than an α-helix. To include this difference in the 3_{10}-helix theory, one of the α-helical initiating residues (i.e., the central unit of either the $h_\alpha h_\alpha c$ or the $ch_\alpha h_\alpha$ triplet) must become a 3_{10}-helix-propagating residue. We arbitrarily chose to assign the propagating statistical weight, w_τ, to the central unit of the $h_\tau h_\tau c$ triplet such that helix propagating unit i is associated with the hydrogen bond formed between the CO of peptide $i - 2$ and the NH of peptide $i + 1$ and τ indicates a 3_{10}-helix conformation. The remainder of the statistical weights applicable to the α-helix–coil theory are maintained.

The models described above for the α-helix–coil and 3_{10}-helix–coil transitions can be combined to describe an equilibrium including pure α-helices, pure 3_{10}-helices, and mixed α-/310-helices (48). In this model, three conformational states are possible, 3_{10}-helical (h_τ), α-helical (h_α), and coil (c). Stretches of residues in h_α conformation are treated as in the pure α-helix model and stretches of residues in h_τ conformation are treated as in the pure 3_{10}-helix model. Mixed helices consist regions of α- and 3_{10}-helical structure, and transitions between the two types of helices. We defined two additional parameters, t_N and t_C, to describe the junction from 3_{10}- to α-helix and from α- to 3_{10}-helix, respectively. The pure 3_{10}-helix and the mixed α-/3_{10}-helix models were subsequently extended to include side-chain interactions (156). 3_{10}-Helices have only $(i, i + 3)$ side-chain interactions, while both $(i, i + 3)$ and $(i, i + 4)$ are possible in mixed helices.

Sheinerman and Brooks (128) independently produced a model for the α/3_{10}/coil equilibrium, based on the ZB formalism, rather than the LR model. They similarly extend the classification of conformations from α/coil to α/3_{10}/coil. Their model differs from ours primarily in that it does not include additional parameters for junctions between α- and 3_{10}-helical segments and that it allows a 3_{10}-helix to extend only from the C-terminus of an α-helix. N-terminal 3_{10}-helical extensions to α-helices are often observed in crystal structures, however (4, 41).

In a π-helix, formation of an $(i, i + 5)$ hydrogen bond requires that four units be constrained to the π-helical conformation, h_π. The π subscript designates the conformation and weights describing the π-helix, whose dihedral angles are distinct from α- and 3_{10}-helices. Assigning statistical weights to individual units requires consideration of the conformations of the unit itself and its three nearest neighbors (48). The initiating statistical weight, v_π, is assigned to a helical unit when one or more of its two N-terminal and nearest C-terminal neighbors are in the coil conformation. The definition of helix initiating units as the two N-terminal and one C-terminal units of each helical stretch is again arbitrary. Units in a π-helical conformation with three helical neighbors are assigned the propagating statistical weight, w_π. A π-helix propagating residue, i, is thus associated with the hydrogen bond between the NH of residue $i + 2$ and the CO of residue $i - 3$.

K. AGADIR

AGADIR is a LR-based helix–coil model developed by Serrano, Muñoz, and coworkers. The original model (157) included parameters for helix propensities excluding backbone hydrogen bonds (attributed to conformational entropy), backbone hydrogen bond enthalpy, side-chain interactions and a term for coil weights at the end of helical sequences (i.e., caps). The single-sequence approximation was used. The original partition function assumed that many helical conformations did not exist, as all conformations in which the residue of interest is not part of a helix were excluded (150, 157). These were corrected in a later version, AGADIRms, which considers all possible conformations (158). If AGADIR and LR models are both applied to the same data, to determine a side-chain interaction energy, for example, the results are similar, showing that the models are now not significantly different (158, 159). The treatment of the helix–coil equilibrium differs in a number of respects from the

ZB and LR models and these have been discussed in detail in by Muñoz and Serrano (158). The minimal helix length in AGADIR is four residues in an h conformation, rather than three. The effect of this assumption is to exclude all helices which contain a single hydrogen bond; only helices with two or more hydrogen bonds are allowed. In practice, this probably makes little difference as *chhhc* conformations are usually unfavorable and hence have low populations. Early versions of AGADIR considered that residues following an acetyl at the N-terminus or preceding an amide at the C-terminus were always in a c conformation; this was changed to allow these to be helical (160).

The current version of AGADIR, AGADIR1s-2 (160), includes terms for electrostatics (160), the helix dipole (160, 161), pH dependence (161), temperature (161), ionic strength (160), N1, N2, and N3 preferences (162) and capping motifs such as the capping box, hydrophobic staple, Schellman motif and Pro-capping motif (160). The free energy of a helical segment, $\Delta G_{helical\ segment}$, is given by $\Delta G_{helical segment} = \Delta G_{Int} + \Delta G_{Hbond} + \Delta G_{SD} + \Delta G_{dipole} + \Delta G_{nonH} + \Delta G_{electrost}$, which are terms for the energy required to fix a residue in helical angles (with separate terms for N1, N2, N3, and N4), backbone hydrogen bonding, side-chain interactions excluding those between charged groups, capping, and helix dipole interactions, respectively. Electrostatic interactions are calculated with Coulomb's equation. Helix dipole interactions were all electrostatic interactions between the helix dipole or free N- and C-termini and groups in the helix. Interactions of the helix dipole with charged groups located outside the helical segment were also included. pH dependence calculations considered a different parameter set for charged and uncharged side chains and their pK_a values. The single-sequence approximation (see above) is used again, unlike in AGADIRms. This means that it must not be used for full protein sequences, though this has been done, even if they do not have any tertiary interactions.

AGADIR is at present the only model that can give a prediction of helix content for any peptide sequence, thus making it very useful. It can also predict NMR chemical shifts and coupling constants. In order to do this it must include estimates of all the terms that contribute to helix stability, notably the 400 possible $(i, i + 4)$ side-chain interactions. Since only a few of these interactions have been measured accurately, the terms used cannot be precise. Further determination of energetic contributions to helix stability is therefore still needed.

L. Lomize–Mosberg Model

Lomize and Mosberg (163) also developed a thermodynamic model for calculating the stability of helices in solution. Interestingly, they extended it to consider helices in micelles or a uniform nonpolar droplet to model a protein core environment. Helix stability in water is calculated as the sum of main chain interactions, which is the free energy change for transferring Ala from coil to helix, the difference in energy when replacing an Ala with another residue, hydrogen bonding and electrostatic interactions between polar side chains and hydrophobic side-chain interactions. An entropic nucleation penalty of two residues per helix is included. Different energies are included for N-cap, N1–N3, C1–C3, C-cap, hydrophobic staples, Schellman motifs, and polar side-chain interactions, based on known empirical data at the time (1996). Hydrophobic interactions were calculated from decreases in nonpolar surface area when they are brought in contact. Helix stability in micelles or nonpolar droplets are found by calculating the stability in water then adding a transfer energy to the nonpolar environment.

M. Extension of the Zimm–Bragg Model

Following the discovery of short peptides that form isolated helices in aqueous solution, Vásquez and Scheraga (164) extended the ZB model to include helix dipole and side-chain interactions. The model is very general as it can include interactions of any spacing within a single helix. It was applied to determine $(i, i + 4)$ and $(i, i + 8)$ interactions. Long-range interactions, beyond the scope of LR models, can thus be included. Roberts (165) and Gans et al. (84) also refined the ZB model to include side-chain interactions.

N. Tertiary Interactions

A challenging problem is to extend helix–coil theory to include interactions between helices. Qian (166) developed a model for coiled coils that included a parameter for the interaction between two helices. Hausrath (167, 168) included a parameter for coupling between two helices in a helix-turn-helix structure.

IV. Forces Affecting α-Helix Stability

A. Helix Interior

Since the advent of crystal structures of proteins it had been noticed that some amino acids appeared frequently in α-helices and others less frequently (169, 170). For example, alanine and leucine are abundant, whereas proline and glycine appear rarely. As more of this kind of information became available a helix propensity scale was derived (171) and eventually allowed prediction of the location of α-helices (and other structures) in folded proteins from their sequence (172).

Different approaches have been used in order to determine the helical propensity or preference of individual amino acids. Scheraga and coworkers used a host–guest strategy (see above) to derive values for the helical preference of various amino acid residues. The host–guest system uses long random copolymers of a water soluble, nonionic guest (PHPG or PHBG), together with a low (10–50%) content of the guest residue. The Zimm–Bragg model s and σ-values of the host homopolymer are used to compute those for the guest (*173, 174*). This has been carried out for all 20 naturally occurring amino acids (*175*). This work has been criticized as the host side chains can interact with each other (*176*). The introduction of a guest residue thus removes host–host interactions and replaces them with PHBG–guest or PHPG–guest side-chain interactions that may obscure the intrinsic helix propensities. We believe that helix propensities are best evaluated in an alanine background, where no side-chain interactions can affect the helix stability.

Rohl *et al.* (*97*) used many alanine-based peptides with the general sequences Ac–(AAKAA)$_m$Y–NH$_2$ (or with Q instead of K) to measure interior helix propensities. Substitutions in the helix interior and subsequent measures of helicity using CD spectroscopy in both water and 40% (v/v) trifluoroethanol (TFE) allowed both the calculation of the Lifson–Roig w parameter and stabilization energy for all 20 amino acids (see Table II). Kallenbach and coworkers (*83*) also used synthetic peptides of the form succinyl–YSEEEEKAKKAXAEEAEKKKK–NH$_2$, where substitutions at X allowed determination of helix stabilizing energies for common amino acids. Stellwagen and coworkers (*80*) made substitutions in position 9 of Ac–Y(EAAAK)$_3$A–NH$_2$. They agree well with the alanine-based peptide work described previously (*97, 177*).

Other groups have investigated helical propensities and stabilization using whole protein methods. Blaber *et al.* (*178–180*) used mutagenesis in the helices of phage T4 lysozyme to study the structural effects of substitutions of amino acids. With the exception of substituting proline, they found that no substitutions significantly distorted the helix backbone. $\Delta\Delta G$ values correlated well (71–93%) with model peptide studies and with studies on the frequency of amino acid occurrence in protein structures (*172*). Fersht and coworkers (*181*) used a similar method with barnase to study the effect of replacing Ala32 of the second helix in this protein with the other 19 naturally common amino acids. They used reversible urea determination to measure free energies of unfolding.

O'Neil and DeGrado (*182*) used substitutions into an α-helical two-stranded coiled-coil system to deduce helix-forming tendencies of common amino acids, through the design of a peptide that forms a noncovalent α-helical dimer, which is in equilibrium with a randomly coiled monomeric state. The α-helices in the dimer contain a single solvent-exposed site that is surrounded by small, neutral amino acid side chains. Each of the commonly occurring amino

TABLE II
HELIX PROPAGATION PROPENSITIES AND FREE ENERGIES OF AMINO ACIDS IN WATER (FROM (97))

Amino acid	$\Delta G°$ (helix) (kcal mol^{-1})	w-value
Ala	−0.27	1.70
Arg	−0.052	1.14
Leu	0.095	0.87
Met	0.25	0.65
Lys	0.019	1.00
Gln	0.28	0.62
Glu	0.21	0.70
Ile	0.44	0.46
Trp	0.69	0.29
Ser	0.52	0.40
Tyr	0.42	0.48
Phe	0.73	0.27
Val	0.77	0.25
His	0.57	0.36
Asn	0.69	0.29
Thr	0.95	0.18
Cys	0.64	0.32
Asp	0.52	0.40
Gly	1.7	0.048
Pro	>3.8	<0.001

acids was substituted into this guest site, and the resulting equilibrium constants for the monomer–dimer equilibrium were determined to provide a list of free energy difference ($\Delta\Delta G°$) values. Again these values show good agreement with those of other groups working in different model systems.

In 1998, Pace and Scholtz (*183*) gathered information from many different sources and derived a scale for the propensity of each amino acid in the helix interior. This is summarized in Table III. The values are in $\Delta(\Delta G)$ relative to alanine as zero. Alanine was taken as zero as it is generally (though not universally) agreed that this amino acid has the highest helical propensity. In particular, host–guest analysis showing alanine to be effectively helix-neutral has been supported by data from some other groups, notably the templated helices of Kemp and coworkers (*103*). Several efforts have been made to try and explain this discrepancy, including implication of the charged groups used to solubilize the alanine-based peptides (*184, 185*). The use of template-nucleated helices has been criticized by Rohl et al. (*186*) who argued that the

TABLE III
SUMMARY OF OTHER EXPERIMENTAL HELIX PROPENSITIES (RELATIVE TO ALANINE)

Amino acid	Helix propensity ($\Delta\Delta G$) (kcal mol^{-1}) (taken from (183))
Ala	0.00
Arg	0.21
Leu	0.21
Met	0.24
Lys	0.26
Gln	0.39
Glu	0.40
Ile	0.41
Trp	0.49
Ser	0.50
Tyr	0.53
Phe	0.54
Val	0.61
His	0.61
Asn	0.65
Thr	0.66
Cys	0.68
Asp	0.69
Gly	1.00
Pro	3.16

low apparent helix propensity of alanine is a consequence of properties of the template-helix junction. However, recent work by Kemp and coworkers (108, 187) may settle the controversy. Using templates to investigate the helix-forming tendency of polyalanine, these workers extended the length of the polyalanine beyond the previous limit of six residues. Below six residues, both this group and Scheraga's had low helix propensities for alanine (see above) but when the limit of six was exceeded, a dramatic increase in helix propensity. For chains with less than six alanines, $w = 1.03$, in agreement with both Kemp and Scheraga's earlier experimental results. For chains with 6–9 alanines, $w = 1.15$ and for more than 10 alanines, $w = 1.26$. This indicates that there is a length-dependent term in the helicity of polyalanine and that the charged groups are not having the effect previously ascribed to them. These values for longer polyalanine sequences are also much more consistent with values published by other groups.

While experimental studies agree that Ala is favored in helices over Gly, the origins of this seemingly simple substitution are less clear. Contributing factors have been suggested to be a difference in backbone conformational entropy in the denatured state, burial of hydrophobic surfaces on folding, and disruption of hydrogen bonding to water (*178, 188–190*). Aromatic side-chain preferences may result from interactions with the backbone (*191*).

B. Caps

Serrano and Fersht (*22*) explored the capping preferences at the N-cap by mutating Thr residues at the N-cap of two helices in barnase. They found that negatively charged residues were favored at the N-cap with a rank order of Asp > Thr > Glu > Ser > Asn > Gly > Gln > Ala > Val. Their result conflicted with the statistical survey result that Asn is one of the most frequently found N-caps in proteins (*12*). Experimentally, Asn destabilized the helices by 1.3 kcal mol^{-1} relative to Thr. The rank order of amino acids N-cap preferences is T4 lysozyme was found to be Thr > Ser > Asn > Asp > Val = Ala > Gly (*23*). They suggested that Asn can be inherently as good an N-cap as Ser or Thr, but it requires a change in backbone dihedral angles of N-cap residues which might be altered in native proteins as the results of tertiary contacts. Indeed, Asn is the most stabilizing residue at N-cap in a peptide model in the absence of tertiary contacts and other side-chain interactions (see below).

The Kallenbach group (*192*) substituted several amino acids at the N-cap position in peptide models in the presence of a capping box. They found that Ser and Arg are the most stabilizing residues with $\Delta\Delta G$ relative to Ala of −0.74 and −0.58 kcal mol^{-1}, respectively, whilst Gly and Ala are less stabilizing. The results are in agreement with the results of Forood *et al.* (*25*), who found that the trend in α-helix inducing ability at the N-cap is Asp > Asn > Ser > Glu > Gln > Ala. A more comprehensive work to determine the preferences for all 20 amino acids at the N-cap position used peptides with a sequence of NH$_2$-XAKAAAAKAAAAKAAGY-CONH$_2$ (*24, 91, 96*). N-Capping free energies ranged from Asn (best) to Gln (worst) (Table IV).

We have used a similar approach using peptide models to probe the preferences at N1 (*92*), N2 (*93*), and N3 (*94*) using peptides with sequences of CH$_3$CO–XA$_4$QA$_4$QAAGY–CONH$_2$, CH$_3$CO–AXA$_4$KA$_4$KAAGY–CONH$_2$, and CH$_3$CO–AAXA$_4$KA$_4$KAGY–CONH$_2$, respectively. The results have given N1, N2, and N3 preferences for most amino acids for these positions (Table IV) and these agree well with preferences seen in protein structures, with the interesting exception of Pro at N1. Petukhov *et al.* (*162, 193*) similarly obtained N1, N2, and N3 preferences for nonpolar and uncharged polar residues by applying AGADIR to experimental helical peptide data, and found almost identical results. The complete sequences of peptides used can be seen in

TABLE IV
AMINO ACID PROPENSITIES AT N- AND C-TERMINAL POSITIONS OF THE HELIX

ΔΔG relative to Ala for transition from coil to the position (kcal mol^{-1})

Residue	N-cap (91)[a]	N-cap (92)[b]	N1 (162), (193)[c]	N2 (93)[d]	N2 (162), (193)[e]	N3 (94)[f]	N3 (162), (193)[g]	C3 (195)[h]	C2 (196)[i]	C2 (195)[j]	C1 (195)[k]	C-cap (91)[l]	C' (273)[m]
A	0	0	0	0	0	0	0	0	0	0	0	0	0
C°												0.2	
C⁻	−1.4	1.0		0.9		–							
D°		0.5		0.7		–						0.2	0.3
D⁻	−1.6	0		−0.2		1.1							
E°		1.0		−0.2		–						−0.4	0.3
E⁻	−0.7	0.1		−0.4		0.6						−0.5	
F	−0.7	1.4		0.9		1.3							0.1
G	−1.2	1.0	0.7	–	0.4	–	0.8	2.1	0.6			0.1	−1.1
H°	−0.7	0.7		0.8		2.6			1.0	0.6	0.4		
H⁺		–		–		–						−0.2	−0.9
I	−0.5	0.5	0.4	0.6	0.5	0.7	0.5	0.2	0.2	0.4	0.5	−0.1	1.5
K⁺	0.1	0.7		0.9		0.9						−0.1	−0.1

	a	b	c	d	e	f	g	h	i	j	k	l	m
L	−0.7	0.4	0.2	0.5	0.5	0.8	0.4		0.1			−0.1	0.1
M	−0.3	0.5	0.1	0.7	0.3	0.7	0.4		0.1			−0.3	−0.4
N	−1.7	—	0.6	1.7	0.7	—	0.7	0.5	0.7	0.4	0.3	0.1	1.2
P	−0.4	0.6	0.5	—									
Q	2.5	0.5	0.3	0.5	0.3	1.2	0.2	0.2	−0.02	0.2	0.05	−0.5	−0.1
R$^+$	−0.1	0.7		0.8		—						−0.4	−0.2
S	−1.2	0.4	0.4	0.7	0.5	1.1	0.6	0.6	0.5	0.7	0.5	0.8	0.3
T	−0.7	0.5	0.5	0.5	0.5	1.2	0.6	0.8	0.6	0.5	0.8		1.1
V	−0.1	0.6		0.5	0.4	—	0.4	0.3	0.4	0.7	0.6	0.9	1.6
W	−1.3	0.4		0.8		4.0							0.7
Y	−0.9	—		—		1.2						−2.2	

[a] NH$_2$–XAKAAAAKAAAAKAAGY–CONH$_2$.
[b] CH$_3$CO–XAAAAQAAAAQAAGY–CONH$_2$.
[c] CH$_3$CO–XAAAAAARAAAARGGY–NH$_2$.
[d] CH$_3$CO–AYAAAAKAAAAKAAGY–CONH$_2$.
[e] CH$_3$CO–AXAAAAARAAAARGGY–NH$_2$.
[f] CH$_3$CO–AAXAAAAKAAAAKAGY–CONH$_2$.
[g] CH$_3$CO–AAXAAAAARAAAARGGY–NH$_2$.
[h] NH$_2$–YGGSAKEAAARAAAAXAA–CONH$_2$.
[i] Substitution of residue 32 (C2 position) of α-helix of ubiquitin.
[j] NH$_2$–YGGSAKEAAARAAAAAXA–CONH$_2$.
[k] NH$_2$–YGGSAKEAAARAAAAAAX–CONH$_2$.
[l] CH$_3$CO–YGAAKAAAAKAAAAKAX–COOH.
[m] Substitution of residue 35 (C′ position) of α-helix of ubiquitin.

the table footnote. In general, at N1, N2, and N3, Asp and Glu as well as Ala are preferred, presumably because negative side chains interact favorably with the helix dipole or NH groups while Ala has the strongest interior helix preference.

Preferences at the C-cap position differ from those at the N-cap. At the N-terminus, the helix geometry favors side chain–backbone hydrogen bonding, so polar residues are preferred (15, 20). At the C-terminus unsatisfied backbone hydrogen bonds are fulfilled by interactions with backbone groups upstream of the helix. Zhou et al. (194) found that Asn is the most favored residue at the C-cap followed by Gln > Ser ∼ Ala > Gly ∼ Thr. Forood et al. (25) tested a limited number of amino acids the C-terminus (C1) finding a rank order of Arg > Lys > Ala. Doig and Baldwin (91) determined the C-capping preferences for all 20 amino acids in α-helical peptides. The thermodynamic propensities of some amino acids at C′, C-cap, C1, C2, and C3 are also included in Table IV (195, 196). C-cap motifs usually include a Gly. The substitution of a Gly C-cap in ubiquitin for D- and L-amino acids has shown that the preference for Gly can be attributed to its ability to adopt a left-handed conformation, rather than better solvation (197, 198). Replacement of this Gly with a Pro causes large structural changes, particularly in the last two turns of the α-helix, though cooperative two-state folding is retained (199).

C. Phosphorylation

Phosphoserine is destabilizing compared to serine at interior helix positions (200, 201). We investigated the effect of placing phosphoserine at the N-cap, N1, N2, N3, and interior position in alanine-based α-helical peptides, studying both the -1 and -2 phosphoserine charge states (202). Phosphoserine stabilizes at the N-terminal positions by as much as 2.3 kcal mol^{-1}, while destabilizes in the helix interior by 1.2 kcal mol^{-1}, relative to serine. The rank order of free energies relative to serine at each position is N2 > N3 > N1 > N-cap > interior. Moreover, -2 phosphoserine is the most preferred residue known at each of these N-terminal positions. Experimental pK_a values for the -1 to -2 phosphoserine transition are in the order N2 < N-Cap < N1 < N3 < interior. Phosphoserine can form stabilizing salt bridges to arginine (201).

D. Noncovalent Side-Chain Interactions

Many studies have been performed on the stabilizing effects of interactions between amino acid side chains in α-helices. These studies have identified a number of types of interaction that stabilize the helix including salt bridges (79, 82, 84, 151, 203–207), hydrogen bonds (151, 207–209), hydrophobic interactions (150, 210–212), basic–aromatic interactions (102, 213) and polar/nonpolar interactions (214). The stabilizing energies of many pairs in these categories have been measured, though some have only been analyzed qualitatively. As described earlier, residue side chains spaced $(i, i + 3)$ and $(i, i + 4)$ are on the same face of the α-helix, though it is the $(i, i + 4)$ spacing that receives most attention in the literature, as these are stronger. A summary of stabilizing energies for side-chain interactions is given in Table V. We give only those that have been measured in helical peptides with the side-chain interaction energies determined by applying helix–coil theory. Almost all are attractive, with the sole exception of the Lys–Lys repulsion.

E. Covalent Side-Chain Interactions

Lactam (amide) bonds formed between NH_3^+ and CO_2^- side chains can stabilize a helix, acting in a similar way to disulfide bridges in a protein by constraining the side chains to be close, reducing the entropy of nonhelical states (215). Lactam bridges between Lys–Asp, Lys–Glu, and Glu–Orn spaced $(i, i + 4)$ have been introduced into analogues of human growth hormone releasing factor (216), and proved to be stabilizing with Lys–Asp most effective. The same Lys–Asp $(i, i + 4)$ lactam was stabilizing in other helical peptide systems (217–220), while Lys–Glu $(i, i + 4)$ lactam bridges were less effective (218). Two overlapping Lys–Asp lactams were even more stabilizing (221). The effect of the ring size formed by the lactam was investigated by replacing Lys with Ornithine or (S)-diaminopropionic acid. A ring size of 21 or 22 atoms was most stabilizing (a Lys–Asp $(i, i + 4)$ lactam is 20 atoms) (216). Lactams between side chains spaced $(i, i + 7)$ (222, 223) or $(i, i + 3)$ (223, 224), spanning two or one turns of the helix have also been reported. $(i, i + 7)$ Disulfide bonds have been introduced into alanine-based peptides, using (D)- and (L)-2-amino-6-mercaptohexanoic acid derivatives (225). Strongly stabilizing effects were observed. Zhang and Morikis showed that an $(i, i + 8)$ hydrophobic interaction could induce bending in a designed helical peptide (226). Disulfide bonds are rare within helices, though they are allowed when the Cys residues are at the N-cap and N3 positions. The CXXC motif at the helix N-terminus is characteristic of the thiol–disulfide oxidoreductase superfamily and stabilizes the helix by 0.5 kcal mol^{-1} (227).

Helix formation can be reversibly photoregulated. Two cysteine residues are crosslinked by an azobenzene derivative which can be photoisomerized from *trans* to *cis*, causing a large increase or decrease in the helix content of the peptide, depending on its spacing (228–230).

F. Capping Motifs

Although the N-terminal capping box sequence stabilizes helices by inhibiting N-terminal fraying, it does not necessary promote elongation unless accompanied by favorable hydrophobic interactions as in a "hydrophobic

TABLE V
Summary of Side-Chain Interaction Energies from Literature

Interaction	$\Delta\Delta G$ (kcal mol^{-1})	References
Ile–Lys $(i, i+4)$	−0.22	(214)
Val–Lys $(i, i+4)$	−0.25	(214)
Ile–Arg $(i, i+4)$	−0.22	(214)
Phe–Met $(i, i+4)$	−0.8	(150)
Met–Phe $(i, i+4)$	−0.5	(150)
Gln–Asn $(i, i+4)$	−0.5	(209)
Asn–Gln $(i, i+4)$	−0.1	(209)
Phe–Lys $(i, i+4)$	−0.14	(102)
Lys–Phe $(i, i+4)$	−0.10	(102)
Phe–Arg $(i, i+4)$	−0.18	(102)
Phe–Orn $(i, i+4)$	−0.4	(213)
Arg–Phe $(i, i+4)$	−0.1	(102)
Tyr–Lys $(i, i+4)$	−0.22	(102)
Glu–Phe $(i, i+4)$	−0.5	(274)
Asp–Lys $(i, i+3)$	−0.12	(238)
Asp–Lys $(i, i+4)$	−0.24	(238)
Asp–His $(i, i+3)$	>−0.63	(275)
Asp–His $(i, i+4)$	>−0.63	(275)
Asp–Arg $(i, i+3)$	−0.8	(276)
Glu–His $(i, i+3)$	−0.23	(238)
Glu–His $(i, i+4)$	−0.10	(238)
Glu–Lys $(i, i+3)$	−0.38	(151)
Glu–Lys $(i, i+4)$	−0.44	(151)
Phe–His $(i, i+4)$	−1.27	(207)
Phe–Met $(i, i+4)$	−0.7	(212)
His–Asp $(i, i+3)$	−0.53	(207)
His–Asp $(i, i+4)$	−2.38	(262)
His–Glu $(i, i+3)$	−0.45	(238)
His–Glu $(i, i+4)$	−0.54	(238)
Lys–Asp $(i, i+3)$	−0.4	(238)
Lys–Asp $(i, i+4)$	−0.58	(238)
Lys–Glu $(i, i+3)$	−0.38	(238)
Lys–Glu $(i, i+4)$	−0.46	(238)
Lys–Lys $(i, i+4)$	+0.17	(205)

(Continues)

TABLE V (Continued)

Interaction	$\Delta\Delta G$ (kcal mol^{-1})	References
Leu–Tyr $(i, i + 3)$	−0.44	(152)
Leu–Tyr $(i, i + 4)$	−0.65	(152)
Met–Phe $(i, i + 4)$	−0.37	(212)
Gln–Asp $(i, i + 4)$	−0.97	(208)
Gln–Glu $(i, i + 4)$	−0.31	(151)
Trp–Arg $(i, i + 4)$	−0.4	(274)
Trp–His $(i, i + 4)$	−0.8	(159)
Tyr–Leu $(i, i + 3)$	−0.02	(152)
Tyr–Leu $(i, i + 4)$	−0.44	(152)
Tyr–Val $(i, i + 3)$	−0.13	(152)
Tyr–Val $(i, i + 4)$	−0.31	(152)
Arg $(i, i + 4)$ Glu $(i, i + 4)$ Arg	−1.5	(277)
Arg $(i, i + 3)$ Glu $(i, i + 3)$ Arg	−1.0	(277)
Arg $(i, i + 3)$ Glu $(i, i + 4)$ Arg	−0.3	(277)
Arg $(i, i + 4)$ Glu $(i, i + 3)$ Arg	−0.1	(277)
Phosphoserine–Arg $(i, i + 4)$	−0.45	(201)

staple" motif (231, 232). The nature of the capping box stabilizing effect thus not only arises from reciprocal hydrogen bonds between compatible residues, but also from local interactions between side chains, helix macrodipole-charged residue interactions and solvation (233).

Despite statistical analyses revealing that Schellman motifs are observed more frequently that expected at the helix C-terminus, this motif populates only transiently in aqueous solution but it is formed in 30% TFE (234). This might be due to the C-terminus being very frayed and the increase of helical content contributed from this motif is small. Energetically this motif is not very favorable due to the entropic cost of fixing a Gly residue at the position C′. The Schellman motif is believed to be a consequence of helix formation and does not involve α-helix nucleation (235). The α_L motif seems to be more stable than the alternative Schellman motif (232).

G. Ionic Strength

Electrostatic interactions between charged side chains and the helix macrodipole can stabilize the helix (88, 98, 236). The interactions are potentially quite strong, but are alleviated by the screening effects of water, ions, and nearby protein atoms. In theory, increasing ionic strength of the solvent (up to 1.0 M) should stabilize the helix through interactions with α-helix dipole moments by shifting the equilibrium between α-helix and random coil, which has a random orientation of the peptide dipoles (237). The energetic of the interaction between fully charged ion pairs can be diminished by added salt and completely screened at 2.5 M NaCl (206, 238). In peptides containing side-chain to side-chain interactions, the effect of ion pairs and charge-helix dipole interactions cannot be clearly separated. There are, however, indications that the interactions of charged residues with the helix macrodipole are less affected than those between charged side chains (238, 239). In coiled-coil peptides, salt also affect hydrophobic residues by strengthening their interactions at the coiled-coil interface. This can be explained through alterations of the peptide–water interactions at high salt concentration. However, this requires a strong kosmotropic anion to accompany the screening cation (240).

H. Enthalpy Change for the Helix–Coil Transition

Thermal unfolding experiments show that the helix unfolds with increasing temperature (241–243). There is no sign of cold denaturation, as seen with proteins. While most work on amino acid preferences for the helix have naturally concentrated on free energies, these can of course be divided into enthalpic and entropic terms, which can show the effects of temperature on helix formation. Early theoretical and experimental studies on long homopolymers gave estimates of the enthalpy for the helix–coil transition of approximately 1 kcal mol res.$^{-1}$ (244–246). Enthalpy and entropy changes for the helix–coil transition are difficult to determine, as the helix–coil transition is very broad, precluding accurate determination of high- and low-temperature baselines by calorimetry (241). Nevertheless, Scholtz et al. (241) were able to use differential scanning calorimetry on a 50 residue alanine-based helix to find the enthalpy change for the helix–coil transition to be between 0.8 and 1.3 kcal mol res.$^{-1}$. Thirteen peptides based on Ac–Y(XEARA)$_6$–NH$_2$, where X is various combinations of aliphatic side chains, gave values between 0.72 and 1.27 kcal mol res.$^{-1}$ (247, 248). These data interestingly showed that ΔC_p for helix formation is negative, in contrast to protein folding. This can be attributed to hydration of the polar backbone, which is dominant in helix formation, in contrast to protein folding, where hydration of polar groups makes a large, and opposite, contribution ΔC_p (249).

All of these studies have considerable inaccuracy due to the very broad nature of the thermal helix–coil transition. This can be neatly overcome by using metal binding to trigger helix formation and isothermal titration calorimetry in a peptide derived from the Ca-binding loop of calmodulin. Accurate measurements of the enthalpy change for the helix–coil transition using this system were found (250–253), and were in good agreement with previous results. Enthalpy values were found for 12 amino acids. The results can be rationalized by changes in backbone hydration (254). A combination of entropy and enthalpy changes agrees very well with experimental free energies (254).

I. Trifluoroethanol

Peptides with sequences of helices in proteins usually show low helix contents in water. An answer to this problem is to add 2,2,2-trifluoroethanol (TFE) to induce helix formation (255–258). For many peptides, the concentration of TFE used to increase the helix content is only up to 40% (255–257, 259, 260). TFE may act by shielding CO and NH groups from to the water solvent while leading to hydrogen bond formation between them. The conformational equilibrium thus shifts toward more compact structures, such as the α-helical conformation (185). The mechanism involves interaction between TFE and water with several interpretations. One view suggests that TFE indirectly disrupts the solvent shell on α-helices (261, 262). Another view proposes that TFE destabilizes the unfolded species and thereby indirectly enhances the kinetics and thermodynamics of folding of the coiled coil (263). A more compromising view suggests that TFE forms clusters in water solution, which at lower concentration pulls the water molecules from the surface of proteins. At higher concentration, TFE clusters associate with appropriate hydrophobic side chains reducing their conformational entropy and switch the conformation at TFE concentration >40% (264).

The propagation propensities of all amino acids increase variably in 40% TFE relative to water. The propagation propensities of the nonpolar amino acids increase greatly in 40% TFE whilst other amino acids propensity increase uniformly. However, glycine and proline are strong helix breakers in both in water and 40% TFE solvents (97). In addition, 40% TFE dramatically alters electrostatic (and polar) interactions and increases the dependence of helix propensities on the sequence (265).

J. pK_a Values

Evaluation of pK_a values of titrable amino acids in a peptide sequence can be used to analyze the strength of the possible interaction they form in water. pK_a shifts of charged residues at the helix termini are significant because they can potentially interact with unsatisfied hydrogen bonds of the NH groups and CO group at the N-terminus and C-terminus, respectively, or the helix dipole. The pK_a values can be measured accurately from the change in ellipticity across a broad range of pH. The asymptotic values of the ellipticities for the different protonation states are fitted to a Henderson–Hasselbach equation to calculate the pK_a.

In general, the pK_a values of Glu and His at N1–N3 is normal compared to those in model compounds. In contrast, Asp and Cys have shifted pK_a to lower values (*92, 93, 98, 151, 236, 266–269*). An exception for negatively charged residues at the N-cap is that they have a lower pK_a (*91*). This may be because side chains at the N-cap can form strong hydrogen bonds to NH groups of N2 and N3, while the bonds formed by side chains at N1, N2, and N3 are much weaker (*15, 20*). For both Asp and Glu, there are about 0.5 more hydrogen bonds per residue at the N-termini than in other protein regions, which helps to explain their lower pK_a values (*270*).

The negatively charged residues at higher pH destabilize helices when at the C-cap (*91*). The increased pK_a may result from an unfavorable electrostatic interaction with the C-terminal dipole or partial negative charges on the terminal CO groups.

K. Relevance to Proteins

Many of the features studied in peptide helices are also applicable to proteins and can be used to rationally modify protein stability or to design new helical proteins. Helices in proteins are often found on the surface with one face exposed to solvent and the other buried in the protein core. Helix propensities and side-chain interactions measured in peptides are thus directly applicable to the solvent-exposed face. Substitutions at buried positions are much more complex and tertiary interactions also make major contributions to stability. Tertiary interactions at helix termini are rare; nearly all side-chain interactions are local (*15*). Preferences for capping sites and the first and last turn of the helix are therefore applicable to most protein helices. The feature of protein helices of amphiphilicity, reflected in possession of a hydrophobic moment (*271*), is irrelevant to monomeric isolated helices.

Acknowledgments

We thank all our coworkers in this field, namely Avi Chakrabartty, Carol Rohl, Buzz Baldwin, Tod Klingler, Ben Stapley, Jim Andrew, Eleri Hughes, Simon Penel, Duncan Cochran, Nicoleta Kokkoni, Jia Ke Sun, Jim Warwicker, Gareth Jones, Jonathan Hirst, Neil Errington, and Teuku Iqbalsyah.

References

1. Pauling, L., Corey, R. B., and Branson, H. R. (1951). The structure of proteins: Two hydrogen-bonded helical configurations of the polypeptide chain. *Proc. Natl Acad. Sci. USA* **37,** 205–211.
2. Perutz, M. F. (1951). New X-ray evidence on the configuration of polypeptide chains. *Nature* **167,** 1053–1054.

3. Kendrew, J. C., Dickerson, R. E., Strandberg, B. E., Hart, R. G., Davies, D. R., Phillips, D. C., and Shore, V. C. (1960). Structure of myoglobin. *Nature* **185,** 422–427.
4. Barlow, D. J., and Thornton, J. M. (1988). Helix geometry in proteins. *J. Mol. Biol.* **201,** 601–619.
5. Scholtz, J. M., and Baldwin, R. L. (1992). The mechanism of α-helix formation by peptides. *Annu. Rev. Biophys. Biomol. Struct.* **21,** 95–118.
6. Baldwin, R. L. (1995). α-Helix formation by peptides of defined sequence. *Biophys. Chem.* **55,** 127–135.
7. Chakrabartty, A., and Baldwin, R. L. (1995). Stability of α-helices. *Adv. Protein Chem.* **46,** 141–176.
8. Kallenbach, N. R., Lyu, P., and Zhou, H. (1996). CD spectroscopy and the helix–coil transition in peptides and polypeptides. *In* "Circular Dichroism and the Conformational Analysis of Biomolecules" (G. D. Fasman, Ed.), pp. 201–259. Plenum Press, New York.
9. Rohl, C. A., and Baldwin, R. L. (1998). Deciphering rules of helix stability in peptides. *Meth. Enzymol.* **295,** 1–26.
10. Andrews, M. J. I., and Tabor, A. B. (1999). Forming stable helical peptides using natural and artificial amino acids. *Tetrahedron* **55,** 11711–11743.
11. Serrano, L. (2000). The relationship between sequence and structure in elementary folding units. *Adv. Protein Chem.* **53,** 49–85.
12. Richardson, J. S., and Richardson, D. C. (1988). Amino acid preferences for specific locations at the ends of α helices. *Science* **240,** 1648–1652.
13. Fonseca, N. A., Camacho, R., and Magalhaes, A. L. (2007). Amino acid pairing at the N- and C-termini of helical segments in proteins. *Proteins Struct. Funct. Bioinf.* **70,** 188–196.
14. Presta, L. G., and Rose, G. D. (1988). Helix signals in proteins. *Science* **240,** 1632–1641.
15. Doig, A. J., MacArthur, M. W., Stapley, B. J., and Thornton, J. M. (1997). Structures of N-termini of helices in proteins. *Protein Sci.* **6,** 147–155.
16. MacGregor, M. J., Islam, S. A., and Sternberg, M. J. E. (1987). Analysis of the relationship between side-chain conformation and secondary structure in globular proteins. *J. Mol. Biol.* **198,** 295–310.
17. Swindells, M. B., MacArthur, M. W., and Thornton, J. M. (1995). Intrinsic ϕ, ψ propensities of amino acids derived from the coil regions of known structures. *Nat. Struct. Biol.* **2,** 596–603.
18. Dunbrack, R., and Karplus, M. (1994). Conformational analysis of the backbone-dependent rotamer preferences of protein side-chains. *Nat. Struct. Biol.* **1,** 334–339.
19. Dunbrack, R. L. (2002). Rotamer libraries in the 21(st) century. *Curr. Opin. Struct. Biol.* **12,** 431–440.
20. Penel, S., Hughes, E., and Doig, A. J. (1999). Side-chain structures in the first turn of the α-helix. *J. Mol. Biol.* **287,** 127–143.
21. Aurora, R., Srinivasan, R., and Rose, G. D. (1994). Rules for α-helix termination by glycine. *Science* **264,** 1126–1130.
22. Serrano, L., and Fersht, A. R. (1989). Capping and alpha-helix stability. *Nature* **342,** 296–299.
23. Bell, J. A., Becktel, W. J., Sauer, U., Baase, W. A., and Matthews, B. W. (1992). Dissection of helix capping in T4 lysozyme by structural and thermodynamic analysis of six amino acid substitutions at Thr 59. *Biochemistry* **31,** 3590–3596.
24. Chakrabartty, A., Doig, A. J., and Baldwin, R. L. (1993). Helix capping propensities in peptides parallel those in proteins. *Proc. Natl Acad. Sci. USA* **90,** 11332–11336.
25. Forood, B., Feliciano, E. J., and Nambiar, K. P. (1993). Stabilization of alpha-helical structures in short peptides via end capping. *Proc. Natl Acad. Sci. USA* **90,** 838–842.
26. Dasgupta, S., and Bell, J. A. (1993). Design of helix ends. Amino acid preferences, hydrogen bonding and electrostatic interactions. *Int. J. Pept. Protein Res.* **41,** 499–511.

27. Harper, E. T., and Rose, G. D. (1993). Helix stop signals in proteins and peptides: The capping box. *Biochemistry* **32,** 7605–7609.
28. Seale, J. W., Srinivasan, R., and Rose, G. D. (1994). Sequence determinants of the capping box, a stabilizing motif at the N-termini of α-helices. *Protein Sci.* **3,** 1741–1745.
29. Muñoz, V., and Serrano, L. (1995). The hydrophobic-staple motif and a role for loop residues in α-helix stability and protein folding. *Nat. Struct. Biol.* **2,** 380–385.
30. Viguera, A. R., and Serrano, L. (1999). Stable proline box motif at the N-terminal end of α-helices. *Protein Sci.* **8,** 1733–1742.
31. Schellman, C. (1980). The alphaL conformation at the ends of helices. *Protein Folding* 53–61.
32. Aurora, R., and Rose, G. D. (1998). Helix capping. *Protein Sci.* **7,** 21–38.
33. Jernigan, R., Raghunathan, G., and Bahar, I. (1994). Characterisation of interactions and metal-ion binding-sites in proteins. *Curr. Opin. Struct. Biol.* **4,** 256–263.
34. Alberts, I. L., Nadassy, K., and Wodak, S. J. (1998). Analysis of zinc binding sites in protein crystal structures. *Protein Sci.* **7,** 1700–1716.
35. Ghadiri, M. R., and Choi, C. (1990). Secondary structure nucleation in peptides—Transition-metal ion stabilized α-helices. *J. Am. Chem. Soc.* **112,** 1630–1632.
36. Kise, K. J., and Bowler, B. E. (2002). Induction of helical structure in a heptapeptide with a metal cross-link: Modification of the Lifson–Roig Helix–coil theory to account for covalent cross-links. *Biochemistry* **41,** 15826–15837.
37. Ruan, F., Chen, Y., and Hopkins, P. B. (1990). Metal ion enhanced helicity in synthetic peptides containing unnatural, metal-ligating residues. *J. Am. Chem. Soc.* **112,** 9403–9404.
38. Cline, D. J., Thorpe, C., and Schneider, J. P. (2003). Effects of As(III) binding on α-helical structure. *J. Am. Chem. Soc.* **125,** 2923–2929.
39. Matzapetakis, M., and Pecoraro, V. L. (2005). Site-selective metal binding by designed α-helical peptides. *J. Am. Chem. Soc.* **127,** 18229–18233.
40. Karpen, M. E., De Haset, P. L., and Neet, K. E. (1992). Differences in the amino acid distributions of 3_{10}-helices and α-helices. *Protein Sci.* **1,** 1333–1342.
41. Baker, E. N., and Hubbard, R. E. (1984). Hydrogen bonding in globular proteins. *Prog. Biophys. Mol. Biol.* **44,** 97–179.
42. Némethy, G., Phillips, D. C., Leach, S. J., and Scheraga, H. A. (1967). A second right-handed helical structure with the parameters of the Pauling–Corey α-helix. *Nature* **214,** 363–365.
43. Millhauser, G. L. (1995). Views of helical peptides—A proposal for the position of 3_{10}-helix along the thermodynamic folding pathway. *Biochemistry* **34,** 3872–3877.
44. Bolin, K. A., and Millhauser, G. L. (1999). α and 3_{10}: The split personality of polypeptide helices. *Acc. Chem. Res.* **32,** 1027–1033.
45. Toniolo, C., and Benedetti, E. (1991). The polypeptide 3_{10}-helix. *Trends Biochem. Sci.* **16,** 350–353.
46. Ramachandran, G. N., and Sasisekharan, V. (1968). Conformation of polypeptides and proteins. *Adv. Protein Chem.* **23,** 283–437.
47. Low, B. W., and Grenville-Wells, H. J. (1953). Generalized mathematical relations for polypeptide chain helixes. The coordinates for the π helix. *Proc. Natl Acad. Sci. USA* **39,** 785–801.
48. Rohl, C. A., and Doig, A. J. (1996). Models for the 3_{10}-helix/coil, π-helix/coil, and α-helix/3_{10}-helix/coil transitions in isolated peptides. *Protein Sci.* **5,** 1687–1696.
49. Shirley, W. A., and Brooks, C. L. (1997). Curious structure in "canonical" alanine based peptides. *Proteins Struct. Funct. Genet.* **28,** 59–71.
50. Lee, K. H., Benson, D. R., and Kuczera, K. (2000). Transitions from α to π helix observed in molecular dynamics simulations of synthetic peptides. *Biochemistry* **39,** 13737–13747.
51. Morgan, D. M., Lynn, D. G., Miller-Auer, H., and Meredith, S. C. (2001). A designed Zn2+-binding amphiphilic polypeptide: Energetic consequences of π-helicity. *Biochemistry* **40,** 14020–14029.

52. Fodje, M. N., and Al-Karadaghi, S. (2002). Occurrence, conformational features and amino acid propensities for the π-helix. *Protein Eng.* **15,** 353–358.
53. Weaver, T. M. (2000). The π-helix translates structure into function. *Protein Sci.* **9,** 201–206.
54. Baldwin, R. L. (2003). In search of the energetic role of peptide hydrogen bonds. *J. Biol. Chem.* **278,** 17581–17588.
55. Epand, R. M., and Scheraga, H. A. (1968). The influence of long-range interactions on the structure of myoglobin. *Biochemistry* **7,** 2864–2872.
56. Taniuchi, J. H., and Anfinsen, C. B. (1969). An experimental approach to the study of the folding of staphylococcal nuclease. *J. Biol. Chem.* **244,** 3864–3875.
57. Spek, E. J., Gong, Y., and Kallenbach, N. R. (1995). Intermolecular interactions in a helical oligo- and poly(L-glutamic acid) at acidic pH. *J. Am. Chem. Soc.* **117,** 10773–10774.
58. Brown, J. E., and Klee, W. A. (1971). Helix–coil transition of the isolated amino terminus of ribonuclease. *Biochemistry* **10,** 470–476.
59. Bierzynski, A., Kim, P. S., and Baldwin, R. L. (1982). A salt bridge stabilizes the helix formed by isolated C-peptide of Ribonuclease A. *Proc. Natl Acad. Sci. USA* **79,** 2470–2474.
60. Kim, P. S., Bierzynski, A., and Baldwin, R. L. (1982). A competing salt-bridge suppresses helix formation by the isolated C-peptide carboxylate of Ribonuclease A. *J. Mol. Biol.* **162,** 187–199.
61. Kim, P. S., and Baldwin, R. L. (1984). A helix stop signal in the isolated S-peptide of Ribonuclease A. *Nature* **307,** 329–334.
62. Shoemaker, K. R., Kim, P. S., Brems, D. N., Marqusee, S., York, E. J., Chaiken, I. M., Stewart, J. M., and Baldwin, R. L. (1985). Nature of the charged-group effect on the stability of the C-peptide helix. *Proc. Natl Acad. Sci. USA* **82,** 2349–2353.
63. Strehlow, K. G., and Baldwin, R. L. (1989). Effect of the substitution Ala–Gly at each of 5 residue positions in the C-peptide helix. *Biochemistry* **28,** 2130–2133.
64. Osterhout, J. J., Baldwin, R. L., York, E. J., Stewart, J. M., Dyson, H. J., and Wright, P. E. (1989). H-1 NMR studies of the solution conformations of an analog of the C-peptide of Ribonuclease-A. *Biochemistry* **29,** 7059–7064.
65. Shoemaker, K. R., Fairman, R., Schultz, D. A., Robertson, A. D., York, E. J., Stewart, J. M., and Baldwin, R. L. (1990). Side-chain interactions in the C-peptide helix: Phe8–His12$^+$. *Biopolymers* **29,** 1–11.
66. Fairman, R., Shoemaker, K. R., York, E. J., Stewart, J. M., and Baldwin, R. L. (1990). The Glu2–Arg10$^+$ side-chain interaction in the C-peptide helix of Ribonuclease A. *Biophys. Chem.* **37,** 107–119.
67. Strehlow, K. G., Robertson, A. D., and Baldwin, R. L. (1991). Proline for alanine substitutions in the C-peptide helix of Ribonuclease A. *Biochemistry* **30,** 5810–5814.
68. Fairman, R., Armstrong, K. M., Shoemaker, K. R., York, E. J., Stewart, J. M., and Baldwin, R. L. (1991). Position effect on apparent helical propensities in the C-peptide helix. *J. Mol. Biol.* **221,** 1395–1401.
69. Blanco, F. J., Jimenez, M. A., Rico, M., Santoro, J., Herranz, J., and Nieto, J. L. (1992). The homologous Angiogenin and Ribonuclease N-terminal fragments fold into very similar helices when isolated. *Biochem. Biophys. Res. Commun.* **182,** 1491–1498.
70. Rico, M., Nieto, J. L., Santoro, J., Bermejo, F. J., and Herranz, J. (1983). H1-NMR parameters of the N-terminal 13-residue C-peptide of ribonuclease in aqueous-solution. *Org. Magn. Res.* **21,** 555–563.
71. Gallego, E., Herranz, J., Nieto, J. L., Rico, M., and Santoro, J. (1983). H1-NMR parameters of the N-terminal 19-residue S-peptide of Ribonuclease in aqueous-solution. *Int. J. Pept. Protein Res.* **21,** 242–253.

72. Rico, M., Nieto, J. L., Santoro, J., Bermejo, F. J., Herranz, J., and Gallego, E. (1983). Low temperature H1-NMR evidence of the folding of isolated Ribonuclease S-peptide. *FEBS Lett.* **162,** 314–319.
73. Rico, M., Gallego, E., Santoro, J., Bermejo, F. J., Nieto, J. L., and Herranz, J. (1984). On the fundamental role of the Glu2$^-$... Arg10$^+$ salt bridge in the folding of isolated Ribonuclease A S-peptide. *Biochem. Biophys. Res. Commun.* **123,** 757–763.
74. Nieto, J. L., Rico, M., Jimenez, M. A., Herranz, J., and Santoro, J. (1985). Amide H1-NMR study of the folding of Ribonuclease C-peptide. *Int. J. Biol. Macromol.* **7,** 66–70.
75. Nieto, J. L., Rico, M., Santoro, J., and Bermejo, J. (1985). NH resonances of Ribonuclease S-peptide in aqueous solution—Low temperature NMR study. *Int. J. Pept. Protein Res.* **25,** 47–55.
76. Santoro, J., Rico, M., Nieto, J. L., Bermejo, J., Herranz, J., and Gallego, E. (1986). C13 NMR spectral assignment of Ribonuclease S-peptide—Some new structural information about its low temperature folding. *J. Mol. Struct.* **141,** 243–248.
77. Rico, M., Herranz, J., Bermejo, J., Nieto, J. L., Santoro, J., Gallego, E., and Jimenez, M. A. (1986). Quantitative interpretation of the helix–coil transition in RNase A S-peptide. *J. Mol. Struct.* **143,** 439–444.
78. Rico, M., Santoro, J., Bermejo, J., Herranz, J., Nieto, J. L., Gallego, E., and Jimenez, M. A. (1986). Thermodynamic parameters for the helix–coil thermal transition of Ribonuclease S-peptide and derivatives from H1 NMR data. *Biopolymers* **25,** 1031–1053.
79. Marqusee, S., and Baldwin, R. L. (1987). Helix stabilization by Glu$^-$... Lys$^+$ salt bridges in short peptides of *de novo* design. *Proc. Natl Acad. Sci. USA* **84,** 8898–8902.
80. Park, S. H., Shalongo, W., and Stellwagen, E. (1993). Residue helix parameters obtained from dichroic analysis of peptides of defined sequence. *Biochemistry* **32,** 7048–7053.
81. Marqusee, S., Robbins, V. H., and Baldwin, R. L. (1989). Unusually stable helix formation in short alanine based peptides. *Proc. Natl Acad. Sci. USA* **86,** 5286–5290.
82. Lyu, P. C., Marky, L. A., and Kallenbach, N. R. (1989). The role of ion-pairs in α-helix stability—2 new designed helical peptides. *J. Am. Chem. Soc.* **111,** 2733–2734.
83. Lyu, P. C., Liff, M. I., Marky, L. A., and Kallenbach, N. R. (1990). Side-chain contributions to the stability of α-helical structure in peptides. *Science* **250,** 669–673.
84. Gans, P. J., Lyu, P. C., Manning, P. C., Woody, R. W., and Kallenbach, N. R. (1991). The helix–coil transition in heterogeneous peptides with specific side chain interactions: Theory and comparison with circular dichroism. *Biopolymers* **31,** 1605–1614.
85. Penel, S., Morrison, R. G., Mortishire-Smith, R. J., and Doig, A. J. (1999). Periodicity in α-helix lengths and C-capping preferences. *J. Mol. Biol.* **293,** 1211–1219.
86. Wada, A. (1976). The α-helix as an electric macro-dipole. *Adv. Biophys.* **9,** 1–63.
87. Hol, W. G. J., van Duijnen, P. T., and Berendsen, H. J. C. (1978). The α-helix dipole and the properties of proteins. *Nature* **273,** 443–446.
88. Aqvist, J., Luecke, H., Quiocho, F. A., and Warshel, A. (1991). Dipoles located at helix termini of proteins stabilize charges. *Proc. Natl Acad. Sci. USA* **88,** 2026–2030.
89. Zhukovsky, E. A., Mulkerrin, M. G., and Presta, L. G. (1994). Contribution to global protein stabilization of the N-capping box in human growth hormone. *Biochemistry* **33,** 9856–9864.
90. Tidor, B. (1994). Helix-capping interaction in λ Cro protein: A free energy simulation analysis. *Proteins Struct. Funct. Genet.* **19,** 310–323.
91. Doig, A. J., and Baldwin, R. L. (1995). N- and C-capping preferences for all 20 amino acids in α-helical peptides. *Protein Sci.* **4,** 1325–1336.
92. Cochran, D. A. E., Penel, S., and Doig, A. J. (2001). Contribution of the N1 amino acid residue to the stability of the α-helix. *Protein Sci.* **10,** 463–470.
93. Cochran, D. A. E., and Doig, A. J. (2001). Effects of the N2 residue on the stability of the α-helix for all 20 amino acids. *Protein Sci.* **10,** 1305–1311.

94. Iqbalsyah, T. M., and Doig, A. J. (2004). Effect of the N3 residue on the stability of the α-helix. *Protein Sci.* **13**, 32–39.
95. Decatur, S. M. (2000). IR spectroscopy of isotope-labeled helical peptides: Probing the effect of N-acetylation on helix stability. *Biopolymers* **54**, 180–185.
96. Doig, A. J., Chakrabartty, A., Klingler, T. M., and Baldwin, R. L. (1994). Determination of free energies of N-capping in α-helices by modification of the Lifson–Roig helix–coil theory to include N- and C-capping. *Biochemistry* **33**, 3396–3403.
97. Rohl, C. A., Chakrabartty, A., and Baldwin, R. L. (1996). Helix propagation and N-cap propensities of the amino acids measured in alanine-based peptides in 40 volume percent trifluoroethanol. *Protein Sci.* **5**, 2623–2637.
98. Huyghues-Despointes, B. M., Scholtz, J. M., and Baldwin, R. L. (1993). Effect of a single aspartate on helix stability at different positions in a neutral alanine-based peptide. *Protein Sci.* **2**, 1604–1611.
99. Brandts, J. R., and Kaplan, K. J. (1973). Derivative spectroscopy applied to tyrosyl chromophores. Studies on ribonuclease, lima bean inhibitor, and pancreatic trypsin inhibitor. *Biochemistry* **10**, 470–476.
100. Edelhoch, H. (1967). Spectroscopic determination of tryptophan and tyrosine in proteins. *Biochemistry* **6**, 1948–1954.
101. Chakrabartty, A., Kortemme, T., Padmanabhan, S., and Baldwin, R. L. (1993). Aromatic side-chain contribution to far-ultraviolet circular dichroism of helical peptides and its effect on measurement of helix propensities. *Biochemistry* **32**, 5560–5565.
102. Andrew, C. D., Bhattacharjee, S., Kokkoni, N., Hirst, J. D., Jones, G. R., and Doig, A. J. (2002). Stabilizing interactions between aromatic and basic side chains in α-helical peptides and proteins. Tyrosine effects on helix circular dichroism. *J. Am. Chem. Soc.* **124**, 12706–12714.
103. Kemp, D. S., Boyd, J. G., and Muendel, C. C. (1991). The helical s-constant for alanine in water derived from template-nucleated helices. *Nature* **352**, 451–454.
104. Kemp, D. S., Allen, T. J., and Oslick, S. L. (1995). The energetics of helix formation by short templated peptides in aqueous solution. 1. Characterization of the reporting helical template Ac–HEl(1). *J. Am. Chem. Soc.* **117**, 6641–6657.
105. Groebke, K., Renold, P., Tsang, K. Y., Allen, T. J., McClure, K. F., and Kemp, D. S. (1996). Template-nucleated alanine–lysine helices are stabilized by position-dependent interactions between the lysine side chain and the helix barrel. *Proc. Natl Acad. Sci. USA* **93**, 4025–4029.
106. Kemp, D. S., Oslick, S. L., and Allen, T. J. (1996). The structure and energetics of helix formation by short templated peptides in aqueous solution. 3. Calculation of the helical propagation constant s from the template stability constants t/c for Ac–Hel(1)–Ala(n)–OH, n = 1–6. *J. Am. Chem. Soc.* **118**, 4249–4255.
107. Kemp, D. S., Allen, T. J., Oslick, S. L., and Boyd, J. G. (1996). The structure and energetics of helix formation by short templated peptides in aqueous solution. 2. Characterization of the helical structure of Ac–Hel(1)–Ala(6)–OH. *J. Am. Chem. Soc.* **118**, 4240–4248.
108. Kennedy, R. J., Walker, S. M., and Kemp, D. S. (2005). Energetic characterization of short helical polyalanine peptides in water: Analysis of 13C = O chemical shift data. *J. Am. Chem. Soc.* **127**, 16961–16968.
109. Austin, R. E., Maplestone, R. A., Sefler, A. M., Liu, K., Hruzewicz, W. N., Liu, C. W., Cho, D. E., Wemmer, D. E., and Bartlett, P. A. (1997). Template for stabilization of a peptide alpha-helix: Synthesis and evaluation of conformational effects by circular dichroism and NMR. *J. Am. Chem. Soc.* **119**, 6461–6472.
110. Mimna, R., Tuchscherer, G., and Mutter, M. (2007). Toward the design of highly efficient, readily accessible peptide N-caps for the induction of helical conformations. *Int. J. Pept. Res. Ther.* **13**, 237–244.

111. Arrhenius, T., and Sattherthwait, A. C. (1989). "Peptides: Chemistry, Structure and Biology." Proceedings of the 11th American Peptide Symposium. Escom, Leiden, Netherlands.
112. Muller, K., Obrecht, D., Knierzinger, A., Stankovic, C., Spiegler, C., Bannwarth, W., Trzeciak, G., Englert, G., Labhardt, A. M., and Schonholzer, P. (1993). "Perspectives in Medicinal Chemistry."pp. 513–531. Helvetica Chimica Acta & VCH, Basel.
113. Gani, D., Lewis, A., Rutherford, T., Wilkie, J., Stirling, I., Jenn, T., and Ryan, M. D. (1998). Design, synthesis, structure and properties of an alpha-helix cap template derived from N-[(2S)-2-chloropropionyl]-(2S)-Pro-(2R)-Ala-(2S,4S)-4-thioPro-OMe which initiates alpha-helical structures. *Tetrahedron* **54,** 15793–15819.
114. Aleman, C. (1997). Conformational properties of α-amino acids disubstituted at the α-carbon. *J. Phys. Chem.* **101,** 5046–5050.
115. Sacca, B., Formaggio, F., Crisma, M., Toniolo, C., and Gennaro, R. (1997). Linear oligopeptides. 401. In search of a peptide 3_{10}-helix in water. *Gazz. Chim. Ital.* **127,** 495–500.
116. Tanaka, M. (2002). Design and conformation of peptides containing α,α-disubstituted alpha-amino acids. *J. Synth. Org. Chem. Jpn.* **60,** 125–136.
117. Lancelot, N., Elbayed, K., Raya, J., Piotto, M., Briand, J. P., Formaggio, F., Toniolo, C., and Bianco, A. (2003). Characterization of the 3_{10}-helix in model peptides by HRMAS NMR spectroscopy. *Chem. Eur. J.* **9,** 1317–1323.
118. Karle, I. L., and Balaram, P. (1990). Structural characteristics of α-helical peptide molecules containing Aib residues. *Biochemistry* **29,** 6747–6756.
119. Banerjee, R., and Basu, G. (2002). A short Aib/Ala-based peptide helix is as stable as an Ala-based peptide helix double its length. *ChemBioChem* **3,** 1263–1266.
120. Yokum, T. S., Gauthier, T. J., Hammer, R. P., and McLaughlin, M. L. (1997). Solvent effects on the 3_{10}-/α-helix equilibrium in short amphipathic peptides rich in α,α-disubstituted amino acids. *J. Am. Chem. Soc.* **119,** 1167–1168.
121. Miick, S. M., Martinez, G. V., Fiori, W. R., Todd, A. P., and Millhauser, G. L. (1992). Short alanine-based peptides may form 3_{10}-helices and not α-helices in aqueous solution. *Nature* **359,** 653–655.
122. Millhauser, G. L., Stenland, C. J., Hanson, P., Bolin, K. A., and van de Ven, F. J. M. (1997). Estimating the relative populations of 3_{10}-helix and α-helix in Ala-rich peptides. A hydrogen exchange and high field NMR study. *J. Mol. Biol.* **267,** 963–974.
123. Fiori, W. R., Lundberg, K. M., and Millhauser, G. L. (1994). A single carboxy-terminal arginine determines the amino-terminal helix conformation of an alanine-based peptide. *Nat. Struct. Biol.* **1,** 374–377.
124. Fiori, W. R., and Millhauser, G. L. (1995). Exploring the peptide 3_{10}-helix/α-helix equilibrium with double label electron spin resonance. *Biopolymers* **37,** 243–250.
125. Yoder, G., Polese, A., Silva, R. A. G. D., Formaggio, F., Crisma, M., Broxterman, Q. B., Kamphuis, C., Toniolo, C., and Keiderling, T. A. (1997). Conformation characterization of terminally blocked L-(αMe)Val homopeptides using vibrational and electronic circular dichroism. 3_{10}-Helical stabilization by peptide-peptide interaction. *J. Am. Chem. Soc.* **119,** 10278–10285.
126. Kennedy, D. F., Crisma, M., Toniolo, C., and Chapman, D. (1991). Studies of peptides forming 3_{10}- and α-helices and β-bend ribbon structures in organic solution and in model biomembranes by Fourier transform infrared spectroscopy. *Biochemistry* **30,** 6541–6548.
127. Hungerford, G., Martinez-Insua, M., Birch, D. J. S., and Moore, B. D. (1996). A reversible transition between an α-helix and a 3_{10}-helix in a fluorescence-labelled peptide. *Angew. Chem. Int. Ed. Engl.* **35,** 326–329.
128. Sheinerman, F. B., and Brooks, C. L. (1995). 3_{10}-Helices in peptides and proteins as studied by modified Zimm–Bragg theory. *J. Am. Chem. Soc.* **117,** 10098–10103.

129. Poland, D., and Scheraga, H. A. (1970). "Theory of Helix–Coil Transitions in Biopolymers." Academic Press, New York.
130. Qian, H., and Schellman, J. A. (1992). Helix/coil theories: A comparative study for finite length polypeptides. *J. Phys. Chem.* **96,** 3987–3994.
131. Doig, A. J. (2002). Recent advances in helix–coil theory. *Biophys. Chem.* **101–102,** 281–293.
132. Zimm, B. H., and Bragg, J. K. (1959). Theory of the phase transition between helix and random coil in polypeptide chains. *J. Chem. Phys.* **31,** 526–535.
133. Lifson, S., and Roig, A. (1961). On the theory of the helix–coil transition in polypeptides. *J. Chem. Phys.* **34,** 1963–1974.
134. Flory, P. J. (1969). "Statistical Mechanics of Chain Molecules." Wiley, New York.
135. Pappu, R. V., Srinivasan, R., and Rose, G. D. (2000). The Flory isolated-pair hypothesis is not valid for polypeptide chains: Implications for protein folding. *Proc. Natl Acad. Sci. USA* **97,** 12565–12570.
136. Tiffany, M. L., and Krimm, S. (1968). New chain conformations of poly(glutamic acid) and polylysine. *Biopolymers* **6,** 1379–1382.
137. Krimm, S., and Tiffany, M. L. (1974). The circular dichroism spectrum and structure of unordered polypeptides and proteins. *Israel J. Chem.* **12,** 189–200.
138. Woody, R. W. (1992). Circular dichroism and conformations of unordered polypeptides. *Adv. Biophys. Chem.* **2,** 37–79.
139. Barron, L. D., Hecht, L., Blanch, E. W., and Bell, A. F. (2000). Solution structure and dynamics of biomolecules from Raman optical activity. *Prog. Biophys. Mol. Biol.* **73,** 1–49.
140. Blanch, E. W., Morozova-Roche, L. A., Cochran, D. A. E., Doig, A. J., Hecht, L., and Barron, L. D. (2000). Is polyproline II helix the killer conformation? A Raman optical activity study of the amyloidogenic prefibrillar intermediate of human lysozyme *J. Mol. Biol.* **301,** 553–563.
141. Smyth, E., Syme, C. D., Blanch, E. W., Hecht, L., Vasak, M., and Barron, L. D. (2000). Solution structure of native proteins irregular folds from Raman optical activity. *Biopolymers* **58,** 138–151.
142. Syme, C. D., Blanch, E. W., Holt, C., Jakes, R., Goedert, M., Hecht, L., and Barron, L. D. (2002). A Raman optical activity study of the rheomorphism in caseins, synucleins and tau. *Eur. J. Biochem.* **269,** 148–156.
143. Rucker, A. L., and Creamer, T. P. (2002). Polyproline II helical structure in protein unfolded states: Lysine peptides revisited. *Protein Sci.* **11,** 980–985.
144. Shi, Z. S., Woody, R. W., and Kallenbach, N. R. (2002). Is polyproline II a major backbone conformation in unfolded proteins? *Adv. Protein Chem.* **62,** 163–240.
145. Pappu, R. V., and Rose, G. D. (2002). A simple model for polyproline II structure in unfolded states of alanine-based peptides. *Protein Sci.* **11,** 2437–2255.
146. Shi, Z., Olson, C. A., Rose, G. D., Baldwin, R. L., and Kallenbach, N. R. (2002). Polyproline II structure in a sequence of seven alanine residues. *Proc. Natl Acad. Sci. USA* **99,** 9190–9195.
147. Firestine, A. M., Chellgren, V. M., Rucker, S. J., Lester, T. E., and Creamer, T. P. (2008). Conformational properties of a peptide model for unfolded α-helices. *Biochemistry* **47,** 3216–3224.
148. Rucker, A. L., Pager, C. T., Campbell, M. N., Qualis, J. E., and Creamer, T. P. (2003). Host–guest scale of left-handed polyproline II helix formation. *Proteins Struct. Funct. Genet.* **53,** 68–75.
149. Andersen, N. H., and Tong, H. (1997). Empirical parameterization of a model for predicting peptide helix/coil equilibrium populations. *Protein Sci.* **6,** 1920–1936.
150. Stapley, B. J., Rohl, C. A., and Doig, A. J. (1995). Addition of side chain interactions to modified Lifson–Roig helix–coil theory: Application to energetics of phenylalanine–methionine interactions. *Protein Sci.* **4,** 2383–2391.

151. Scholtz, J. M., Qian, H., Robbins, V. H., and Baldwin, R. L. (1993). The energetics of ion-pair and hydrogen-bonding interactions in a helical peptide. *Biochemistry* **32**, 9668–9676.
152. Shalongo, W., and Stellwagen, E. (1995). Incorporation of pairwise interactions into the Lifson–Roig model for helix prediction. *Protein Sci.* **4**, 1161–1166.
153. Argos, P., and Palau, J. (1982). Amino acid distribution in protein secondary structures. *Int. J. Pept. Protein Res.* **19**, 380–393.
154. Kumar, S., and Bansal, M. (1998). Dissecting α-helices: Position-specific analysis of α-helices in globular proteins. *Proteins* **31**, 460–476.
155. Sun, J. K., Penel, S., and Doig, A. J. (2000). Determination of alpha-helix N1 energies by addition of N1, N2 and N3 preferences to helix–coil theory. *Protein Sci.* **9**, 750–754.
156. Sun, J. K., and Doig, A. J. (1998). Addition of side-chain interactions to 3_{10}-helix/coil and alpha-helix/3_{10}-helix/coil theory. *Protein Sci.* **7**, 2374–2383.
157. Muñoz, V., and Serrano, L. (1994). Elucidating the folding problem of helical peptides using empirical parameters. *Nat. Struct. Biol.* **1**, 399–409.
158. Muñoz, V., and Serrano, L. (1997). Development of the multiple sequence approximation within the AGADIR model of α-helix formation: Comparison with Zimm–Bragg and Lifson–Roig formalisms. *Biopolymers* **41**, 495–509.
159. Fernández-Recio, J., Vásquez, A., Civera, C., Sevilla, P., and Sancho, J. (1997). The tryptophan/histidine interaction in α-helices. *J. Mol. Biol.* **267**, 184–197.
160. Lacroix, E., Viguera, A. R., and Serrano, L. (1998). Elucidating the folding problem of α-helices: Local motifs, long-range electrostatics, ionic-strength dependence and prediction of NMR parameters. *J. Mol. Biol.* **284**, 173–191.
161. Muñoz, V., and Serrano, L. (1995). Elucidating the folding problem of helical peptides using empirical parameters. II. Helix macrodipole effects and rational modification of the helical content of natural peptides. *J. Mol. Biol.* **245**, 275–296.
162. Petukhov, M., Muñoz, V., Yumoto, N., Yoshikawa, S., and Serrano, L. (1998). Position dependence of non-polar amino acid intrinsic helical propensities. *J. Mol. Biol.* **278**, 279–289.
163. Lomize, A. L., and Mosberg, H. I. (1997). Thermodynamic model of secondary structure for α-helical peptides and proteins. *Biopolymers* **42**, 239–269.
164. Vásquez, M., and Scheraga, H. A. (1988). Effect of sequence-specific interactions on the stability of helical conformations in polypeptides. *Biopolymers* **27**, 41–58.
165. Roberts, C. H. (1990). A hierarchical nesting approach to describe the stability of α helices with side chain interactions. *Biopolymers* **30**, 335–347.
166. Qian, H. (1994). A thermodynamic model for the helix–coil transition coupled to dimerization of short coiled-coil peptides. *Biophys. J.* **67**, 349–355.
167. Hausrath, A. C. (2006). A model for the coupling of α-helix and tertiary contact formation. *Protein Sci.* **15**, 2051–2061.
168. Xian, W., Connolly, P. J., Oslin, M., Hausrath, A. C., and Osterhout, J. J. (2006). Fundamental processes of protein folding: Measuring the energetic balance between helix formation and hydrophobic interactions. *Protein Sci.* **15**, 2062–2070.
169. Guzzo, A. V. (1965). The influence of amino acid sequence on protein structure. *Biophys. J.* **5**, 809–822.
170. Davies, D. R. (1964). A correlation between amino acid composition and protein structure. *J. Mol. Biol.* **9**, 605–609.
171. Chou, P. Y., and Fasman, G. D. (1974). Conformational parameters for amino acids in helical, b-sheet and random coil regions calculated from proteins. *Biochemistry* **13**, 211–221.
172. Chou, P. Y., and Fasman, G. D. (1978). Empirical predictions of protein conformation. *Annu. Rev. Biochem.* **47**, 251–276.

173. Von Dreele, P. H., Poland, D., and Scheraga, H. A. (1971). Helix–coil stability constants for the naturally occurring amino acids in water. I. Properties of copolymers and approximate theories. *Macromolecules* **4**, 396–407.
174. Von Dreele, P. H., Lotan, N., Ananthanarayanan, V. S., Andreatta, R. H., Poland, D., and Scheraga, H. A. (1971). Helix–coil stability constants for the naturally occurring amino acids in water. II. Characterization of the host polymers and application of the host–guest technique to random poly-(hydroxypropylglutamine-co-hydroxybutylglutamine). *Macromolecules* **4**, 408–417.
175. Wójcik, J., Altman, K. H., and Scheraga, H. A. (1990). Helix–coil stability constants for the naturally occurring amino acids in water. XXIV. Half-cystine parameters from random poly (hydroxybutylglutamine-co-S-methylthio-L-cysteine). *Biopolymers* **30**, 121–134.
176. Padmanabhan, S., York, E. J., Gera, L., Stewart, J. M., and Baldwin, R. L. (1994). Helix-forming tendencies of amino acids in short (hydroxybutyl)-L-glutamine peptides: An evaluation of the contradictory results from host–guest studies and short alanine-based peptides. *Biochemistry* **33**, 8604–8609.
177. Chakrabartty, A., Kortemme, T., and Baldwin, R. L. (1994). Helix propensities of the amino acids measured in alanine-based peptides without helix-stabilizing side-chain interactions. *Protein Sci.* **3**, 843–852.
178. Blaber, M., Zhang, X. J., and Matthews, B. W. (1993). Structural basis of amino acid alpha-helix propensity. *Science* **260**, 1637–1640.
179. Blaber, M. W., Baase, W. A., Gassner, N., and Matthews, B. W. (1993). Structural basis of amino acid α-helix propensity. *Science* **260**, 1637–1640.
180. Blaber, M., Zhang, X. J., Lindstrom, J. D., Pepiot, S. D., Baase, W. A., and Matthews, B. W. (1994). Determination of alpha-helix propensity within the context of a folded protein: Sites 44 and 131 in bacteriophage-T4 lysozyme. *J. Mol. Biol.* **235**, 600–624.
181. Horovitz, A., Matthews, J. M., and Fersht, A. R. (1992). α-Helix stability in proteins. II. Factors that influence stability at an internal position. *J. Mol. Biol.* **227**, 560–568.
182. O'Neil, K. T., and DeGrado, W. F. (1990). A thermodynamic scale for the helix-forming tendencies of the commonly occurring amino acids. *Science* **250**, 646–651.
183. Pace, C. N., and Scholtz, J. M. (1998). A helix propensity scale based on experimental studies of peptides and proteins. *Biophys. J.* **75**, 422–427.
184. Vila, J., Williams, R. L., Grant, J. A., Wójcik, J., and Scheraga, H. A. (1992). The intrinsic helix-forming tendency of L-alanine. *Proc. Natl Acad. Sci. USA* **89**, 7821–7825.
185. Vila, J. A., Ripoll, D. R., and Scheraga, H. A. (2000). Physical reasons for the unusual α-helix stabilization afforded by charged or neutral polar residues in alanine-rich peptides. *Proc. Natl Acad. Sci. USA* **97**, 13075–13079.
186. Rohl, C. A., Fiori, W., and Baldwin, R. L. (1999). Alanine is helix-stabilizing in both template-nucleated and standard peptide helices. *Proc. Natl Acad. Sci. USA* **96**, 3682–3687.
187. Kennedy, R. J., Tsang, K. Y., and Kemp, D. S. (2002). Consistent helicities from CD and template t/c data for N-templated polyalanines: Progress toward resolution of the alanine helicity problem. *J. Am. Chem. Soc.* **124**, 934–944.
188. Serrano, L., Neira, J. L., Sancho, J., and Fersht, A. R. (1992). Effect of alanine versus glycine in alpha-helices on protein stability. *Nature* **356**, 453–455.
189. Creamer, T. P., and Rose, G. D. (1992). Side-chain entropy opposes alpha-helix formation but rationalizes experimentally determined helix-forming propensities. *Proc. Natl Acad. Sci. USA* **89**, 5937–5941.
190. Scott, K. A., Alonso, D. O. V., Sato, S., Fersht, A. R., and Daggett, V. (2007). Conformational entropy of alanine versus glycine in protein denatured states. *Proc. Natl Acad. Sci. USA* **104**, 2661–2666.

191. Palermo, N. Y., Csontos, J., Owen, M. C., Murphy, R. F., and Lovas, S. (2007). Aromatic-backbone interactions in model α-helical peptides. *J. Comp. Chem.* **28**, 1208–1214.
192. Lyu, P. C., Zhou, H. X. X., Jelveh, N., Wemmer, D. E., and Kallenbach, N. R. (1992). Position-dependent stabilizing effects in α-helices—N-terminal capping in synthetic model peptides. *J. Am. Chem. Soc.* **114**, 6560–6562.
193. Petukhov, M., Uegaki, K., Yumoto, N., Yoshikawa, S., and Serrano, L. (1999). Position dependence of amino acid intrinsic helical propensities II: Non-charged polar residues: Ser, Thr, Asn, and Gln. *Protein Sci.* **8**, 2144–2150.
194. Zhou, H. X. X., Lyu, P. C. C., Wemmer, D. E., and Kallenbach, N. R. (1994). Structure of C-terminal α-helix cap in a synthetic peptide. *J. Am. Chem. Soc.* **116**, 1139–1140.
195. Petukhov, M., Uegaki, K., Yumoto, N., and Serrano, L. (2002). Amino acid intrinsic α-helical propensities III: Positional dependence at several positions of C-terminus. *Protein Sci.* **11**, 766–777.
196. Ermolenko, D. N., Richardson, J. M., and Makhatadze, G. I. (2003). Noncharged amino acid residues at the solvent-exposed positions in the middle and at the C terminus of the α-helix have the same helical propensity. *Protein Sci.* **12**, 1169–1176.
197. Bang, D., Gribenko, A. V., Tereshko, V., Kosiakoff, A. A., Kent, S. B. H., and Makhatadze, G. I. (2006). Dissecting the energetics of α-helix C-cap termination through chemical protein synthesis. *Nat. Chem. Biol.* **2**, 139–143.
198. Rose, G. D. (2006). Lifting the lid on helix capping. *Nat. Chem. Biol.* **2**, 123–124.
199. Ermolenko, D. N., Dangi, B., Gvritishvili, A., Gronenborn, A. M., and Makhatadze, G. I. (2007). Elimination of the C-cap in ubiquitin—Structure, dynamics and thermodynamic consequences. *Biophys. Chem.* **126**, 25–35.
200. Szalik, L., Moitra, J., Krylov, D., and Vinson, C. (1997). Phosphorylation destabilizes α-helices. *Nat. Struct. Biol.* **4**, 112–114.
201. Liehr, S., and Chenault, H. K. (1999). A comparison of the α-helix forming propensities and hydrogen bonding properties of serine phosphate and α-amino-γ-phosphonobutyric acid. *Bioorg. Med. Chem. Lett.* **9**, 2759–2762.
202. Andrew, C. D., Warwicker, J., Jones, G. R., and Doig, A. J. (2002). Effect of phosphorylation on α-helix stability as a function of position. *Biochemistry* **41**, 1897–1905.
203. Horovitz, A., Serrano, L., Avron, B., Bycroft, M., and Fersht, A. R. (1990). Strength and co-operativity of contributions of surface salt bridges to protein stability. *J. Mol. Biol.* **216**, 1031–1044.
204. Merutka, G., and Stellwagen, E. (1991). Effect of amino acid ion pairs on peptide helicity. *Biochemistry* **30**, 1591–1594.
205. Stellwagen, E., Park, S. H., Shalongo, W., and Jain, A. (1992). The contribution of residue ion pairs to the helical stability of a model peptide. *Biopolymers* **32**, 1193–1200.
206. Huyghues-Despointes, B. M., Scholtz, J. M., and Baldwin, R. L. (1993). Helical peptides with three pairs of Asp–Arg and Glu–Arg residues in different orientations and spacings. *Protein Sci.* **2**, 80–85.
207. Huyghues-Despointes, B. M., and Baldwin, R. L. (1997). Ion-pair and charged hydrogen-bond interactions between histidine and aspartate in a peptide helix. *Biochemistry* **36**, 1965–1970.
208. Huyghues-Despointes, B. M., Klingler, T. M., and Baldwin, R. L. (1995). Measuring the strength of side-chain hydrogen bonds in peptide helices: The Gln·Asp (i, $i + 4$) interaction. *Biochemistry* **34**, 13267–13271.
209. Stapley, B. J., and Doig, A. J. (1997). Hydrogen bonding interactions between Glutamine and Asparagine in α-helical peptides. *J. Mol. Biol.* **272**, 465–473.
210. Padmanabhan, S., and Baldwin, R. L. (1994). Tests for helix-stabilizing interactions between various nonpolar side chains in alanine-based peptides. *Protein Sci.* **3**, 1992–1997.

211. Padmanabhan, S., and Baldwin, R. L. (1994). Helix-stabilizing interaction between tyrosine and leucine or valine when the spacing is i, $i + 4$. *J. Mol. Biol.* **241**, 706–713.
212. Viguera, A. R., and Serrano, L. (1995). Side-chain interactions between sulfur-containing amino acids and phenylalanine in α-helices. *Biochemistry* **34**, 8771–8779.
213. Tsou, L. K., Tatko, C. D., and Waters, M. L. (2002). Simple cation-p interaction between a phenyl ring and a protonated amine stabilizes an α-helix in water. *J. Am. Chem. Soc.* **124**, 14917–14921.
214. Andrew, C. D., Penel, S., Jones, G. R., and Doig, A. J. (2001). Stabilizing nonpolar/polar side-chain interactions in the α-helix. *Proteins Struct. Funct. Genet.* **45**, 449–455.
215. Taylor, J. W. (2002). The synthesis and study of side-chain lactam-bridged peptides. *Biopolymers* **66**, 49–75.
216. Campbell, R. M., Bongers, J., and Felxi, A. M. (1995). Rational design, synthesis, and biological evaluation of novel growth-hormone releasing-factor analogs. *Biopolymers* **37**, 67–68.
217. Chorev, M., Roubini, E., McKee, R. L., Gibbons, S. W., Goldman, M. E., Caulfield, M. P., and Rosenblatt, M. (1991). Cyclic parathyroid-hormone related protein antagonists—Lysine 13 to Aspartic Acid 17 [I to (I + 40)] side-chain to side-chain lactamization. *Biochemistry* **30**, 5968–5974.
218. Osapay, G., and Taylor, J. W. (1990). Multicyclic polypeptide model compounds. 1. Synthesis of a tricyclic amphiphilic α-helical peptide using an oxime resin, segment-condensation approach. *J. Am. Chem. Soc.* **112**, 6046–6051.
219. Bouvier, M., and Taylor, J. W. (1992). Probing the functional conformation of neuropeptide-T through the design and study of cyclic analogs. *J. Med. Chem.* **35**, 1145–1155.
220. Kapurniotu, A., and Taylor, J. W. (1995). Structural and conformational requirements for human calcitonin activity—Design, synthesis, and study of lactam-bridged analogs. *J. Med. Chem.* **38**, 836–847.
221. Bracken, C., Gulyas, J., Taylor, J. W., and Baum, J. (1994). Synthesis and nuclear-magnetic-resonance structure determination of an α-helical, bicyclic, lactam-bridged hexapeptide. *J. Am. Chem. Soc.* **116**, 6431–6432.
222. Chen, S. T., Chen, H. J., Yu, H. M., and Wang, K. T. (1993). Facile synthesis of a short peptide with a side-chain-constrained structure. *J. Chem. Res. S* **6**, 228–229.
223. Zhang, W. T., and Taylor, J. W. (1996). Efficient solid-phase synthesis of peptides with tripodal side-chain bridges and optimization of the solvent conditions for solid-phase cyclizations. *Tetrahedron Lett.* **37**, 2173–2176.
224. Luo, P. Z., Braddock, D. T., Subramanian, R. M., Meredith, S. C., and Lynn, D. G. (1994). Structural and thermodynamic characterization of a bioactive peptide model of apolipoprotein-E—Side-chain lactam bridges to constrain the conformation. *Biochemistry* **33**, 12367–12377.
225. Jackson, D. Y., King, D. S., Chmielewski, J., Singh, S., and Schultz, P. G. (1991). General approach to the synthesis of short α-helical peptides. *J. Am. Chem. Soc.* **113**, 9391–9392.
226. Zhang, L., and Morikis, D. (2007). Solution structure of a bent α-helix. *Biochemistry* **46**, 12959–12967.
227. Iqbalsyah, T., Moutevelis, E., Warwicker, J., Errington, N., and Doig, A. J. (2006). The CXXC motif at the N terminus of an α-helical peptide. *Protein Sci.* **15**, 1945–1950.
228. Kumita, J. R., Smart, O. S., and Woolley, G. A. (2000). Photo-control of helix content in a short peptide. *Proc. Natl Acad. Sci. USA* **97**, 3803–3808.
229. Flint, D. G., Kumita, J. R., Smart, O. S., and Wolley, G. A. (2002). Using an azobenzene cross-linker to either increase or decrease peptide helix content upon *trans*-to-*cis* photoisomerization. *Chem. Biol.* **9**, 391–397.

230. Kumita, J. R., Flint, D. G., Smart, O. S., and Woolley, G. A. (2002). Photo-control of peptide helix content by an azobenzene cross-linker: Steric interactions with underlying residues are not critical. *Protein Eng.* **15,** 561–569.
231. Jiménez, M. A., Muñoz, V., Rico, M., and Serrano, L. (1994). Helix stop and start signals in peptides and proteins. The capping box does not necessarily prevent helix elongation. *J. Mol. Biol.* **242,** 487–496.
232. Kallenbach, N. R., and Gong, Y. X. (1999). C-terminal capping motifs in model helical peptides. *Bioorg. Med. Chem.* **7,** 143–151.
233. Petukhov, M., Yumoto, N., Murase, S., Onmura, R., and Yoshikawa, S. (1996). Factors that affect the stabilization of α-helices in short peptides by a capping box. *Biochemistry* **35,** 387–397.
234. Viguera, A. R., and Serrano, L. (1995). Experimental analysis of the Schellman motif. *J. Mol. Biol.* **251,** 150–160.
235. Gong, Y., Zhou, H. X., Guo, M., and Kallenbach, N. R. (1995). Structural analysis of the N-and C-termini in a peptide with consensus sequence. *Protein Sci.* **4,** 1446–1456.
236. Armstrong, K. M., and Baldwin, R. L. (1993). Charged histidine affects α-helix stability at all positions in the helix by interacting with the backbone charges. *Proc. Natl Acad. Sci. USA* **90,** 11337–11340.
237. Scholtz, J. M., York, E. J., Stewart, J. M., and Baldwin, R. L. (1991). A neutral water-soluble, α-helical peptide: The effect of ionic strength on the helix–coil equilibrium. *J. Am. Chem. Soc.* **113,** 5102–5104.
238. Smith, J. S., and Scholtz, J. M. (1998). Energetics of polar side-chain interactions in helical peptides: Salt effects on ion pairs and hydrogen bonds. *Biochemistry* **37,** 33–40.
239. Lockhart, D. J., and Kim, P. S. (1993). Electrostatic screening of charge and dipole interactions with the helix backbone. *Science* **260,** 198–202.
240. Jelesarov, I., Durr, E., Thomas, R. M., and Bosshard, H. B. (1998). Salt effects on hydrophobic interaction and charge screening in the folding of a negatively charged peptide to a coiled coil (leucine zipper). *Biochemistry* **37,** 7539–7550.
241. Scholtz, J. M., Marqusee, S., Baldwin, R. L., York, E. J., Stewart, J. M., Santoro, M., and Bolen, D. W. (1991). Calorimetric determination of the enthalpy change for the alpha-helix to coil transition of an alanine peptide in water. *Proc. Natl Acad. Sci. USA* **88,** 2854–2858.
242. Yoder, G., Pancoska, P., and Keiderling, T. A. (1997). Characterization of alanine-rich peptides, Ac-(AAKAA)n-GY-NH2 ($n = 1$–4), using vibrational circular dichroism and Fourier transform infrared conformational determination and thermal unfolding. *Biochemistry* **36,** 5123–5133.
243. Huang, C. Y., Klemke, J. W., Getahun, Z., DeGrado, W. F., and Gai, F. (2001). Temperature-dependent helix–coil transition of an alanine based peptide. *J. Am. Chem. Soc.* **123,** 9235–9238.
244. Schellman, J. A. (1955). The stability of hydrogen-bonded peptide structures in aqueous solution. *C. R. Trav. Lab. Carlsberg Ser. Chim.* **29,** 230–259.
245. Rialdi, G., and Hermans, J. (1966). Calorimetric heat of the helix–coil transition of poly-L-glutamic acid. *J. Am. Chem. Soc.* **88,** 5719–5720.
246. Chou, P. Y., and Scheraga, H. A. (1971). Calorimetric measurement of enthalpy change in the isothermal helix–coil transition of poly-L-lysine in aqueous solution. *Biopolymers* **10,** 657–680.
247. Richardson, J. M., and Makhatadze, G. I. (2004). Temperature dependence of the thermodynamics of helix–coil transition. *J. Mol. Biol.* **335,** 1029–1037.
248. Richardson, J. M., McMahon, K. W., MacDonald, C. C., and Makhatadze, G. I. (1999). MEARA sequence repeat of human CstF-64 polyadenylation factor is helical solution: A spectroscopic and calorimetric study. *Biochemistry* **38,** 1029–1037.

249. Makhatadze, G. I., and Privalov, P. L. (1990). Heat capacity of proteins. I. Partial molar heat capacity of individual amino acid residues in aqueous solution: Hydration effect. *J. Mol. Biol.* **213,** 375–384.
250. Siedlecka, M., Goch, G., Ejchart, A., Sticht, H., and Bierzynski, A. (1999). α-Helix nucleation by calcium-binding peptide loop. *Proc. Natl Acad. Sci. USA* **96,** 903–908.
251. Goch, G., Maciejczyk, M., Oleszczuk, M., Stachowiak, D., Malicka, J., and Bierzynski, A. (2003). Experimental investigation of initial steps of helix propagation in model peptides. *Biochemistry* **42,** 6840–6847.
252. Lopez, M. M., Chin, D. H., Baldwin, R. L., and Makhatadze, G. I. (2002). The enthalpy of the alanine peptide helix measured by isothermal titration calorimetry using metal-binding to induce helix formation. *Proc. Natl Acad. Sci. USA* **99,** 1298–1302.
253. Richardson, J. M., Lopez, M. M., and Makhatadze, G. I. (2005). Enthalpy of helix–coil transition: Missing link in rationalizing the thermodynamics of helix-forming propensities of the amino acid residues. *Proc. Natl Acad. Sci. USA* **102,** 1413–1418.
254. Makhatadze, G. I. (2006). Thermodynamics of α-helix formation. *Adv. Protein Chem.* **72,** 199–226.
255. Nelson, J. W., and Kallenbach, N. R. (1986). Stabilization of the ribonuclease S-peptide α-helix by trifluoroethanol. *Proteins Struct. Funct. Genet.* **1,** 211–217.
256. Nelson, J. W., and Kallenbach, N. R. (1989). Persistence of the α-helix stop signal in the S-peptide in trifluoroethanol solutions. *Biochemistry* **28,** 5256–5261.
257. Sonnichsen, F. D., Van Eyk, J. E., Hodges, R. S., and Sykes, B. D. (1992). Effect of trifluoroethanol on protein secondary structure: An NMR and CD study using a synthetic actin peptide. *Biochemistry* **31,** 8790–8798.
258. Waterhous, D. V., and Johnson, W. C. (1994). Importance of environment in determining secondary structure in proteins. *Biochemistry* **33,** 2121–2128.
259. Jasanoff, A., and Fersht, A. R. (1984). Quantitative determination of helical propensities. Analysis of data from trifluoroethanol titration curves. *Biochemistry* **33,** 2129–2135.
260. Albert, J. S., and Hamilton, A. D. (1995). Stabilization of helical domains in short peptides using hydrophobic interactions. *Biochemistry* **34,** 984–990.
261. Walgers, R., Lee, T. C., and Cammers-Goodwin, A. (1998). An indirect chaotropic mechanism for the stabilization of helix conformation of peptides in aqueous trifluoroethanol and hexafluoro-2-propanol. *J. Am. Chem. Soc.* **120,** 5073–5079.
262. Luo, P., and Baldwin, R. L. (1999). Interaction between water and polar groups of the helix backbone: An important determinant of helix propensities. *Proc. Natl Acad. Sci. USA* **96,** 4930–4935.
263. Kentsis, A., and Sosnick, T. R. (1998). Trifluoroethanol promotes helix formation by destabilizing backbone exposure: Desolvation rather than native hydrogen bonding defines the kinetic pathway of dimeric coiled coil folding. *Biochemistry* **37,** 14613–14622.
264. Reiersen, H., and Rees, A. R. (2000). Trifluoroethanol may form a solvent matrix for assisted hydrophobic interactions between peptide side chains. *Protein Eng.* **13,** 739–743.
265. Myers, J. K., Pace, N., and Scholtz, J. M. (1998). Trifluoroethanol effects on helix propensity and electrostatic interactions in the helical peptide from ribonuclease T1. *Protein Sci.* **7,** 383–388.
266. Nozaki, Y., and Tanford, C. (1967). Intrinsic dissociation constants of aspartyl and glutamyl carboxyl groups. *J. Biol. Chem.* **242,** 4731–4735.
267. Kyte, J. (1995). "Structure in Protein Chemistry." Garland Publishing, Inc., New York.
268. Kortemme, T., and Creighton, T. E. (1995). Ionisation of cysteine residues at the termini of model α-helical peptides. Relevance to unusual thiol pKa values in proteins of the thioredoxin family. *J. Mol. Biol.* **253,** 799–812.

269. Miranda, J. J. (2003). Position-dependent interactions between cysteine and the helix dipole. *Protein Sci.* **12,** 73–81.
270. Porter, M. A., Hall, J. R., Locke, J. C., Jensen, J. H., and Molina, P. A. (2006). Hydrogen bonding is the prime determinant of carboxyl pKa values at the N-termini of α-helices. *Proteins Struct. Funct. Bioinf.* **63,** 621–635.
271. Eisenberg, D., Weiss, R. M., and Terwilliger, T. C. (1982). The helical hydrophobic moment: A measure of the amphiphilicity of a helix. *Nature* **299,** 371–374.
272. Muñoz, V., and Serrano, L. (1995). Analysis of $i, i + 5$ and $i, i + 8$ hydrophobic interactions in a helical model peptide bearing the hydrophobic staple motif. *Biochemistry* **34,** 15301–15306.
273. Thomas, S. T., Loladze, V. V., and Makhatadze, G. I. (2001). Hydration of the peptide backbone largely defines the thermodynamic propensity scale of residues at the C' position of the C-capping box of α-helices. *Proc. Natl Acad. Sci. USA* **98,** 10670–10675.
274. Shi, Z., Olson, C. A., Bell, A. J., and Kallenbach, N. R. (2002). Non-classical helix-stabilizing interactions: C–H...O–H-bonding between Phe and Glu side chains in α-helical peptides. *Biophys. Chem.* **101–102,** 267–279.
275. Luo, R., David, L., Hung, H., Devaney, J., and Gilson, M. K. (1999). Strength of solvent-exposed salt-bridges. *J. Phys. Chem. B* **103,** 366–380.
276. Marqusee, S., and Sauer, R. T. (1994). Contributions of a hydrogen bond/salt bridge network to the stability of secondary and tertiary structure in lambda repressor. *Protein Sci.* **3,** 2217–2225.
277. Shi, Z., Olson, C. A., Bell, A. J., and Kallenbach, N. R. (2001). Stabilization of α-helix structure by polar side-chain interactions: Complex salt bridges, cation π interactions and C–H...O–H bonds. *Biopolymers* **60,** 366–380.

Folding and Wrapping Soluble Proteins: Exploring the Molecular Basis of Cooperativity and Aggregation

ARIEL FERNÁNDEZ, XI ZHANG, AND JIANPING CHEN

Department of Bioengineering and Program in Applied Physics, Rice University, Houston, Texas 77005

I. Folding, Cooperativity, and Wrapping of Soluble Proteins: An Overview 54
II. Wrapping the Folded Structure 57
III. Exploring the Molecular Basis of Folding Cooperativity 62
    A. Understanding Cooperativity Through Wrapping 62
    B. Generating Wrapping/Folding Trajectories 66
    C. Evolving Wrapping Patterns and Fixed Wrapping Motifs Along Folding Trajectories 71
    D. Solvation Nanoscale Model of Wrapping and Cooperativity 76
IV. Protein Under-Wrapping, Misfolding, and Aggregation 80
    A. Pathogenically Under-Wrapped Proteins 80
    B. Under-Wrapped Proteins and Epigenetic Polymorphism 83
    References 85

In this chapter, we survey recent advances in the understanding of the folding of soluble proteins, mostly focusing on the molecular basis of cooperativity. In this regard, we explore the concept of protein wrapping and its bearing on the expediency of the folding process for natural proteins. Wrapping refers to the environmental modulation or protection of intramolecular electrostatic interactions through an exclusion of surrounding water that takes place as the chain folds onto itself. Thus, a special many-body picture of the folding process is shown to emerge where the folding chain not only interacts with itself but also shapes the microenvironments that stabilize or destabilize the interactions. After describing reported results on the dynamics of folding, wrapping and water exclusion, we examine the endpoint of the folding process. In particular, we classify folds according to their extent of wrapping and explore relationships between severely under-wrapped proteins, aggregation, and epigenetic polymorphism.

I. Folding, Cooperativity, and Wrapping of Soluble Proteins: An Overview

In spite of burgeoning effort, the physical underpinnings to the protein folding process remain elusive or, rather, difficult to cast in an operational or useful form enabling *ab initio* structure prediction (1–10). Probably for these same reasons, the possibility of inferring the folding pathway of a soluble protein solely from first principles continues to elude all major research efforts.

While most scientists would agree that the dominant intramolecular forces driving the folding process are essentially electrostatic and hydrophobic in nature, not all seem aware that such forces are modulated by an equally important factor: the shaping or framing of the solvent microenvironments where they become operational (10–12). This last aspect is not a minor detail. It suffices to recall that an electrostatic interaction occurring in bulk water would be 78 times weaker than the same interaction occurring in vacuum (12, 13). Thus, as a peptide chain folds onto itself, it shapes the microenvironments that surround each pairwise interaction, and hence the strength and stability of such interactions need to be rescaled according to the extent to which they become "wrapped" or surrounded by other parts of the chain (2, 10). Thus, pairwise interactions between different parts of the peptide chain not only entail the units directly engaged in the interaction but also the units involved in shaping their microenvironment, and the latter are just as important as they determine either the persistence or the ephemeral nature of such interactions and ultimately the integrity of the protein structure. This fact makes the folding problem essentially a many-body problem and points to the heart of cooperativity, a pivotal attribute of the folding process (4, 6).

To delineate the physical underpinnings of cooperativity, we need to examine the folding process rather closely and from a rigorous physicochemical perspective: With an amide and carbonyl group per residue, the backbone of the protein chain is highly polar and this feature poses severe constraints on the nature of the hydrophobic collapse and on the chain composition of proteins capable of sustaining such a collapse (14, 15). Thus, the hydrophobic collapse entails the dehydration of backbone amides and carbonyls and such a process would be thermodynamically disfavored if it were not for the possibility of amides and carbonyls to engage in hydrogen bonding with each other (9). Hence not every hydrophobic collapse qualifies as being conducive to folding the protein chain: only a collapse that ensures the formation and protection of backbone hydrogen bonds is likely to ensure an expedient folding of the chain (2). On the other hand, polar-group hydration competes with intramolecular hydrogen bonds, compromising the structural integrity of proteins with a deficiently wrapped backbone (9). Thus, the need for formation and protection of intramolecular hydrogen bonds from water attack imposes constraints on the chain composition of an efficient folder capable of sustaining a reproducible and expedient hydrophobic collapse (2). These assertions require rigorous demonstration and hence call for further physicochemical investigation of the nanoscale-level dehydration of backbone hydrogen bonds (12–14).

In accord with this picture, it has been postulated that as water-soluble proteins fold, the hydrogen-bond pairing of backbone amides and carbonyls is concurrent with the hydrophobic collapse of the chain (9, 15, 16). This fact has been rationalized taking into account that the thermodynamic cost associated with the dehydration of unpaired polar groups is relatively high, and that the hydrophobic collapse hinders the backbone hydration by shielding it from water (2). On the other hand, the strength and stability of hydrogen bonds clearly depend on the microenvironment where they occur: the proximity of nonpolar groups to a hydrogen bond enhances the electrostatic interaction by descreening the partial charges, and stabilizes it by hindering the hydration of the polar groups in the nonbonded state. Thus, to guarantee the integrity of soluble protein structure and the expediency of the folding process, most intramolecular hydrogen bonds, however transient or long-lived, must be surrounded or "wrapped" by nonpolar groups fairly thoroughly as to become significantly dehydrated (2, 14). This observation has implications at an ensemble-average level accessible to experimentalists (16–18). It may help understand the fact that single-domain proteins are likely to be two-state folders (17, 18), with a single kinetic barrier dominating the folding process at the ensemble-average level (17).

Taken together, the hydration propensity of amide and carbonyl and the dehydration-induced strengthening of their electrostatic association represent two conflictive tendencies, suggesting that there must be a crossover point in the dehydration propensity of a backbone hydrogen bond. If the bond is poorly wrapped by a few nonpolar groups that cluster around it, then hydration of the paired amide and carbonyl is likely to be favorable, but as the hydrogen bond becomes better wrapped intramolecularly, the surrounding water loses too many hydrogen bonding partnerships and thus, further removal of surrounding water is promoted (2). This switch-over to a dehydration propensity signals a commitment to fold into a compact structure in which most backbone hydrogen bonds are thoroughly dehydrated (2). In turn, this inference demands rigorous justification that may be achieved through a survey of experimental and theoretical results.

The crossover point in hydrogen-bond dehydration propensity may be regarded as representing a local characterization of the folding transition state if we adopt the backbone hydrogen-bond dehydration as a generic folding coordinate (9). Once the folding process has progressed beyond the crossover point, further dehydration of the backbone is favored in consonance with the downhill nature of the folding process beyond the transition state (17). Thus, a transition-state conformation commits the chain to fold partly because the partially wrapped hydrogen bonds trigger their further desolvation, in turn fostering further chain compaction (2, 17). This compaction is essential to augment the number of nonpolar groups within the hydrogen-bond microenvironments, thus protecting the bonds from water attack.

However compelling, the folding scenario described above requires a rigorous justification that entails an understanding of the cross-over behavior at a local level. This understanding prompts us to focus on the solvent environment of an individual hydrogen bond (12, 19, 20) in order to address the following question: How can we demonstrate that a partially wrapped hydrogen bond is likely to attract nonpolar groups in accord with its purported propensity to promote its further dehydration? Previous reported research effectively addressed this question by introducing an experimental platform to measure the adsorption of proteins with wrapping defects onto a hydrophobic "wrapping" surface (19). The experiments exploited a set-up for evanescent-field spectroscopy adapted to correlate adsorption uptake with extent of protein under-wrapping. Hence the attractive drag exerted by under-wrapped hydrogen bonds on test hydrophobes became accessible (19).

This study was motivated by the earlier observations that under-wrapped hydrogen bonds in native structure, the so-called *dehydrons* (12, 14, 21–26), play a pivotal role in driving protein associations, as such associations contribute intermolecularly to the wrapping of the preformed structure (21). In consistency with current terminology, the force stemming from the dehydration propensity of the partially wrapped hydrogen bond is hereby termed *dehydronic*. The dehydronic force arises as a nonpolar group approaches a dehydron with a net effect of immobilizing and ultimately removing surrounding water molecules. This displacement lowers the polarizability of the microenvironment which, in turn, deshields the paired charges (12, 13). Thus, a net attractive force is exerted by the dehydron on a nonpolar group. Since the water molecules solvating an amide and carbonyl paired by a dehydron are necessarily depleted of some hydrogen-bonding partners, the work required for their ultimate removal from the bond surroundings is expected to be minimal (13, 19, 20). The dehydronic force, denoted $\Phi(\mathbf{R})$, is orthogonal to the Coulomb field generated by the polar pair, and may be described within a quasicontinuous treatment by the equation:

$$\Phi(\mathbf{R}) = -\nabla_R[4\pi\varepsilon(\mathbf{R})]^{-1}qq'/r, \tag{1}$$

where $\mathbf{R}$ represents the position vector of the hydrophobe or nonpolar group with respect to the center of mass of the hydrogen-bonded polar pair, ∇_R is the gradient taken with respect to this vector, r is the distance between the charges of magnitude q and q' paired by the hydrogen bond, and the local permittivity coefficient $\varepsilon = \varepsilon(\mathbf{R})$ subsumes the polarizability of the microenvironment, which is generically dependent on the position of the test hydrophobe (13). An appropriate expression for $\varepsilon(\mathbf{R})$ valid at the nanoscales is unavailable at present, because of the discreteness of the dielectric medium and the need to include individual solvent dipole correlations. Although a mean-field dielectric description is unsatisfactory (13), it is still possible to assert that $\Phi(\mathbf{R})$ is an attractive force since a decrease in $\|\mathbf{R}\|$ entails a decrease in local polarization which, in turn, enhances the Coulomb attraction.

Building on this analysis, we may quantify the net hydrophobicity η of a hydrogen bond by taking into account the surface flux of the dehydronic field $\omega^{-1}\Phi(\mathbf{R})$ (ω = volume of test hydrophobe) generated by the hydrogen bond. Thus, in accord with Gauss theorem we obtain:

$$\eta = \iint_\Sigma \Phi(\mathbf{R}) d\sigma(\mathbf{R}), \qquad (2)$$

where Σ is the closed surface of the dehydration domain of the hydrogen bond and $d\sigma(\mathbf{R})$ is the differential surface area vector at position $\mathbf{R}$.

This hydrophobicity (or equivalently, the dehydronic force it generates) is the key to guarantee that the backbone hydrogen bonds concurrent with the hydrophobic collapse of the protein chain will trigger further compaction, ultimately leading to the formation of the native fold. The following sections will survey published work supporting this physical picture of wrapping-related phenomena.

II. Wrapping the Folded Structure

A first-principle survey of the dynamics of protein folding leads us to introduce the concept of wrapping as a means to better understand the phenomenon. However, while wrapping is a well defined category (21–26), the wrapping dynamics concurrent with the folding process can only be captured by folding algorithms (2) but so far not directly probed. This state of affairs prompts us to focus on the endpoints of the folding process, that is the native folds, and examine the wrapping of such sustainable protein conformations.

As indicated above, the structural integrity of a soluble protein is contingent on its capacity to exclude water from its amide–carbonyl hydrogen bonds (2, 14). Water-exposed intramolecular hydrogen bonds in native folds, the so-called dehydrons constitute structural singularities representing wrapping or packing defects that have been recently characterized (12–14). In turn, these defects favor the removal of surrounding water as a means to strengthen and stabilize the underlying electrostatic interaction (19), and thus are implicated in protein associations (21) and macromolecular recognition (27, 28). The strength and stability of dehydrons may be modulated by an external agent. More precisely, intramolecular hydrogen bonds which are not "wrapped" by a sufficient number of nonpolar groups may become stabilized and strengthened by the attachment of a ligand or binding partner that further contributes to their dehydration (21). The net gain in Coulomb energy associated with wrapping a dehydron has been experimentally determined to be ∼4 kJ/mol (19). The adhesive force exerted by a dehydron on a hydrophobe at 6 Å distance is approximately 7.8 pN, a magnitude comparable to the hydrophobic attraction between two nonpolar moieties that frame unfavorable interfaces with water. Furthermore, dehydrons are decisive factors driving association in 38% of the PDB complexes and constitute significant factors (interfacial dehydron density larger than average on individual partners) in about 95% of all complexes reported in the PDB (21).

These discrete effects relating to local water structuring around packing defects cannot be captured properly by existing continuous models of the interfacial electrostatics. Such models are based on mean-force potential approximations to the solvation interactions where solvent degrees of freedom are averaged out. Thus, continuous models should be adapted to deal with local dielectric modulations promoted by the interfacial regions.

Dehydrons may be identified from atomic coordinates of proteins with reported structure, as illustrated in Fig. 1. Thus, we need to introduce an auxiliary quantity, the extent of hydrogen-bond wrapping, ρ, indicating the number of nonpolar groups contained within a "desolvation domain" around the bond. This domain is typically defined as two intersecting balls of fixed radius (approximately thickness of three water layers) centered at the α-carbons of the residues paired by the amide–carbonyl hydrogen bond. In structures of soluble proteins, at least two thirds of the backbone hydrogen bonds are wrapped on average by ρ = 26.6±7.5 nonpolar groups for a desolvation ball radius 6 Å. Dehydrons lie in the tails of the distribution, that is, their microenvironment contains 19 or fewer nonpolar groups, so their ρ-value is below the mean (ρ = 26.6) minus one standard deviation (σ = 7.5) (22–26).

Thus, the overall wrapping of a protein may be assessed by determining the percentage of intramolecular hydrogen bonds with $\rho \leq 19$, that is, the percentage of dehydrons in its structure. An illustration on the representation of the wrapping of a protein is given in Fig. 2.

Under-wrapped or dehydron-rich regions in soluble proteins are typically molecular markers for protein associations because of their propensity towards further dehydration (21). Thus, specific residues of the binding partner contribute to the desolvation of some of the dehydrons, as they enter the

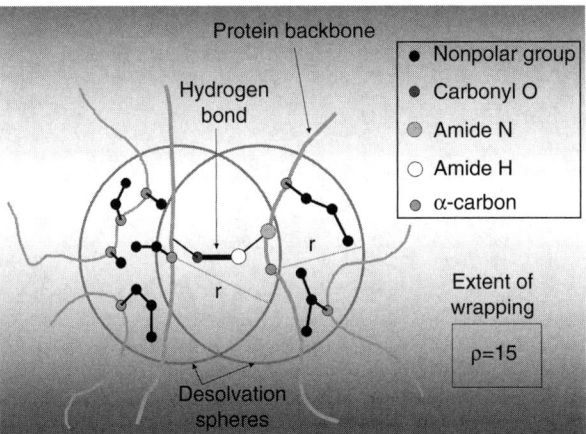

Fig. 1. Wrapping of a backbone hydrogen bond in a soluble protein. Intramolecular hydrogen bonds in soluble proteins prevail only if they are protected from water attack. Thus, their extent of intramolecular wrapping by nonpolar groups (black balls) becomes central to define their stability and strength (12). The extent of hydrogen-bond wrapping indicates the number of nonpolar groups contained within a desolvation domain defined as two intersecting balls of fixed radius centered at the α-carbons of the residues paired by the hydrogen bond. While the wrapping statistics on hydrogen bonds vary with this value, the tails of the distribution remain invariant, thus enabling a unique identification of under-wrapped hydrogen bonds (dehydrons). Dehydrons are packing defects that become stabilized upon removal of surrounding water through protein associations. Thus, dehydrons may be regarded as structural features defining the gene sensitivity to its interactive context and its reliance on binding partnerships to maintain structural integrity. (See Color Insert.)

desolvation domain of intramolecular hydrogen bonds upon association. This intermolecular wrapping is illustrated in Fig. 3, displaying the colicin + cognate immunity protein IM9 complex. The interface region of colicin (chain contour region 70–100) contains 13 dehydrons in the free (uncomplexed) molecule, making it vulnerable to water attack. Upon association, specific residues of the binding partner contribute to the desolvation of some of the dehydrons, as they enter the desolvation domain of the intramolecular hydrogen bonds of colicin. This intermolecular wrapping reduces the vulnerability of colicin, which has seven dehydrons in the interface upon complexation, instead of the original 13, in its free form. The highly vulnerable interface of colicin and the large difference in the number of dehydrons upon complexation is indicative of a severe dosage imbalance if colicin were expressed without its binding partner, be it the immunity protein IM9 or an alterative partner capable of providing comparable level of structural protection.

While wrapping defects in protein structure are clear markers of protein interactivity (21) and node centrality in the protein-interaction network (26), a protein with an excessively under-wrapped native fold is prone to misfolding

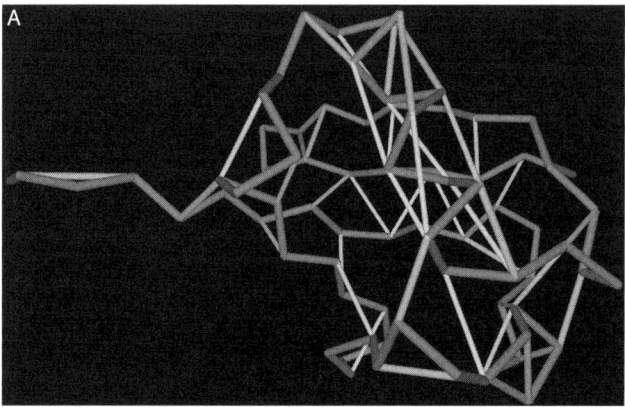

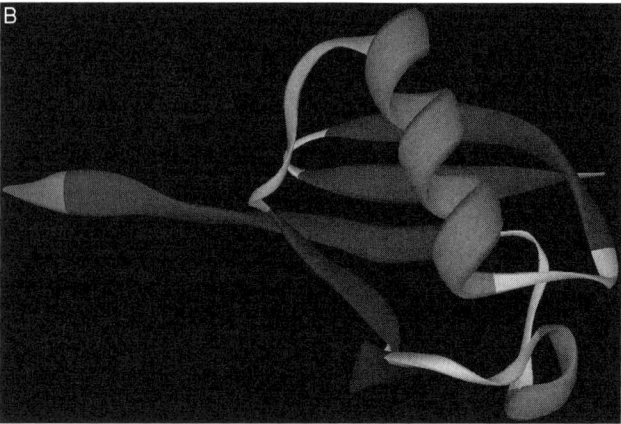

Fig. 2. Illustration of the under-wrapping of protein structure. (A) Dehydron pattern for human ubiquitin (PDB accession code 1UBI). Dehydrons are indicated as green segments joining the α-carbons of the paired units, well-wrapped hydrogen bonds ($\rho > 19$) are shown in light grey, and the protein backbone is conventionally shown as blue virtual bonds joining the α-carbons of consecutive amino acid units. The displayed structure has 64 backbone hydrogen bonds, out of which 16 are dehydrons. Thus, the extent of under-wrapping for this protein is 16/64 = 25%. (B) The ubiquitin structure is also displayed in ribbon representation for easy visualization. (See Color Insert.)

and self-aggregation, a fact that has been extensively documented (11, 13, 22). The misfolding propensity of a severely under-wrapped protein is expected based on its vulnerability to water attack. Thus, an excess of 50% dehydrons in the structure of a soluble protein is indicative of a possible misfolder with prion-like functionalities that may lead to aberrant aggregation (11), or possibly

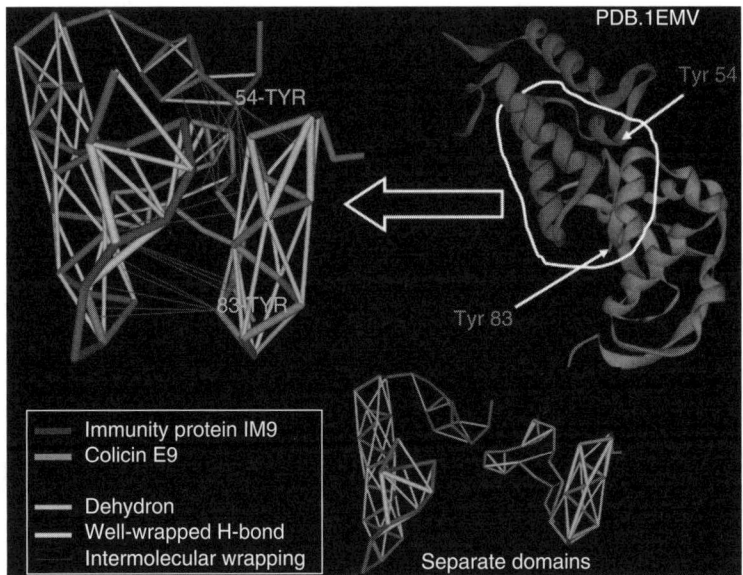

Fig. 3. Intermolecular wrapping in the complex colicin + cognate immunity protein IM9 (PDB.1EMV) as a means of protecting colicin structure. Dehydrons are indicated as green segments joining the α-carbons of the paired units, well-wrapped hydrogen bonds are shown in light grey, and the protein backbone is conventionally shown as virtual bonds joining the α-carbons of consecutive amino acid units. Complexation reduces the structural vulnerability of colicin by reducing the number of dehydrons at the interface from 13 to 7. Intermolecular wrapping is depicted by thin blue lines from the α-carbon of the wrapping residue to the middle of the hydrogen bond that is intermolecularly protected. Thus, an intermolecularly wrapping residue contributes with nonpolar groups to the dehydration of the preformed hydrogen bond from the binding partner.

even to epigenetic phenotype polymorphism. The latter possibility may arise if the putative prion is implicated in regulation of the gene expression, and hence its aggregation sequesters it from the regulatory context, altering the phenotype.

The extent of under-wrapping in PDB-reported proteins ranges from 10% to 60%. Thus, a typical enzyme, like human dehydrofolate reductase (DHF) has 24% dehydrons (Fig. 4), while the cellular form of the human prion PrPC has 55% dehydrons (Fig. 5). Prions possess an extent of under-wrapping comparable to membrane proteins, where under-wrapping, of course, does not represent a structural vulnerability as there is no surrounding water. This observation leads to consider the enticing possibility that prions may have been transmembrane proteins displaced to a cytosolic space through evolution.

III. Exploring the Molecular Basis of Folding Cooperativity

A. Understanding Cooperativity Through Wrapping

The wrapping dynamics concurrent with the folding process cannot be probed directly at this time, much like the conformational exploration defining the folding process still remains opaque to experimentalists (*16–18*). Hence, to make progress in our understanding of the wrapping dynamics, we need to capture this process through *ab initio* folding algorithms independently benchmarked and validated against PDB-reported structures and experimental data on the folding kinetics (*2*).

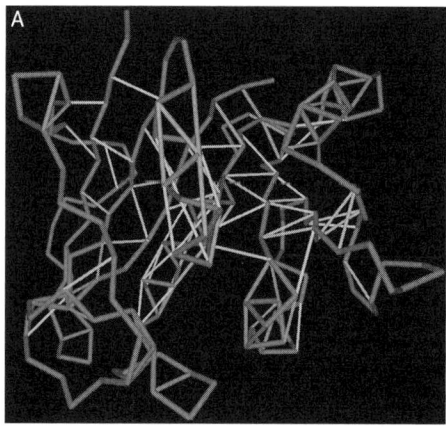

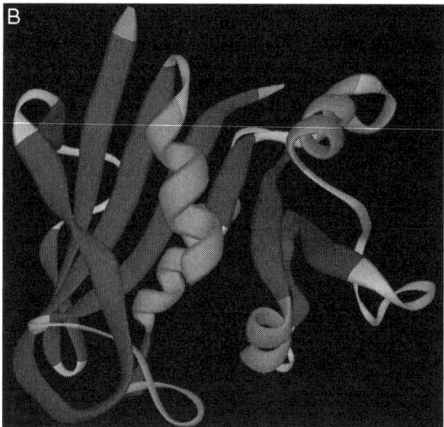

FIG. 4. Dehydron and hydrogen-bond pattern (A) for the enzyme dehydrofolate reductase (human ortholog) (DHF) (PDB.1HFP). (B) Ribbon representation is provided to ease visualization.

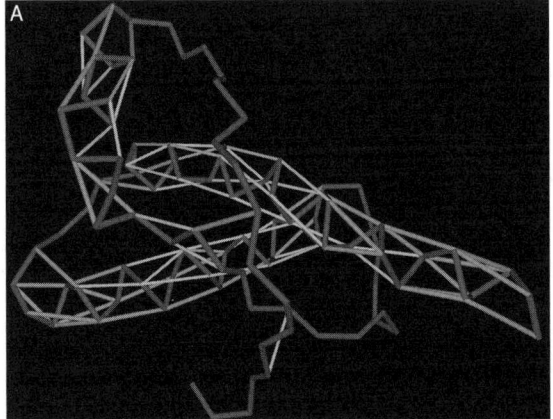
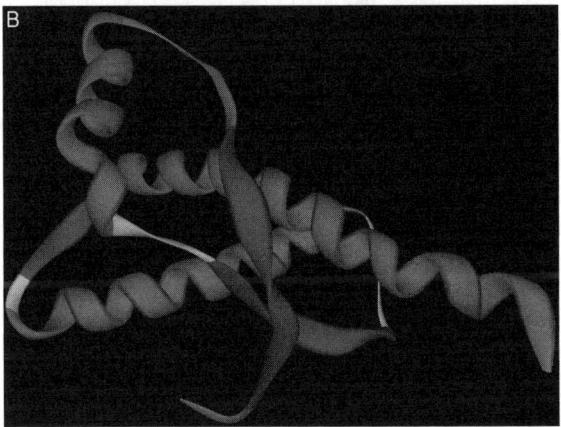

Fig. 5. Dehydron and hydrogen-bond pattern (A) for the human prion PrP^C in cellular form (PDB.1QM0). (B) Ribbon representation is provided to ease visualization.

The investigations into the basis of folding cooperativity prompted us to implement a program to algorithmically reproduce protein folding dynamics, which would capture the essential features of this attribute by incorporating the wrapping concept (2, 9, 12, 13). To access realistic folding timescales beyond the μs range, the program builds on an implicit solvent model of cooperativity (6) and exploits a coarse-grained representation of the torsional Ramachandran (Φ and Ψ)-state of each residue (10). The implicit treatment of the solvent dictates that the program assess at each step the microenvironments of intramolecular hydrogen bonds that form as byproducts of the chain attempts at achieving a sustainable compactification.

On the other hand, the microenvironmental dielectric or, equivalently, the hydrogen-bond hydrophobicity (cf. Eq. (2)), needs to be computed in a coarse manner (simplifying the assessment given in Fig. 1) at each step of the simulations in order to reduce computation time and access relevant timescales. The strength and stability of intramolecular hydrogen bonds are rescaled with each computational step in consonance with the degree of dryness or dehydration of the bond determined by the overall folding state of the chain (2, 6, 10). Thus, the extent of bond dehydration defines its net hydrophobicity (Eq. (2)), which in turn promotes further wrapping begetting further compaction of the chain.

To the best of our knowledge, this is the only *ab initio* folding algorithm that captures cooperativity while possessing predictive potential in regards to both structure and pathway. Furthermore, the algorithm appears to reproduce a basic feature of the folding kinetics of single-domain proteins: its purported two-stateness observed at the ensemble-average level of experimentation (16, 17). This dynamic feature is likely to arise as an overall consequence of the cross-over behavior in hydrogen-bond dehydration propensity, although a "hard proof" of this statement has not been produced to this day, remaining a challenging open problem.

In spite of the difficulties inherent to a direct probing, the folding of a soluble protein is believed to be accompanied by a progressive structuring, immobilization, and eventual exclusion of water surrounding backbone hydrogen bonds (9, 15). This process turns hydrogen bonds into major determinants of the folding pathway and compensates for the thermodynamic penalties associated with desolvation of the backbone polar groups (29). Thus, the need to wrap hydrogen bonds as a means to ensure their integrity has been shown to determine a constraint and hence define a constant of motion in the long-time limit behavior of coarse-grained *ab initio* simulations (2, 29). In this regard, an examination of one of the longest available (1 μs) all-atom simulations with explicit solvent (30) reveals that the average extent of wrapping is indeed a constant of motion for the folding trajectory (2). It is well documented that such a stabilization is best achieved by clustering five residues with nonpolar side-chain groups around the backbone hydrogen bonds, an arrangement that yields an optimal stabilization of the intramolecular hydrogen bond with a minimal conformational-entropy cost (2). The reported results thus support and clarify the view that hydrophobic surface burial should be commensurate with hydrogen-bond formation (16) and enabled us to introduce a basic wrapping motif inherent to the folding dynamics of soluble proteins (2).

The solvent dynamics concurrent with protein folding remains as elusive as the folding process itself. Yet, some progress may be achieved by exploiting molecular dynamic simulations, particularly if such simulations are endowed with some predictive value and are able to capture the essential and independently identified kinetic features of the folding process (2, 17). Thus, we have conjectured and shown that the progressive structuring, immobilization, and ultimate removal of water surrounding amide–carbonyl hydrogen bonds turn the latter into determinants of the folding pathway. This is so because hydrophobic collapse triggers or is concurrent with secondary structure formation (9), although this remains a highly controversial issue (31). The connection between hydrophobic interactions, water structuring, and the strength of hydrogen bonds was first delineated in pioneering work by Scheraga and co-workers (32). Thus, the inaccessibility of hydrogen bonds to solvent takes place as the protein strategically places hydrophobes around its backbone polar groups. This process induces hydrogen-bond formation as a means to compensate for the unfavorable burial of the backbone polar groups. In this regard, natural questions arise and have been addressed through the examination of wrapping dynamics along the folding process (2):

1. Does the wrapping or protection of backbone hydrogen bonds promote an expedient folding and if so, how?
2. How can we identify the conformation or conformational ensemble that commits the chain to fold expeditiously into its native structure?
3. Can we establish whether hydrophobic collapse precedes or is concurrent with secondary structure formation?

We have approached these problems by establishing a relationship between the modulation of electrostatic shielding and the wrapping of the protein conformation along generated folding trajectories. This link has been cemented on statistical information on hydrophobic clustering around native backbone hydrogen bonds, information that has lead us to establish a pervasive wrapping motif in native structures (33). Second, we have established the dynamic relevance of the formation of such hydrophobic-/hydrogen-bond assemblies during folding (2). This analysis was carried out exploiting both a coarse-grained *ab initio* folding algorithm based on an implicit solvent model (29) and a long (1 μs) all-atom molecular dynamics simulation (30) of the type pioneered by Karplus (34), Levitt (35), and Brooks (36), among many others. Both algorithmic approaches reveal a nearly constant average extent of hydrogen-bond wrapping along the folding trajectories with relatively small dispersion, suggesting the existence of a constant of motion. Third, we have provided a theoretical treatment to rationalize the recurrent configurations of hydrophobic clusters in folding protein chains by computing the optimization of their wrapping roles (2, 9, 33).

Put together, these results provide evidence supporting a dynamic picture of folding in which surface burial is commensurate with hydrogen-bond formation or, phrased differently, hydrophobic collapse is concurrent with the formation of secondary structure (9, 16, 37). Furthermore, the results added a new element to the picture: hydrophobic collapse occurs productively in so far as hydrophobes can also exert a protective role (37), wrapping the backbone hydrogen bonds, and they do so to the same extent throughout the folding pathway, irrespective of the number of hydrogen bonds present at a given time. This last assertion enabled us to postulate a constant of motion underlying the folding process.

B. Generating Wrapping/Folding Trajectories

To validate the previous assertions, the dynamics of backbone desolvation were studied using the so-called folding machine (FM), an *ab initio* wrapping-based algorithm designed to generate low-resolution folding pathways (2, 29), and contrasted against wrapping information extracted from all-atom explicit-solvent simulations (30).

The trustworthiness of the FM-generated folding pathways (29) is guaranteed by independent corroboration of the predictive value of the FM. This algorithm has been successful at predicting crucial dynamic features of complex folders that do not follow the two-state kinetic picture, such as β-lactoglobulin (8). The native-like and nonnative conformations occurring along the folding pathway in this presumed nonhierarchical folder as well as the productive role of nonnative interactions in preventing misfolding or incorrect structure condensation were predicted through the FM and subsequently validated experimentally (38). This corroboration added significant leverage to the wrapping-based FM algorithm, making it a powerful tool to study folding cooperativity.

As described in several publications (2, 6–9, 29, 39), rather than encompassing all the structural detail for each step, the FM focuses on the time-evolution of backbone torsional constraints imposed by steric clashes with side-chains and rescales the chain's intramolecular potential according to the wrapping microenvironments around pairwise dielectric-dependent interactions. Thus, each pairwise interaction of the peptide chain, be it hydrophobic or electrostatic, is effectively enhanced or weakened according to the extent of burial of the paired groups. The net decrease of free energy associated with the backbone hydrogen bond desolvation is typically only in the range 0.5–1.2 kcal/mol, due to the opposing increase in the solvation free energy of the polar amide and carbonyl groups. The implicit treatment of the solvent required that we introduce three-body correlations to characterize the wrapping of pre-formed intramolecular interactions, and accordingly, rescale the internal energy terms with each iteration; that is, every time the pattern of three-body correlations around an interaction has changed. Both the reduction in structural resolution and the implicit-solvent treatment are essential to make realistic folding time-scales ($>1~\mu$s) accessible to the FM computations. Thus, the FM *ab initio* approach is geared to generate folding pathways with a coarse structural resolution needed to make folding time-scales computationally accessible. The FM algorithm utilizes no *a priori* information on target folds (unlike the so-called Go models which use the native fold as input for the simulator (40–42)), nor does it incorporate any energetic biases. The model focuses on the torsional Φ and Ψ constraints that are applied to backbone torsional state due to the steric hindrances imposed by the side-chains. Each residue is assigned to a region, or basin, in the Ramachandran map, and changes in configuration occur by hopping to a new basin. By dealing with the evolution of constraints (i.e., Ramachandran basins) rather than the backbone torsional coordinates themselves, the dynamics are judiciously simplified (43). The algorithm consists of a stochastic simulation of the coarsely resolved dynamics, simplified to the level of time-evolving Ramachandran-basin assignments. An operational premise is that steric restrictions imposed by the side-chains on the backbone may be subsumed into the basin-hopping dynamics. The side-chain constraints define regions in the Ramachandran map that can be explored in order to obtain an optimized pattern of nonbonded interactions.

The basin location of each residue coarsely defines the topology of the protein conformation. This string of basin locations, termed the local topology matrix or LTM(t), reflects the inherent geometrical constraints of a real polypeptide chain. The precise coordinates of the chain (i.e., the physical realization of an LTM) are defined by explicit Φ and Ψ angles determined by an optimization process that is turned on every 10 hopping steps. To maintain structural continuity during a folding trajectory, the explicit dihedral angles are retained for each residue from one time-step to the next until that residue Ramachandran basin is scheduled to change, according to a dynamic criterion (43).

Thus, to make torsional moves in 3-D space, translating the *"modulo-basin topology,"* a conformation is generated with a set of explicit Φ and Ψ angles compatible with the basin string. This explicit realization is used to identify the extent of structural involvement of each residue. Thus, as expected, the more structurally involved the residue, the less likely it may be engaged in a basin hopping, and the algorithmic rules do in fact reflect this fact. The degree of structural involvement is quantified energetically with a semiempirical potential. This potential is used to determine which residues change their Ramachandran basin in the next step. Upon a basin transition, the new structure is energetically minimized by changing Φ and Ψ angles within the chosen basins.

The basin-hopping probability is dependent on the extent of structural engagement of the residue, which is defined by the energetic cost associated with the virtual move of changing basin, higher the more structurally engaged the residue is. On the other hand, the probability of hoping to a target basin (given that a hoping move is scheduled to occur) depends on the target-basin lake area or its microcanonical entropy. To fit experimental folding measurements (*16, 17, 44*), a free residue is assigned a basin-hopping rate fixed at 10^9 Hz (*2, 6, 29*). The basic tenets governing interbasin hopping in the FM algorithm are: (i) interbasin hopping is slower than intrabasin exploration. This "adiabatic tenet" warrants a subordination of the backbone (Φ and Ψ) search to the LTM evolution or "*modulo-basin dynamics*" (*43*); (ii) side-chain torsional exploration occurs on a faster times-scale than backbone LTM dynamics. The last premise introduces a second adiabatic approximation (*43*), justifying the averaging of side-chain torsional motions in the stages of folding that precede a final side-chain fine tuning on the native backbone fold. This simplification is adequate to represent early stages of compaction and hydrogen-bond wrapping.

The FM captures the molecular basis of folding cooperativity by introducing an effective enhancement of dielectric-dependent two-body interactions according to the extent of wrapping of the interaction (Fig. 1), which translates as a rescaling of the zero-order (in-bulk) pairwise contributions depending on the number of wrapping side-chain groups. The weakening of hydrophobic attractions depending on the extent of hydrophobic burial of the paired nonpolar groups is treated in a similar manner.

The three main representations of the folding state of a chain captured by the FM, modulo-basin torsional state (LTM), 3-D, and wrapping, are illustrated in Figs. 6A–C. This figure shows the endpoint conformation of a representative simulation for the thermophilic variant of protein G (PDB code: 1gb4) performed at 313 K, pH 7, and consisting of 10^6 steps (*2*). The endpoint was largely reproduced in 66 of 91 runs and has RMSD ~ 4 Å from PDB entry 1GB4.

The wrapping model subsumed into the FM algorithm reflects the fact that hydrogen bonds are extremely context-sensitive. The algorithm, however, treats the solvent implicitly. This simplification requires that we introduce three-body correlations involving the wrapping residues (Fig. 6) as an operational means to incorporate rescalings of the intramolecular potential according to the microenvironmental modulations that take place during the course of folding. Explicitly, these correlations rescale the "zeroth-order" pairwise interactions by determining their extent of desolvation. For consistency, the wrapping of a hydrogen bond is also introduced in a coarse-grained manner in this analysis. Thus, in contrast with the detailed wrapping assessment (cf. Fig. 1), here the ρ-parameter indicates the number of residues contributing with nonpolar side-chain groups to the dehydration of the hydrogen bond. Each residue contributing to the dehydration of a hydrogen bond determines a three-body correlation.

The wrapping effect may also be cast in thermodynamic terms: due to their destabilizing effect on the nonbonded state, the hydrophobes surrounding a dielectric-dependent interactive pair become enhancers of the interaction.

A

	1	2	3	4	5	6	7	8	9	10
Aminoacid	MET-M	THR-T	THR-T	PHE-F	LYS-K	LEU-L	ILE-I	ILE-I	ASN-N	GLY-G
R-basin										
Phi-angle	-60.48	-53.47	-111.35	-146.29	-77.71	-142.19	-60.03	-136.53	-146.28	-83.41
Psi-angle	-50.21	-46.87	114.70	166.19	134.23	132.85	117.82	139.68	137.50	0.57
Omega-angle	180.00	180.00	180.00	180.00	180.00	180.00	180.00	180.00	180.00	180.00

	11	12	13	14	15	16	17	18	19	20
Aminoacid	LYS-K	THR-T	LEU-L	LYS-K	GLY-G	GLU-E	ILE-I	THR-T	ILE-I	GLU-E
R-basin										
Phi-angle	-77.16	-130.78	-61.56	-110.69	-76.55	-128.50	-141.75	-129.74	-131.06	-57.29
Psi-angle	2.03	132.05	-25.00	104.99	86.44	106.57	175.45	137.67	129.69	-32.23
Omega-angle	180.00	180.00	180.00	180.00	180.00	180.00	180.00	180.00	180.00	180.00

	21	22	23	24	25	26	27	28	29	30
Aminoacid	ALA-A	VAL-V	ASP-D	ALA-A	ALA-A	GLU-E	ALA-A	GLU-E	LYS-K	ILE-I
R-basin										
Phi-angle	-63.64	-60.87	-62.93	-144.65	-61.10	-59.75	-61.47	-61.16	-54.21	-61.01
Psi-angle	-32.49	-47.50	-33.90	136.37	-43.25	-27.33	-50.01	-46.13	-43.25	-46.07
Omega-angle	180.00	180.00	180.00	180.00	180.00	180.00	180.00	180.00	180.00	180.00

	31	32	33	34	35	36	37	38	39	40
Aminoacid	PHE-F	LYS-K	GLN-Q	TYR-Y	ALA-A	ASN-N	ASP-D	ASN-N	GLY-G	ILE-I
R-basin										
Phi-angle	-63.08	-54.65	-63.38	-58.81	-118.93	-67.19	-88.71	-78.66	80.98	-61.82
Psi-angle	-45.96	-44.88	-49.83	-27.40	114.72	-53.04	0.80	138.34	-94.77	-33.78
Omega-angle	180.00	180.00	180.00	180.00	180.00	180.00	180.00	180.00	180.00	180.00

	41	42	43	44	45	46	47	48	49	50
Aminoacid	ASP-D	GLY-G	GLU-E	TRP-W	THR-T	TYR-Y	ASP-D	ASP-D	ALA-A	THR-T
R-basin										
Phi-angle	-65.78	84.02	-66.69	-133.65	-64.78	-145.61	-76.24	-132.24	-68.87	-127.87
Psi-angle	-48.66	-3.46	-52.64	174.31	124.11	171.01	137.33	143.91	-54.06	114.38
Omega-angle	180.00	180.00	180.00	180.00	180.00	180.00	180.00	180.00	180.00	180.00

	51	52	53	54	55	56	57
Aminoacid	LYS-K	THR-T	PHE-F	THR-T	VAL-V	THR-T	GLU-E
R-basin							
Phi-angle	62.05	-81.78	-73.09	-129.16	-141.35	-62.94	59.51
Psi-angle	58.19	131.81	136.63	137.30	132.79	120.79	55.61
Omega-angle	180.00	180.00	180.00	180.00	180.00	180.00	180.00

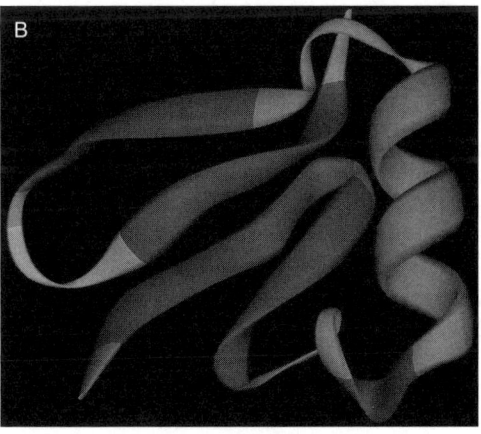

FIG. 6. (Continued)

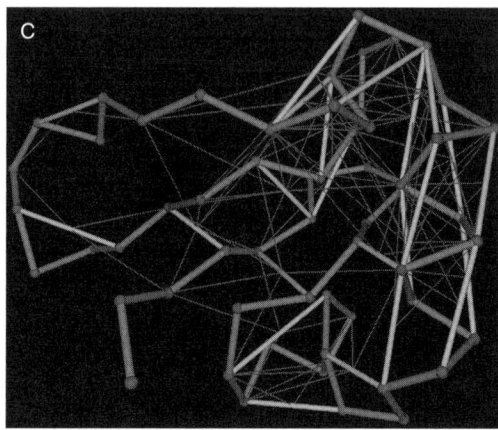

FIG. 6. Three representations of the conformational state of thermophilic variant of protein G (PDB code: 1GB4) obtained after 10^6 FM iterations. (A) LTM or backbone torsional state represented "modulo Ramachandran basin." Each basin is indicated by the quadrant in the (Φ, Ψ)-torus where it occurs. Thus, blue stands for the extended residue conformations including the β-strand states, red corresponds to a manifold of conformations containing the right-handed α-helix, green denotes the basin containing the left-handed helical conformation, while grey represents the basin in the lower right quadrant which is only accessible to Gly. (B) 3-D ribbon representation of the endpoint chain conformation. (C) Wrapping state of the chain in the endpoint conformation. The chain backbone is shown as virtual bonds joining α-carbons depicted in pink, hydrogen bonds are shown as grey segments joining α-carbons, and three-body "wrapping" correlations (cf. Fig. 1) are shown as thin blue lines joining the α-carbon of the wrapping residue with the center of the wrapped hydrogen bond. A wrapping residue is defined as a contributor of nonpolar groups to the hydrogen-bond microenvironment. Thus, the FM keeps track of the pairwise interactions as well as of the evolving microenvironments of such interactions determined by the evolving chain conformation.

As folding progresses, the effective hydrophobic energy contribution becomes progressively transferred into an effective energy of backbone desolvation in which the amide–carbonyl hydrogen bonds become determinants of protein structure and compensate for the rise in self-energy of the paired groups.

To take into account this effect, we have incorporated a phenomenological approach to wrapping electrostatics (13) as a means of accounting for changes in permittivity associated with changes in the conformation-dependent environments that affect pairwise interactions. This procedure entails a rescaling of the intramolecular potential terms as folding progresses, an operation requiring keeping track of vicinal hydrophobe positions with each FM iteration (cf. Fig. 6C).

C. Evolving Wrapping Patterns and Fixed Wrapping Motifs Along Folding Trajectories

To take into account the effect of wrapping on dielectric-dependent pairwise interactions, and assess its role in protein folding and cooperativity, we have adopted a semiempirical procedure to algorithmically keep track of the conformation-dependent environments that affect pairwise intrachain interactions. Thus, the in-bulk potential energy contributions are regarded as zero-order terms, while cooperative effects arise due to the wrapping of favorable interactions brought about by hydrophobic third-body participation (cf. Fig. 6C).

The basic question addressed by such endeavors (2, 29) becomes what is the dynamic relevance of optimal wrapping vis-à-vis the protein's commitment to fold? To tackle this question, we analyze a representative simulation for the thermophilic variant of protein G. This simulation performed at 313 K, pH 7, consists of 10^6 steps, and was essentially reproduced in 66 of 91 FM runs. All the runs generated a stationary fold within RMSD ~ 4 Å from the native structure (PDB entry 1GB44) and a dramatic decrease in potential energy around 0.6 ms (Fig. 7).

In accord with experimental tenets, direct examination of the time-dependent behavior of contact order suggests a nucleation process, whereby a sustainable large-scale organization is achieved only at 8×10^{-4} s after a relatively lengthy trial-and-error process (0–0.63 ms), followed by a critical regime (0.63–0.8 ms). Direct inspection of Fig. 8 reveals that the transition from local to large-scale organization is actually defined by a sudden burst in the number of three-body correlations starting at 6.0×10^{-4}–7×10^{-4} s.

The nucleation picture revealed by Fig. 8 has been further confirmed by examination of total internal energy of the peptide chain and solvent-exposed area plots. The energy experiences a sudden decrease in the 6.3×10^{-4}–8.0×10^{-4} s region concurrent with a dramatic decrease in the solvent-exposed area. The point at which the protein is actually committed to fold can be inferred by performing runs with different starting conformations extracted from the 6.3×10^{-4}–8.0×10^{-4} s time window. This commitment arises when a sustainable number of three-body correlations (native or nonnative) equal to or larger than the final almost stationary number are reached (Fig. 9). In the case of protein G, the burst time window is 6.3×10^{-4}–8.0×10^{-4} s and a sustainable population of three-body correlations is maintained in the region $7.0(\pm 0.2) \times 10^{-4}$ s.

A similar computation was carried out for ubiquitin using the best FM runs described in (45). The exposed surface area at the transition state is estimated to be 7200 Å^2, while the random coil conformation exposes approximately 10,800 Å^2. Thus, we find that the transition state buries 3600 Å^2, approximately 60% of the total area buried in the native fold, in good agreement with the experimental results (16).

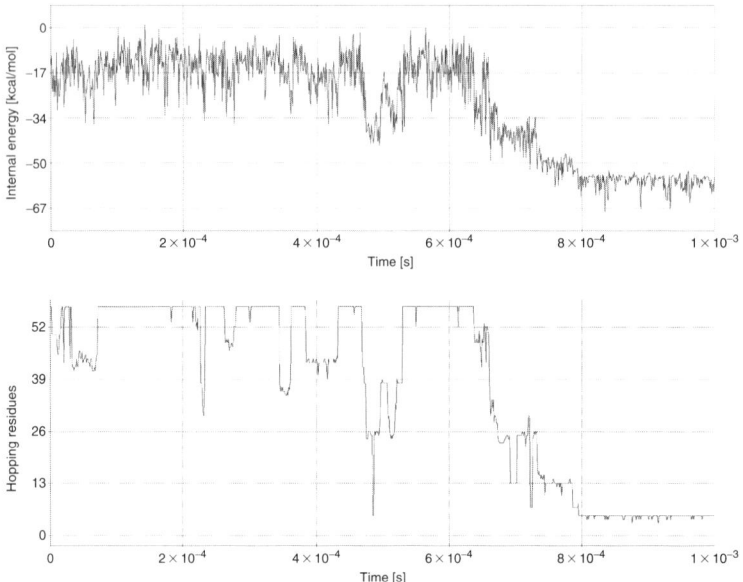

FIG. 7. Internal energy and number of hopping residues along a reproducible representative FM trajectory for the thermophilic variant of protein G. A hopping residue is defined as tagged for a Ramachandran basin transition in the coarse-grained stochastic process that underlies the FM torsional dynamics (10).

The dominant folding pathway for the variant of protein G is coarsely displayed in Fig. 8, where the abscissas denote residue numbers and the ordinates, the time axis. The Ramachandran basin assignment for each residue is given as a function of time and the topology of the entire chain is designated by a row in the histogram. The blue color (basin 1) designates the large basin containing the β strand and PP2 conformations, the red color (basin 2) designates the basin containing the right-handed helix conformation, green (basin 3) designates the basin containing the left-handed helix coordinates, and gray (basin 4) corresponds to the fourth basin present only in glycine. The figure clearly reveals the emergence of a stable large-scale organization that prevails after a critical period located at the 6.3×10^{-4}–8.0×10^{-4} s interval.

We have also reported the extent of protection on backbone hydrogen bonds along the folding pathway. Figure 10 shows that the average extent of hydrogen-bond protection, $\rho(t)$, converges to the value $\rho = 5$ in the long-time limit that starts right after the trial-and-error period; that is, at the sharp burst in $C_3 = C_3(t)$. This regime is associated with the region $C_3 > 60$. The stationary native-like population of $15(\pm 1)$ backbone hydrogen bonds are protected by $75(\pm 5)$ three-body correlations. We see that the $\rho = 5$ value becomes an

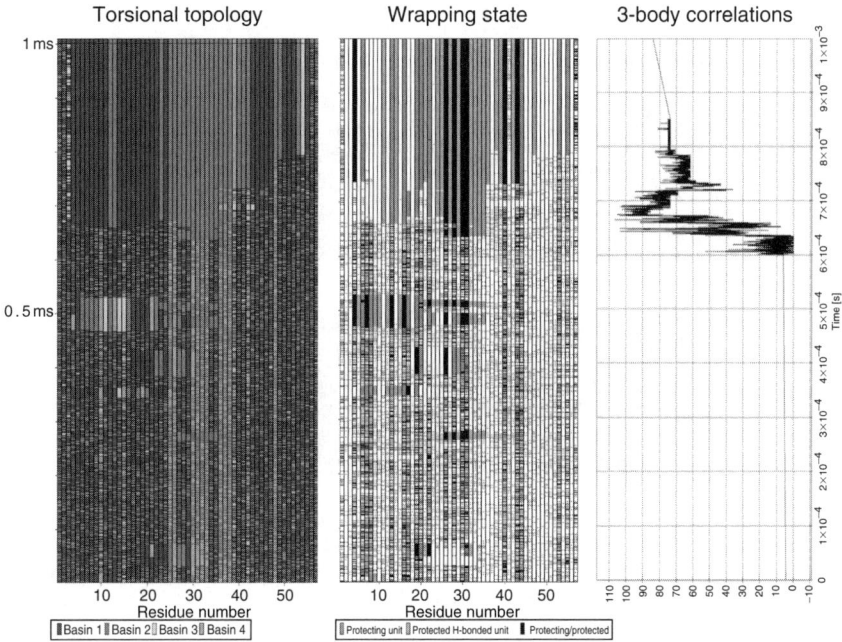

FIG. 8. Three views of the wrapping/folding dynamics for the thermophilic variant of protein G obtained from a representative expeditious FM trajectory. The left panel represents the time evolution of the local topology matrix (LTM) with the abscissas denoting residue numbers and the ordinates, the time axis. The Ramachandran basin assignment for each residue is given as a function of time and the topology of the entire chain is designated by a row in the histogram. Color convention is consistent with Fig. 6A. The middle panel represents the different roles exerted by residues along the chain at different times. Thus, a pink residue-time entry indicates a residue exerting a protecting or wrapping role at the particular time, a grey entry indicates a residue engaged in pairwise interaction which is being protected or wrapped by other hydrophobes, and a black entry indicates a dual role as protector or wrapper and also engagement in a hydrogen bond which is being exogenously protected. The right panel indicates the total number of three-body correlations representing the wrapping dynamics concurrent with the expeditious folding process. Notice that the burst phase in three-body correlations coincides with the region of transition from a trial-and-error phase to a sustainable structure.

approximate constant of motion in the critical region and beyond, that is for $C_3 > 60$. A similar result holds for ubiquitin (2, 45): The native-like stationary population of $28(\pm 3)$ backbone hydrogen bonds is now wrapped by $140(\pm 5)$ three-body correlations.

Thus, the wrapping results (2, 45) proved to be more specific and clear-cut than earlier attempts at establishing whether buried surface area is commensurate with hydrogen-bond formation (16, 17, 44). It is difficult to infer from such studies whether hydrophobic collapse triggers hydrogen-bond formation,

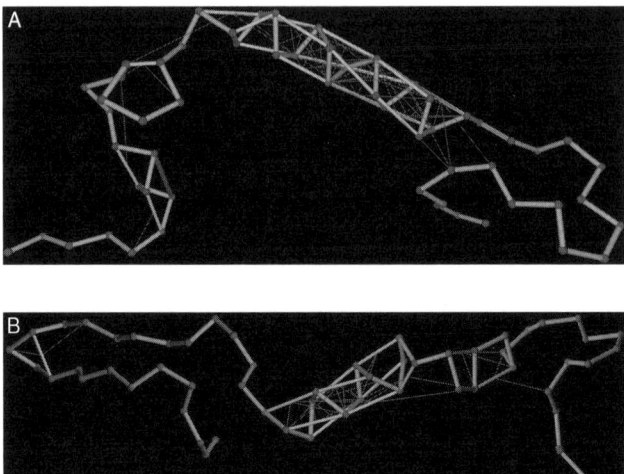

Fig. 9. Wrapping patterns for chain conformations occurring during the burst phase (cf. Fig. 8) of protein-G variant generated at 6.4×10^{-4}s (A) and 6.5×10^{-4}s (B). These conformations commit the chain to fold, are part of the "transition state ensemble," and do not contain exclusively native interactions, as it becomes apparent in panel A.

or vice versa. However, Figs. 8 and 10 reveal that the productive build-up in hydrogen-bond formation beyond the "trial-and-error folding regime" is invariably assisted by the nearly constant $\rho = 5$ wrapping value.

To further validate our conclusions by contrasting them against all-atom molecular dynamics simulations, we have analyzed one of the longest all-atom simulations with explicit solvent: the 1 µs simulation of the villin headpiece by Duan and Kollman (30). Their trajectory was examined using the FM's representation and visualization tools, and provides valuable information on the wrapping dynamics in the trial and error regime. The analysis revealed an almost constant proportionality between C_3 and Q along the entire trajectory (2). Taken together, our reported results (2, 45) reveal that $\rho = 5$ is likely to be a constant of motion for the folding trajectory. Testing the universal validity of this constant of motion may prove to be a daunting task, as the wrapping of all or most good folders would need to be dynamically investigated. Nevertheless, the preceding findings instill confidence in the universal validity of this hypothesis.

A theoretical analysis based on a nanoscale treatment of the solvent dielectrics further supports this hypothesis. Thus, in (2) we have proven that a hydrogen bond is embedded in the lowest dielectric when surrounded by five average-shaped hydrophobic residues, and this optimal wrapping arrangement represents a compromise between crowding and proximity to the hydrogen

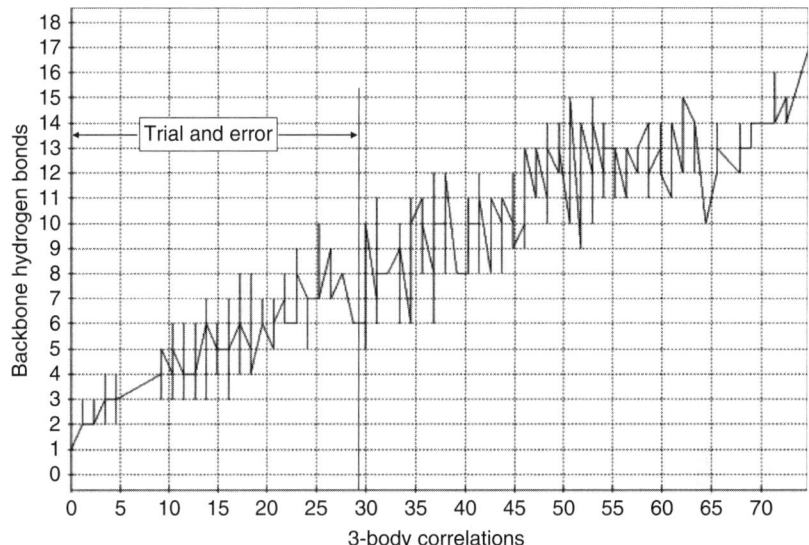

FIG. 10. Number of backbone hydrogen bonds plotted against number of three-body correlations extracted from representative FM trajectory capturing the wrapping/folding dynamics for the thermophilic variant of protein G.

bond subject to the steric constraints determined by a minimum approach distance. This motif is realized only approximately in real folded proteins due to the diversity of shapes and sizes of the wrapping side-chains. This approximate law holds for both native structure (33) and folding dynamics (2). In this regard, this wrapping motif may be regarded as a structural element that captures the basic component of energy transduction from hydrophobic association to structure formation. Furthermore, it implies that a fundamental constraint in protein architecture applicable to native structures applies also throughout the folding trajectory.

Kinetic amide isotope studies by the Sosnick group (16, 44) imply that helical hydrogen-bond formation and surface area burial form to the same degree in the transition state for single-domain proteins that fold in a two-state manner. Extensive helix formation does not occur prior to significant hydrophobic association at the limiting step. The surveyed results on individual folding trajectories indicate that commensurate burial occurs both on the way up to the limiting barrier (prenucleation) as well as afterwards, in consistency with the ensemble-average evidence stemming from the kinetic folding experiments.

D. Solvation Nanoscale Model of Wrapping and Cooperativity

The aim of this section is to describe a semiempirical model of nanoscale solvation that captures the dielectric modulation brought about by the approach of a hydrophobe to a protein hydrogen bond (2, 12, 13). In essence, the model captures the solvent-ordering effect promoted by the hydrophobe and quantifies the effect of this induced organization on the electrostatics of a preformed amide–carbonyl hydrogen bond (Fig. 11). This model reproduces the crossover point in hydrogen-bond dehydration propensity that characterizes the folding transition state if we adopt the extent of backbone hydrogen-bond dehydration as generic reaction coordinate for the folding process (9).

In view of the dynamic results presented in the previous section, the description of the nano-solvation model can be made more specific. Here, the aim is to report on a rationalization of the $n = 5$ coarse-wrapping motif (33) through an implicit-solvent model that reproduces the modulation of the dielectric environment in which intramolecular hydrogen bonds are formed (46). We defined a Cartesian coordinate system by placing the carbonyl oxygen atom effective charge q at the center of coordinates, define the x-axis as that along the carbonyl–amide hydrogen bond, and place the amide hydrogen atom at position $\mathbf{r}$, 1.4–2.1 Å away along the positive x-axis. We assume the hydrogen bond to be surrounded by a discrete number of identical spherical hydrophobic units of radius $d/2$ (the parameter d is defined below) centered at fixed positions $\mathbf{r}_j, j = 1, 2, \ldots, n$. This is an idealized picture but one that can be dealt with analytically.

Previously reported implicit-solvent approaches (46) take into account the solvent structuring induced by the solvent–hydrophobe interface (cf. Fig. 11), translate this effect into a distance-dependence permittivity, $\varepsilon(\mathbf{r})$, and quantify the effect on the coulomb screening. A more heuristic, practical, and phenomenological approach has been reported earlier (2) and is rooted in two pivotal components: (i) perturbation of the diffraction structure of bulk water as hydrophobes are incorporated at fixed positions; and (ii) recovery of their solvent-structuring effect by inverse Fourier transforming the previous result given in frequency space.

To propagate the solvent-structuring effect induced by the presence of the hydrophobic spheres, we replaced the position-dependent dielectric by an integral kernel convoluted with the electric field at position $\mathbf{r}$ to represent the correlations with the field at neighboring positions $\mathbf{r}'$. This prompts us to replace the classical Poisson equation by the heuristic relation:

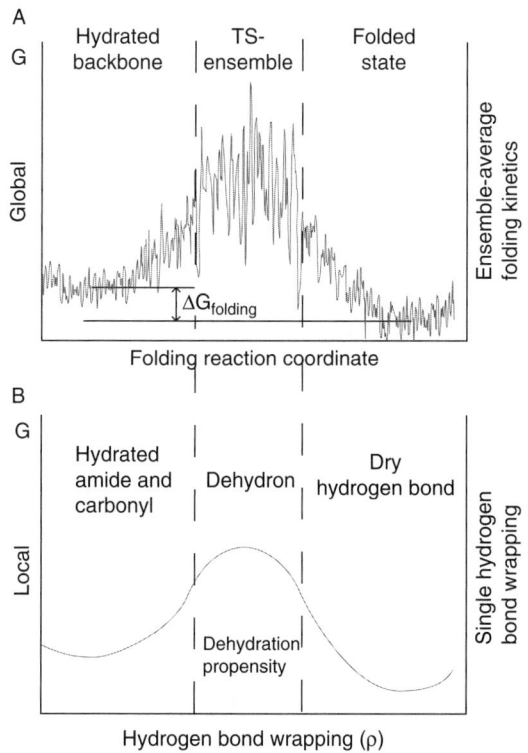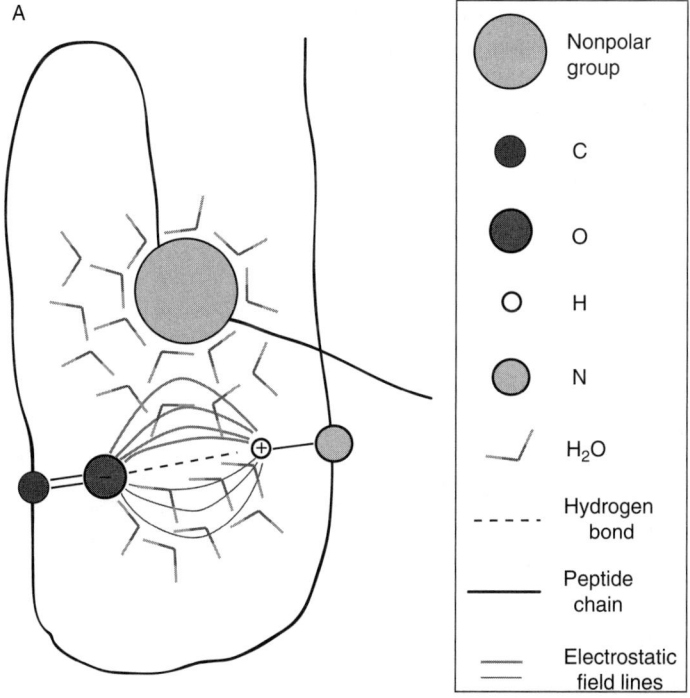

$$\text{div}\left[\int K(\mathbf{r},\mathbf{r}',\{\mathbf{r}_j\})E(\mathbf{r}')d\mathbf{r}'\right] = 4\pi q\delta(\mathbf{r}) \quad (3)$$

where the kernel $K(\mathbf{r},\mathbf{r}',\{\mathbf{r}_j\})$ is parametrically dependent on the fixed hydrophobe positions. In the absence of vicinal hydrophobic units, the correlations decay as $\exp(-\|\mathbf{r}-\mathbf{r}'\|/\xi)$ (ξ = characteristic correlation length). In the limit $\xi \to 0$, we get: $K(\mathbf{r},\mathbf{r}') \sim \delta(\mathbf{r}'-\mathbf{r})$, yielding the standard Poisson equation.

The correlation kernel reflects the relationship between diffraction and dielectric. For bulk water, we get $K(\mathbf{r},\mathbf{r}') = K(\mathbf{r}-\mathbf{r}')$ by inverse transforming its frequency $\mathbf{k}$-vector representation:

$$K(\mathbf{r}-\mathbf{r}') = \int \exp[i\mathbf{k}\cdot(\mathbf{r}-\mathbf{r}')]L(\mathbf{k})d\mathbf{k} \quad (4)$$

Here $L(\mathbf{k}) = [(\varepsilon_w - \varepsilon_0)/(1 + \varepsilon_w\|\mathbf{k}\|^2\xi^2/\varepsilon_0) + \varepsilon_0]$, $\xi \approx 5$ Å denotes the characteristic length, and ε_w and ε_0 are the permittivities of water and vacuum, respectively. To obtain the correlation kernel with n hydrophobic units, we need to incorporate their solvent-structuring effect:

$$K(\mathbf{r},\mathbf{r}',\{\mathbf{r}_j\}) = \left[\int \exp[i\mathbf{k}\cdot(\mathbf{r}-\mathbf{r}')]L(\mathbf{k})d\mathbf{k}\right] \times \left[1 + \sum_{j=1,\ldots,n}\Gamma_j(\mathbf{r},\mathbf{r}')\right] \quad (5)$$

On phenomenological grounds we get (2, 46): $\Gamma_j(\mathbf{r},\mathbf{r}')\sim\exp[-(\|\mathbf{r}-\mathbf{r}_j\| + \|\mathbf{r}'-\mathbf{r}_j\|)/\Lambda]$, for $\|\mathbf{r}-\mathbf{r}_j\|$ and $\|\mathbf{r}'-\mathbf{r}_j\| > d/2$ with characteristic length Λ subsuming the range of the water-structuring effect. This parameter is typically fixed at 1.8 Å, the effective thickness of a single water layer (cf. Fig. 11).

As previously reported, we have solved Eqs. (3) and (5) by Fourier transformation (2). We obtain the electric field $E(\mathbf{r})$ by inverse Fourier transformation of the solution to Eq. (3) in k-representation:

$$\int E(\mathbf{r})d\mathbf{r} = -(4\pi q)\int \exp(i\mathbf{k}\cdot\mathbf{r})\|\mathbf{k}\|^{-2}[K(\mathbf{k},\{\mathbf{k}_j\})]^{-1}d\mathbf{k} \quad (6)$$

FIG. 11. Basic tenets of nanoscale solvation model. (A) The model captures the cross-over behavior in hydrogen-bond wrapping that translates at the folding-kinetics level into a commitment of the chain to fold. (B) The solvent ordering promoted by the approaching hydrophobe enhances the electrostatics, an effect that must be captured by the model. The induced organization decreases the polarizability of the environment, preventing water dipoles from aligning with the electrostatic field lines. The thicker lines represent a stronger field. By contrast, the region exposed to bulk water facilitates dipole organization along the field lines, weakening the electrostatic field (thin lines).

Direct residue evaluation at the first-order poles $k = \pm i(\varepsilon_0/\varepsilon_w)^{1/2}\xi^{-1}$ ($k = \|\mathbf{k}\|$) and $\mathbf{k} = \mathbf{k}_j \pm i\Lambda^{-1}$, yields the electric field $E(\mathbf{r})$ by retaining only the real part in the residue calculation:

$$E(\mathbf{r}) = (q/r^2)[(\varepsilon_0^{-1} - \varepsilon_w^{-1})\Omega(\{\mathbf{r}_j\})(1 + r/\xi)\exp(-r/\xi) + \varepsilon_w^{-1}] \quad (7)$$

where:

$$\Omega(\{\mathbf{r}_j\}) = \prod_{j=1,\ldots,n} [1 + \exp(-\|\mathbf{r}_j\|/\Lambda)] \times [(1 + \exp(-\|\mathbf{r} - \mathbf{r}_j\|/\Lambda)] \quad (8)$$

and $r = \|\mathbf{r}\|$. Equations (7) and (8) describe the net effect of the wrapping hydrophobic arrangement on the electric field. In accord with Eqs. (7) and (8), the wrapping effect can be captured by replacing the permittivity constant ε_w for bulk water by an effective permittivity $\varepsilon = \varepsilon(\mathbf{r},\{\mathbf{r}_j\})$ defined as:

$$\varepsilon = [(\varepsilon_0^{-1} - \varepsilon_w^{-1})\Omega(\{\mathbf{r}_j\})(1 + r/\xi)\exp(-r/\xi) + \varepsilon_w^{-1}] \quad (9)$$

This quantity tends to the bulk-limit for long interaction distances ($r/\xi \to \infty$).

At this point, we may turn to the problem of finding the optimal wrapping arrangement and contrasting our implicit-solvent result with the phenomenological result emerging from the simulations (cf. Figs. 8 and 10). Since $\varepsilon_0^{-1} \gg \varepsilon_w^{-1}$, finding the wrapping cluster with the lowest dielectric in its interior is tantamount to finding the arrangement $\{\mathbf{r}_j\}$ that maximizes the function $\Omega(\{\mathbf{r}_j\})$, in accord with Eq. (9). We thus reported (2) the optimal arrangement $\{\mathbf{r}_j\}$ of hydrophobes that yields the maximum value Ω^* for $\Omega(\{\mathbf{r}_j\})$. First, we computed the maximum $\Omega^\circ(n)$ of $\Omega(\{\mathbf{r}_j\})$ for each fixed n subject to the constraint of preserving a minimum distance d between any two hydrophobes. The d was taken to be 5 Å in accord with typical minimal distances between α carbon atoms in tertiary structure (33). Our results are qualitatively invariant in the range 4.5 Å $\leq d \leq$ 6 Å. Using the Lagrange multipliers method to minimize the effective permittivity, we find that the optimal arrangement is invariably obtained by fixing $n-2$ hydrophobes at distance d from each other and equidistantly from the O and H atoms, and placing the remaining two along the x-axis at distance $(\Lambda + \eta)(1 - n^{-2})$ (to first approximation) away from the C and N atoms, with η = C–O distance in the carbonyl group. This gives for $n = 4$ (tetrahedron): $\Omega^\circ(4) = 3.419$; for $n = 5$ (trigonal bipyramid): $\Omega^\circ(5) = 4.144$; for $n = 6$ (square bipyramid), $\Omega^\circ(6) = 3.952$; and for $n = 7$ (pentagonal bipyramid), $\Omega^\circ(7) = 3.421$. Similar calculations for all n allow us to establish the following order relations:

$$\Omega^*(3) < \Omega^*(4) < \Omega^*(5) > \Omega^*(6) > \Omega^*(7) > \ldots \quad (10)$$

Thus, $\Omega°(n)$ has a single maximum at $n = 5$. This maximum is expected on the basis of the two conflictive tendencies in the stabilization of a hydrogen bond: (i) bringing close to the hydrogen bond as many hydrophobes as possible and (ii) bringing them as close to the hydrogen bond as possible. However, both demands start becoming mutually incompatible due to the steric hindrances implicit in the Lennard-Jones repulsive terms.

Thus, in full agreement with the result described in Fig. 10, a hydrogen bond is embedded in the lowest dielectric when surrounded by five wrapping residues, the optimal compromise between crowding and proximity to the hydrogen bond. In practice, this wrapping motif is realized only approximately due to the diversity of shapes and sizes of the amino acid side-chains.

IV. Protein Under-Wrapping, Misfolding, and Aggregation

A. Pathogenically Under-Wrapped Proteins

C. M. Dobson provided significant evidence supporting the view that amyloidogenic aggregation, an often pathogenic state of proteins, is a generic phase of peptide chains (47–49). The term generic phase refers to a way of organizing in three dimensions dominated by main-chain interactions and essentially oblivious of the information encoded in the primary sequence. Such intermolecular associations appear to be dominated by a basic structural motif: the cross-β structure (50), an intermolecular sheet-pleated pattern ubiquitous in the fibrous state of aggregation. We should emphasize that this assertion remains conjectural, as no crystal of the fiber for natural prions has been obtained as yet. On the other hand, the alternative folded phase of the peptide chain is relatively well understood: natural soluble proteins tend to adopt single-molecule conformations of marginal stability, often requiring binding partnerships or complexation to preserve the integrity of the so-called native fold (21).

While the folding process and its final stable outcome are very much dependent on the amino-acid composition of the chain (1), the amyloid state appears to be fairly insensitive to the information encoded in the side chains: At first sight, amyloidogenic aggregation does not seem to require an "aggregation code." But further analysis reveals that it must place severe constraints on the primary sequence, as some proteins tend to be relatively prone to aggregate even under physiological conditions (11), while others require extreme conditions to do so or simply do not aggregate reproducibly (47, 51). In addition, negative-design features of the folded state have been recognized as responsible for averting aggregation (52). Thus, it is not entirely correct to characterize the aberrant aggregation as a "polymer physics phase," shared by polypeptides with arbitrary, suboptimal, or random sequence, in contrast with the folded state, whose existence and integrity is determined typically unambiguously by the primary sequence.

Clearly, a selection pressure operates to optimize the primary sequence, so it can render a good folder, that is, an expeditious structure seeker, and a stable soluble structure. This optimization is needed to prevent the functionally competent fold from reverting to a primeval amyloid phase. On the other hand, certain sequences are better optimized to escape aggregation than others even under conditions known to sustain the native fold (*11*).

While amyloidogenic aggregation has been shown to be always plausible provided sufficiently stringent denaturation conditions are applied (*47–51*), a marked amyloidogenic propensity has been detected on a number of proteins under physiological or near-physiological conditions, particularly if the monomeric folding domain is deprived of its natural interacting partners (*22*). Such findings imply that not all soluble structures have been optimized to the same degree in order to avert aggregation, and that the more reliant the structure is on binding partnerships or complexations, the more vulnerable it becomes in regards to reverting to the primeval phase. Thus, an over-expression of a folding domain with high complexation requirements *in vivo*, or the modification of its binding partners as a result of genetic accident, or any factor that distorts its natural interactive context are likely to bolster a transition to an amyloidogenic state (*11*).

These observations lead us to the following question: What type of deficiency in the native fold constitutes a signal for aberrant aggregation? A recent assessment of the *wrapping* of soluble structure (*12–14*) might prove critical to address this problem. For highly under-wrapped proteins (~50% dehydrons), densities higher than 4 dehydrons per 1000 Å^2 on the protein surface become inducers of protein aggregation (*11*). This observation turns the wrapping analysis into a powerful diagnosis tool.

Thus, the condition of "keeping the structure dry in water" becomes a requirement to preserve the structural integrity of soluble proteins and imposes a severe building constraint (and thereby an evolutionary pressure) on such proteins. Thus, it is expected that the optimization of the structures resulting from this type of evolutionary constraint would be uneven over a range of soluble proteins, resulting in marked differences in aggregation propensity.

All in all, this analysis of the backbone desolvation of the native state supports and clarifies the physical picture put forth by Dobson (*47, 48*), in which amyloidogenic propensity depends crucially on the fact that main-chain interactions become dominant in detriment of the amino-acid sequence that encodes the folded state. Precisely, main-chain interactions may dominate as the main chain of the folded state is not properly protected from water attack (*11*). It is instructive to compare this statement with the local analysis of Avbelj and

Baldwin (15) in the sense that backbone solvation is a determinant of β-sheet propensity. Thus, an over-exposed backbone hydrogen bond in the native fold is an indicator of a failure in folding cooperativity, as it reveals an inability to remove water from an interactive polar pair by means of a many-body correlation, and at the same time, it is a signal enabling the diagnosis of amyloidogenic propensity.

Often, the inability to properly wrap a structure intramolecularly is compensated by protein complexation. This clarifies the physical picture suggesting that the more dependent the folding domain is on its interactive partnerships to preserve its structure, the more likely it is to be prone to revert to its primeval aggregated phase. In essence, we could regard these competing structural alternatives as reflecting a struggle for the survival of backbone hydrogen bonds. Thus, the wrapping concept enables us to discern why some soluble proteins may have been better optimized to avoid amyloidogenic aggregation than others.

Direct inspection of the pattern of desolvation of the main chain clearly reveals that the cellular fold of the human prion (Fig. 5) is too vulnerable to water attack and the same time too sticky to avert aggregation. Clearly, its sequence has not been optimized to "keep the backbone hydrogen bonds dry" in the folded state. In fact, their extent of exposure of backbone hydrogen bonds is the highest among soluble proteins in the entire PDB, with the sole exception of some toxins whose stable fold is held together by a profusion of disulfide bridges (21).

It is suggestive that an inability to protect the main chain is precisely the type of deficiency that best correlates with a propensity to reverse to a primeval aggregation phase determined by main-chain interactions. The actual mechanism by which such defects induce or nucleate the transition is still opaque, although the inherent adhesiveness of packing defects obviously plays a role.

Recently, an atomic-detail structure of a fibrillogenic aggregate, with its β-sheets parallel to the main axis and the strands perpendicular to it, was reported and revealed a tight packing of β-sheets (53). The cross-β spine of the fibrillogenic peptide GNNQQNY reveals a double parallel β-sheet with tight packing of side chains leading to the full dehydration of intrasheet backbone–backbone and side-chain–side-chain hydrogen bonds. However, there is not a single pair-wise interaction between the β–sheets, no hydrogen bond and no hydrophobic interaction. Instead, a direct examination of the crystal structure reveals that the association is driven by the dehydration propensity of preformed intrasheet dehydrons, as depicted in Fig. 12.

To conclude, we may include some evolutionary remarks. A paradigmatic discovery in biology revealed that folds are conserved across species to perform specific functions. However, the wrapping of such folds is clearly not conserved (24, 25). This fact suggests how complex physiologies may be achieved without dramatically expanding genome size, a standing problem in biology. Considerable

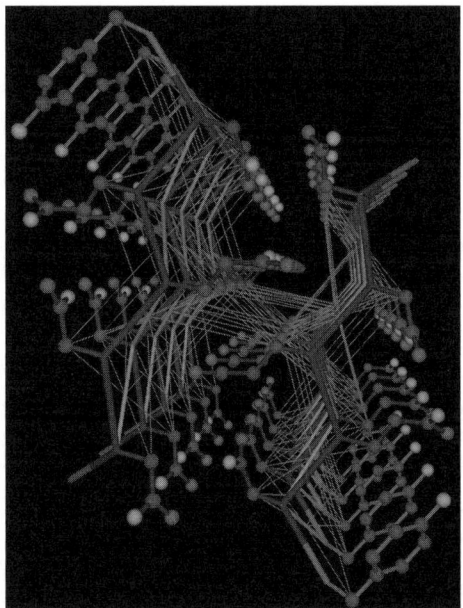

FIG. 12. Pattern of intersheet wrapping of backbone–backbone hydrogen bonds in the fibrillogenic state of peptide, GNNQQNY. The strand backbone is represented as virtual bonds (blue) joining α-carbons, and backbone hydrogen bonds are displayed as light grey lines joining the α-carbons of the paired residues. A thin blue line from the β-carbon of a residue to the baricenter of a hydrogen bond indicates wrapping of the bond by the residue: at least one nonpolar group from the residue is contained in the desolvation sphere of the bond.

network complexity may be achieved by actually fostering a higher level of complexation or binding partnership, as promoted by a more precarious wrapping of the isolated protein domains. According to our previous analysis, such complex design entails an inherent danger: the reversal of highly underwrapped folding domains to an amyloidogenic phase even under physiological conditions. Amyloidosis is, thus, likely to be a consequence of high complexity in proteomic connectivity, as dictated by the structural fragility of highly interactive proteins (13, 26). Thus, the relationships between network centrality, structural wrapping, and propensity for aberrant aggregation merit further investigation.

B. Under-Wrapped Proteins and Epigenetic Polymorphism

Prion-like aggregation has been widely recognized as a causative of pathogenic dysfunction (54), but Susan Lindquist and her co-workers (55) have demonstrated that there may be also epigenetic consequences to prion-like aggregation. These novel insights lead to a significant extension of the prion hypothesis. Furthermore, while a connection between protein under-wrapping and propensity for aberrant amyloidogenic aggregation has been unraveled (*11*), no misfolding inference can be made for cases when the extent of under-wrapping is so severe that no soluble structure can be maintained by the monomeric chain. This case appears to be of biological interest in the light of work by the Lindquist group (*55*). This group has focused on nonpathogenic yeast prions capable of promoting phenotypic polymorphism through a transmittable conformational change that they regard as having epigenetic consequences, thus defining a protein-based element of inheritance. The yeast prions that this group has studied, especially *PSI* in yeast gene Sup35, are unlikely to sustain a monomeric structure as their biophysical experiments reveal.

From the wrapping perspective, this is so because the prion sequences contain large windows (>30 residues) containing mostly under-wrapping residues (especially G, N, Q, S, A, P). In turn, these residues are under-wrappers because they contain few nonpolar side-chain groups, thereby exposing the backbone to the solvent, while some (i.e., S, N, Q) even prevent other nonpolar groups from clustering around the backbone hydrogen bonds. The other under-wrapping residue, D, is unlikely to be found in such prions as its charge promotes hydration, thus introducing a negative design element for aggregation.

These observations are in accordance with the structural characterization by Krishnan and Lindquist (*55*), which places yeast prions in the twilight between order and disorder, in consonance with the wrapping-based prion characterization (*13*).

Thus, the epigenetic consequences associated with misfolding of severely under-wrapped proteins incapable of sustaining monomeric structure should become mandatory items in the research agenda. The focus of such endeavors are proteins endowed with large windows (>30) of under-wrapping residues. When combined with tyrosine (probably needed for stacking), such motifs are likely promoters of self-aggregation leading to pathology (*11*) or epigenetic prion-based phenotype polymorphism (*55*). Thus, future research efforts should be directed at identifying highly under-wrapped human genes containing the sequence motifs indicated, and assessing the epigenetic consequences of their transformation into sequestered aggregates. A preliminary scanning of circa 16,000 human proteins in Swiss-Prot reveals 13 genes with the severely under-wrapped motif. Among them, RNF12, AF9, MAML2 are implicated in cancer promotion, while seven are involved in transcription regulation, hence with likely epigenetic consequences associated with a conformational switching into sequestered aggregate states.

The research of A. F. is supported by NIH grant R01-GM072614, and by the John and Ann Doerr Fund for Computational Biomedicine.

References

1. Anfinsen, C. B. (1973). Principles that govern the folding of protein chains. *Science* **181**, 223–230.
2. Fernández, A., Sosnick, T. R., and Colubri, A. (2002). Dynamics of hydrogen-bond desolvation in folding proteins. *J. Mol. Biol.* **321**, 659–675.
3. Cieplack, M. (2004). Cooperativity and contact order in protein folding. *Phys. Rev. E* **69**, 031907.
4. Jewett, A., Pande, V. S., and Plaxco, K. W. (2003). Cooperativity, smooth energy landscapes and the origins of topology-dependent protein folding rates. *J. Mol. Biol.* **326**, 247–253.
5. Scalley-Kim, M., and Baker, D. (2004). Characterization of the folding energy landscapes of computer generated proteins suggests high folding free energy barriers and cooperativity may be consequences of natural selection. *J. Mol. Biol.* **338**, 573–583.
6. Fernández, A., Colubri, A., and Berry, R. S. (2002). Three-body correlations in protein folding: The origin of cooperativity. *Physica A* **307**, 235–259.
7. Fernández, A., Kostov, K., and Berry, R. S. (1999). From residue matching patterns to protein folding topographies: General model and bovine pancreatic trypsin inhibitor. *Proc. Natl. Acad. Sci. USA* **96**, 12991–12996.
8. Fernández, A., Colubri, A., and Berry, R. S. (2000). Topology to geometry in protein folding: Beta-lactoglobulin. *Proc. Natl. Acad. Sci. USA* **97**, 14062–14066.
9. Fernández, A., Kardos, J., and Goto, J. (2003). Protein folding: Could hydrophobic collapse be coupled with hydrogen-bond formation. *FEBS Lett.* **536**, 187–192.
10. Fernández, A. (2001). Conformation-dependent environments in folding proteins. *J. Chem. Phys.* **114**, 2489–2502.
11. Fernández, A., Kardos, J., Scott, R., Goto, Y., and Berry, R. S. (2003). Structural defects and the diagnosis of amyloidogenic propensity. *Proc. Natl. Acad. Sci. USA* **100**, 6446–6451.
12. Fernández, A. (2004). Keeping dry and crossing membranes. *Nat. Biotechnol.* **22**, 1081–1084.
13. Pietrosemoli, N., Crespo, A., and Fernández, A. (2007). Dehydration propensity of order-disorder intermediate regions in soluble proteins. *J. Proteome Res.* **6**, 3519–3526.
14. Fernández, A., and Scott, R. (2003). Dehydron: A structure-encoded signal for protein interactions. *Biophys. J.* **85**, 1914–1928.
15. Avbelj, F., and Baldwin, R. L. (2003). Role of backbone solvation and electrostatics in generating preferred peptide backbone conformations: Distributions of phi. *Proc. Natl. Acad. Sci. USA* **100**, 5742–5747.
16. Krantz, B. A., Moran, L. B., Kentsis, A., and Sosnick, T. R. (2000). D/H amide kinetic isotope effects reveal when hydrogen bonds form during protein folding. *Nat. Struct. Biol.* **7**, 62–71.
17. Fersht, A. (2000). Transition-state structure as a unifying basis in protein-folding mechanisms: Contact order, chain topology, stability, and the extended nucleus mechanism. *Proc. Natl. Acad. Sci. USA* **97**, 1525–1929.
18. Plaxco, K. W., Simmons, K. T., and Baker, D. (1998). Contact order, transition state placement and the refolding rates of single domain proteins. *J. Mol. Biol.* **277**, 985–994.
19. Fernández, A., and Scott, L. R. (2003). Adherence of packing defects in soluble proteins. *Phys. Rev. Lett.* **91**, 018102.
20. Fernández, A. (2003). What caliber pore is like a pipe? Nanotubes as modulators of ion gradients *J. Chem. Phys.* **119**, , Communication°°, 5315–5319.
21. Fernández, A., and Scheraga, H. A. (2003). Insufficiently dehydrated hydrogen bonds as determinants of protein interactions. *Proc. Natl. Acad. Sci. USA* **100**, 113–118.

22. Fernández, A., and Berry, R. S. (2003). Proteins with H-bond packing defects are highly interactive with lipid bilayers: Implications for amyloidogenesis. *Proc. Natl. Acad. Sci. USA* **100,** 2391–2396.
23. Fernández, A. (2004). Functionality of wrapping defects in soluble proteins: What cannot be kept dry must be conserved. *J. Mol. Biol.* **337,** 477–483.
24. Fernández, A., Scott, R. L., and Berry, R. S. (2004). The nonconserved wrapping of conserved folds reveals a trend towards increasing connectivity in proteomic networks. *Proc. Natl. Acad. Sci. USA* **101,** 2823–2827.
25. Fernández, A., and Berry, R. S. (2004). Molecular dimension explored in evolution to promote proteomic complexity. *Proc. Natl. Acad. Sci. USA* **101,** 13460–13465.
26. Fernández, A. (2007). Molecular basis for evolving modularity in the yeast protein interaction network. *PLoS Comp. Biol.* **3,** e226.
27. Deremble, C., and Lavery, R. (2005). Macromolecular recognition. *Curr. Opin. Struc. Biol.* **15,** 171–175.
28. Ma, B., Elkayam, T., Wolfson, H., and Nussinov, R. (2003). Protein–protein interactions: Structurally conserved residues distinguish between binding sites and exposed protein surfaces. *Proc. Natl. Acad. Sci. USA* **100,** 5772–5777.
29. Fernández, A., Shen, M., Colubri, A., Sosnick, T. R., and Freed, K. F. (2003). Large-scale context in protein folding: Villin headpiece. *Biochemistry* **42,** 664–671.
30. Duan, Y., and Kollman, P. A. (1998). Pathways to a protein folding intermediate observed in a 1-microsecond simulation in aqueous solution. *Science* **282,** 740–744.
31. Baldwin, R. L. (2002). Making a network of hydrophobic clusters. *Science* **295,** 1657–1658.
32. Nemethy, G., Steinberg, I. Z., and Scheraga, H. A. (1963). The influence of water structure and hydrophobic contacts on the strength of side-chain hydrogen bonds in proteins. *Biopolymers* **1,** 43–69.
33. Fernández, A., and Berry, R. S. (2002). Extent of hydrogen-bond protection in folded proteins: A constraint on packing architectures. *Biophys. J.* **83,** 2475–2481.
34. Novotny, J., Bruccoleri, R., and Karplus, M. (1984). Analysis of incorrectly folded protein models. Implications for structure predictions. *J. Mol. Biol.* **177,** 787–818.
35. Daggett, V., and Levitt, M. (1992). A model of the molten globule state from molecular dynamics simulations. *Proc. Natl. Acad. Sci. USA* **89,** 5142–5146.
36. Brooks, C. L., III, and Case, D. (1993). Simulations of peptide conformational dynamics and thermodynamics. *Chem. Rev.* **93,** 2487–2502.
37. Fernández, A., and Rogale, K. (2004). Sequence-space selection of cooperative model proteins. *J. Phys. A: Math. Gen.* **37,** 197–202.
38. Kuwata, K., Shastry, R., Cheng, H., Hoshino, M., Batt, C. A., Goto, Y., and Roder, H. (2001). Structural and kinetic characterization of early folding events in beta-lactoglobulin. *Nat. Struct. Biol.* **8,** 151–155.
39. Fernández, A., Appignanesi, G., and Colubri, A. (2001). Finding the collapse-inducing nucleus in a folding protein. *J. Chem. Phys.* **114,** 8678–8684.
40. Nymeyer, H., Garcia, A. E., and Onuchic, J. N. (1998). Folding funnels and frustration in off-lattice minimalist protein landscapes. *Proc. Natl. Acad. Sci.* **95,** 5921–5928.
41. Onuchic, J. N., Luthey-Schulten, Z., and Wolynes, P. G. (1997). Theory of protein folding: The energy landscape perspective. *Annu. Rev. Phys. Chem.* **48,** 545–600.
42. Chan, H. S., and Dill, K. A. (1997). From levinthal to pathways to funnels. *Nat. Struct. Biol.* **4,** 10–19.
43. Fernández, A., Colubri, A., and Berry, R. S. (2001). Topologies to geometries in protein folding: Hierarchical and nonhierarchical scenarios. *J. Chem. Phys.* **114,** 5871–5888.
44. Shi, Z., Krantz, B. A., Kallenbach, N., and Sosnick, T. R. (2002). Contribution of hydrogen bonding to protein stability estimated from isotope effects. *Biochemistry* **41,** 2120–2129.

45. Fernández, A. (2002). Time-resolved backbone desolvation and mutational hot spots in folding proteins. *Proteins: Struct. Funct. Genet.* **47,** 447–457.
46. Despa, F., Fernández, A., and Berry, R. S. (2004). Dielectric modulation of biological water. *Phys. Rev. Lett.* **93,** 228104.
47. Dobson, C. M. (1999). Protein misfolding, evolution and disease. *Trends Biochem. Sci.* **24,** 329–332.
48. Dobson, C. M. (2001). The structural basis of protein folding and its links with human disease. *Philos. Trans. R. Soc. Lond. Ser. B* **356,** 133–145.
49. Fändrich, M., and Dobson, C. M. (2002). The behavior of polyamino acids reveals an inverse side chain effect in amyloid structure formation. *EMBO J.* **21,** 5682–5690.
50. Sunde, M., and Blake, C. C. F. (1998). From the globular to the fibrous state: Protein structure and structural conversion in amyloid formation. *Q. Rev. Biophys.* **31,** 1–39.
51. Dobson, C. M. (2002). Protein misfolding diseases: Getting out of shape. *Nature* **418,** 729–730.
52. Richardson, J. S., and Richardson, D. C. (2002). Natural β-sheet proteins use negative design to avoid edge-to-edge aggregation. *Proc. Natl. Acad. Sci. USA* **99,** 2754–2759.
53. Nelson, R., Sawaya, M., Balbirnie, M., Madsen, A., Riekel, C., Grothe, R., and Eisenberg, D. (2005). Structure of the cross-beta spine of amyloid-like fibrils. *Nature* **435,** 773–778.
54. Prusiner, S. B. (1998). Prions. *Proc. Natl. Acad. Sci. USA* **95,** 13363–13383.
55. Krishnan, R., and Lindquist, S. L. (2005). Structural insights into a yeast prion illuminate nucleation and strain diversity. *Nature* **435,** 765–772.

Rescuing Proteins of Low Kinetic Stability by Chaperones and Natural Ligands: Phenylketonuria, a Case Study

Aurora Martinez,
Ana C. Calvo, Knut Teigen,
and Angel L. Pey

Department of Biomedicine, University of Bergen, Jonas Lies vei 91, Bergen 5009, Norway

I. Introduction	90
A. PAH and PKU	92
II. Protein Folding	93
A. Protein Folding and Unfolding	93
B. Thermodynamic and Kinetic Stability of the Native Structure	94
C. Folding and Stability of hPAH	97
III. Misfolding	100
A. Intracellular Control of Misfolding: The Quality Control System	100
B. Protein Folding and Degradation in the Cytosol	101
C. Physiological Processes Leading to Misfolding	103
D. PKU as a Misfolding Disease Caused by Mutations	105
E. Current Therapeutic Approaches to Treat PKU	109
IV. Ligand Binding	110
A. Protein Stabilization by Specific Ligand Binding	110
B. Ligand-Binding Studies of PAH	111
V. Strategies to Correct Misfolding	116
A. The Chaperone Concept	116
B. Chemical Chaperones	117
C. Pharmacological Chaperones	118
D. Natural Chaperone Ligands	119
VI. Concluding Remarks	123
References	124

Misfolding diseases are a group of harmful disorders in which the main molecular mechanism for loss of gene function is either an accelerated degradation of the protein or its aggregation inside or outside the cell. Phenylketonuria (PKU) is a disease caused by deleterious mutations in phenylalanine hydroxylase (PAH) and constitutes a paradigm for misfolding diseases.

In fact, it has been shown that there are substantial overall correlations between the mutational energetic impact on the native state and both *in vitro* residual activities and patient metabolic phenotypes. Recent efforts have concentrated on the characterization of the therapeutic use of the PAH cofactor, tetrahydrobiopterin (BH_4), for the treatment of PKU—mainly mild forms—with methods spanning from biophysical investigations to *in vivo* and clinical studies. Supplementation with BH_4 aids to stimulate PAH activity by raising the concentration of the cofactor to supraphysiological levels but more importantly, BH_4 appears to exert a chaperone effect protecting the PAH mutants against degradation and inactivation. When used therapeutically BH_4 and other *natural chaperone ligands* can be considered particular cases of pharmacological chaperones which can rescue the misfolded proteins by stimulating their correct folding *in vivo* and/or the stabilization of native-like conformations. This finding has encouraged research to attempt the rescue of the majority of PKU mutations by specific compounds of varying chemical nature designed for patient-tailored therapeutics.

I. Introduction

Folding is the process by which a protein reaches a functional and stable native structure, while misfolding can be seen as the failure to attain this fully functional conformation. The field of protein folding has recently experienced major discoveries that have lead to novel interpretations of folding mechanisms, that is, the demonstration of barrierless one-state downhill folding in addition to the classical two or multistate mechanisms (1, 2), the recognition of functional "natively unfolded" proteins (3, 4), the role of protein misfolding in the pathogenesis of human diseases (5), and the involvement of partly folded intermediates in the catalysis of misfolding, aggregation, and amyloid formation (6). Diseases—inherited and acquired—which are caused by the enhanced tendency of mutant proteins to misfold and to either undergo intracellular degradation or deleterious aggregation with formation of amyloid fibers are referred to as misfolding or conformational diseases. Over the last years, it has been shown that the molecular basis of the genetic metabolic disease phenylketonuria (PKU) is a loss of function notably caused by misfolding of mutated phenylalanine hydroxylase (PAH), a cytosolic protein involved in hepatic catabolism of L-Phe.

To avoid pathogenic misfolding, the cell is equipped with protein quality control systems (QCS) mainly including chaperones, the ubiquitin proteasome pathway (UPP) and, in some instances, the aggresome. For cytoplasmatic proteins such as PAH, the operating protective mechanisms are notably the chaperones Hsp70, Hsp90, TRiC, and other associated proteins, which assist in folding and maintenance of stable native protein conformations, and the protein degradative system, notably the UPP *(7, 8)*.

Binding of a ligand to a specific binding site on the native state of a protein will influence the unfolding equilibrium which will be shifted towards the natively folded state, resulting in an increase in protein stability *(9)*. Proteins, and in particular enzymes, often contain divalent cations, metals, and organic cofactors. These natural ligands stabilize the native structure in the *resting* state and also modulate the conformation and functionality of the protein in the *active* states. In fact, some proteins with tendency to adopt functional partially folded conformations *in vivo* usually rely on cofactors to stabilize the most stable native conformation and accelerate folding *(10–12)*, and cofactors appear as regulators of folding kinetics and stability. Therefore, it seems that many proteins require the binding of these *natural chaperone ligands* to adopt functional states, or to display long half-lives *in vivo*. For PAH, the cosubstrate $(6R)$-L-*erythro*-5,6,7,8-tetrahydrobiopterin (BH_4)—usually referred to as cofactor—is known to stabilize PAH *(13–16)* and it has been established that supplementation with the cofactor BH_4 is an effective therapeutic aid for the correction of patients with certain forms of PKU, known as BH_4-responsive PKU *(17, 18)*. Stimulation of activity by increasing the *in vivo* concentration of the cofactor, notably for mutants with defect binding affinity is a probable molecular mechanism leading to the response to BH_4 *(19)*, as has also been proposed for the correction of other genetic diseases which are ameliorated by high-dose vitamin supplementation—vitamins being precursors for many enzyme cofactors *(20)*. We have accumulated solid evidence that, in addition, BH_4 functions as a natural chaperone ligand for PAH *(15, 19)*. During the last years, the concept of rescuing misfolded proteins by stimulating their refolding by small molecules with a structural resemblance to the natural ligands (pharmacological chaperones) has been greatly developed and exploited for successful therapeutic intervention *(21, 22)*. Pharmacological chaperones can also exert their therapeutic function through stabilization of native-like states *(23, 24)*. Understanding the structure–function–stability relationships in the selected enzymes, with *in vivo* studies in animal models are essential to fully apply the natural and pharmacological chaperone concept. Translational methodological approaches integrating biochemical, biophysical, computational techniques, and expression studies in cellular and animal models are necessary to fully develop these therapeutic concepts. Here, we introduce general thermodynamic and kinetic aspects of protein folding, stability, and ligand binding, followed by a more focused discussion of protein misfolding and disease. We then present specific therapeutic approaches aimed to target enzyme misfolding and reduced activity. Notable interest is devoted to PAH and the disease PKU, which constitutes a paradigm for misfolding diseases and an important research motivation for the authors.

A. PAH and PKU

PAH (phenylalanine 4-monooxygenase, EC 1.14.16.1) catalyzes the para-hydroxylation of L-Phe to L-Tyr in the presence of BH_4 as natural cofactor and O_2 as additional substrate (Fig. 1). PAH activity is the rate-limiting step in L-Phe catabolism, consuming about 75% of the L-Phe input from the diet and protein catabolism under physiological conditions (26). A defect in one of the enzymatic steps in this complex L-Phe hydroxylating system usually leads in humans to hyperphenylalaninemia and, more severely, to PKU (OMIM 261600). About 98% of the hyperphenylalaninemic patients present mutations

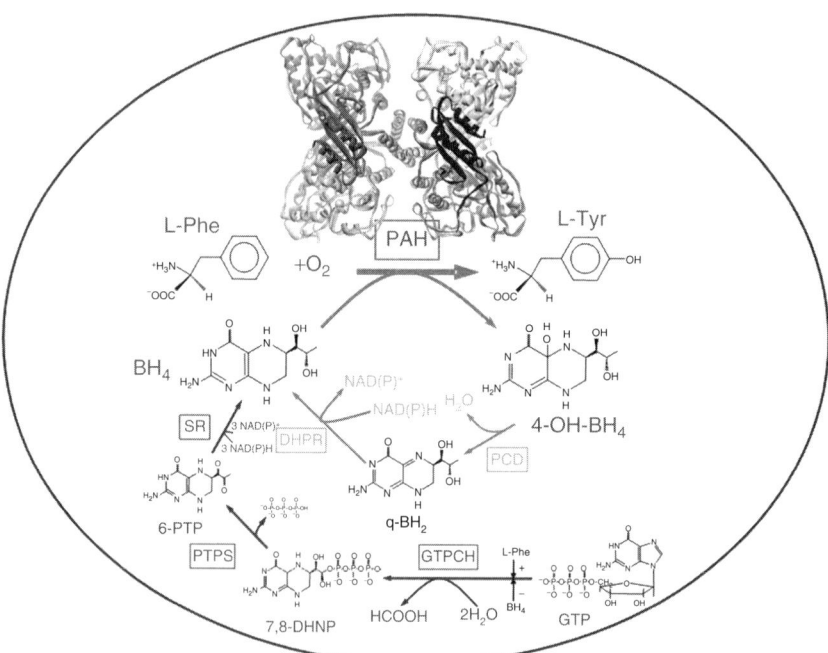

FIG. 1. Hydroxylation of L-Phe by mammalian PAH (red pathway) including the BH_4 regenerating (green) and the *de novo* biosynthetic pathways of BH_4 *in vivo* (blue). The enzymes implicated in these processes are enclosed by frames and the regulatory effects of L-Phe (+) and BH_4 (−) on the rate-limiting step for BH_4 synthesis are also shown. Abbreviations used: BH_4, (6R)-L-*erythro*-5,6,7,8-tetrahydrobiopterin; 4-OH-BH_4, pterin-4a-carbinolamine; q-BH_2, quinoinoid 7,8-dihydrobiopterin; 6-PTP, 6-pyruvoyl-5,6,7,8-tetrahydropterin; 7,8-DHNP, 7,8-dihydroneopterin triphosphate; PCD, pterin 4a-carbinolamine dehydratase; DHPR, dihydropteridine reductase; SR, sepiapterin reductase; PTPS, 6-pyruvoyl-5,6,7,8-tetrahydropterin synthase; GTPCH, GTP cyclohydrolase I. Adapted from (25) with permission from SPS Verlagsgesellschaft MBH. The structural model of tetrameric PAH is also shown; composite model created using the structures PDB 2PHM (rat) and PDB 2PAH (human). (See Color Insert.)

in the *PAH* gene *(26)*. Information on mutations is provided at the PAH locus knowledgebase (PAHdb) *(27)* (http://www.pahdb.mcgill.ca/). PKU was one of the first genetic diseases for which a metabolic explanation was provided *(28)* and the first to exhibit a chemical explanation for mental retardation *(29)*. More recently, PKU has become a paradigm for protein misfolding diseases *(30–32)*. The large genetic heterogeneity of the disease (more than 500 mutations in the same locus *(33)*) and the vast information on expression studies provide a unique opportunity to use PKU as a model system to test and develop new theoretical, experimental, and therapeutical tools for studying misfolding diseases.

Human PAH (hPAH) is a cytosolic homotetrameric enzyme. The determination of the crystal structure of dimeric and tetrameric truncated forms has increased our understanding of the mechanism for catalysis and regulation and genotype/phenotype relationships in PKU *(34–37)*. Mammalian PAH presents a three-domain structure (Fig. 1): (i) the regulatory domain (residues 1–110) includes the phosphorylation site (Ser16) at the autoregulatory sequence (residues 1–30) and is essential for the expression of both positive cooperativity and activation induced by the substrate L-Phe and inhibitory effects caused by the natural cofactor BH_4; (ii) the catalytic domain (residues 111–410) contains the active site, with the non-heme iron center and the binding sites for L-Phe and BH_4; and (iii) the oligomerization domain (residues 411–452) contains an antiparallel β-sheet (411–424) and a coiled-coil motif (428–452) essential for dimer and tetramer formation, respectively. The overall structure of the enzyme consists of a dimer of dimers associated asymmetrically through the tetramerization region *(38)*. Although there is no high resolution structure for the tetrameric full-length PAH protein, composite models have been prepared by combining the structures of several truncated forms solved by X-ray diffraction (Fig. 1).

II. Protein Folding

A. Protein Folding and Unfolding

Protein folding is the process by which proteins are able to rapidly and spontaneously self-assemble into a highly structured conformation with a certain biological function *(39)*. The *classical view* of protein folding established that a protein chain folds from a highly dynamic and globally disordered unfolded state to the functional and biologically more relevant native state through a sequence of individual steps and intermediate states *(40)*. Some of these intermediate states may be significantly populated during folding and may channel the folding reaction towards the native state ("on-pathway" intermediates) while in other cases, they may represent a sort of dead-end conformations that impede or retard the acquisition of the native conformation ("off-pathway" intermediates) *(40, 41)*. On the other hand, in the *new view*, protein folding is envisioned as a statistical search or diffusion of the individual protein molecules through a multidimensional energy surface ("energy landscape"). In this *view*, the concept of "folding pathway" as a sequence of events is changed to the funnel concept of multiple and parallel events *(40)*. In the case of small proteins, this funnel-like energy landscape is smooth and no intermediates are populated, while in larger polypeptides the landscape is expected to be rougher, and both on- and off-pathways might be populated during the folding reaction *(39)*. Both views assume that the native state is placed on a free-energy minima (as a thermodynamically stable state), even though other states with low free energies may also be accessible to the polypeptide sequence (e.g., aggregates and amyloid conformers) *(39)*.

The characterization of the folding mechanism and relative stabilities of different states accessible to a protein (native, partially folded, and unfolded states) is relevant not only to know how and why a protein folds, a problem for physical chemists and biophysicists, but it also has deep implications in the understanding of protein function and regulation from physiological and pathological perspectives *(39)*. It is especially relevant to learn about the stability of native proteins compared to the partially unfolded states, which often turn out to be relatively close to the native state in terms of thermodynamic stability *(42, 43)*, and may determine *in vivo* the relevant protein stability and potential misfolding properties. Moreover, since more than 70% of the eukaryotic proteins are multidomain, it is also necessary to address how the presence of these domains (especially the domain–domain interface contributions) can affect the thermodynamic as well as kinetic stabilities of the individual domains in the overall structure, the cooperativity of their (un) folding (e.g., displaying two or multistate folding) and the misfolding properties inside the cell *(44)*.

B. Thermodynamic and Kinetic Stability of the Native Structure

The concept of *thermodynamic stability* arises from studies where a native protein unfolds and refolds in a reversible manner. The thermodynamic stability is determined by the free-energy difference between the different protein states populated during the unfolding/refolding reaction (Fig. 2). It is widely accepted that most small (<100 residues) and single domain proteins fold in a equilibrium two-state manner, in which the protein exists only in two different macrostates (native (N) and unfolded (U)) without significantly populating intermediate states *(41)* (Fig. 2A). In this situation, the stability of a native

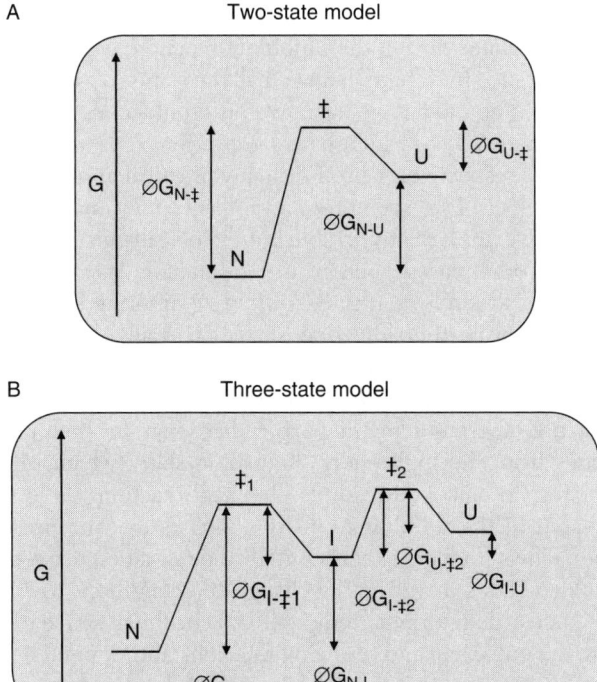

FIG. 2. Simple free-energy representations of a two-state (A) and three-state (B) folding/unfolding equilibria. N, I, and U stand for the native, (on-pathway) intermediate, and unfolded states, respectively, while ‡ stand for transition states. The unfolding free-energy changes between states i and j are indicated as ΔG_{i-j} and the activation free energies for the kinetic unfolding and refolding between the i state and the corresponding transition state (‡) as $\Delta G_{i-\ddagger}$.

protein arises from the free-energy difference between the unfolded and native states ($\Delta G_{N-U} = G_U - G_N = -RT \ln K_u$). The value of ΔG_{N-U} is generally a small number resulting from the almost complete cancellation of large and opposite enthalpic (positive ΔH from breaking bonds) and entropic (positive ΔS from increase in molecular flexibility) contributions to protein stability (45). Any perturbation that preferentially destabilizes the native state, such as temperature, pH, ionic strength, or chaotropic agents like urea or guanidium chloride, will shift the equilibrium toward the unfolded state, resulting in a similar monotonic unfolding transition independently of the variable we attempt to measure (ellipticity by CD, fluorescence, absorbance, hydrodynamic radius, etc.) (46–48). In the case of larger and/or multidomain proteins, we often observe significant deviations from the two-state behavior if unfolding is monitored by different techniques (39, 41, 49). It is therefore required to describe more complex models that include the presence of significantly populated intermediates during the unfolding reaction, which will allow us to estimate the difference in stability between all the states. As seen in Fig. 2B, for a three-state unfolding model, we have two different values for unfolding free energies, ΔG_{N-I} and ΔG_{I-U}, which have been referred to as the relevant and residual stabilities, respectively (42). For many practical applications, the free-energy gap for the N to I step is more relevant since it evaluates the stability of the native protein versus a relative stable but biologically inactive conformation (43). However, in other cases (such as for molten globules, MGs), both N to I and I to U steps are relevant since unfolding intermediates may display multiple functions or partnerships due to their partially disordered state and high-conformational plasticity (50–52).

The concept of *kinetic stability* arises from the assumption that in a folding/unfolding reaction, there is a kinetic barrier that separates both the native and unfolded states from the high energy transition state (which represents the rate-limiting step, in analogy to simple chemical reactions) (Fig. 2), and this barrier is related to the unfolding and refolding rates. The presence of the kinetic barrier allows studying experimentally the relaxation between the native and the unfolded states if we rapidly change the conditions to strongly favor one of these states (denaturants, temperature). In the absence of any kinetic complication, the unfolding and refolding rate constants depend linearly on the denaturant concentration, yielding the classical V-shaped *chevron plot*, which allows to estimate the folding and unfolding rates extrapolated to the absence of denaturant assuming linear dependence on denaturant concentration (41, 53). Deviations from this ideal behavior are typically associated with the presence of kinetic intermediates. In the case of denaturation by temperature, unfolding rates usually display temperature dependences following the Arrhenius equation while the refolding rates often show significant curvatures due to temperature effects on protein stability or oversimplification of the temperature dependence by the Arrhenius approach (54–56).

For many protein systems, it has been observed that the kinetic energy barrier between the native and either partially denatured (intermediates) or highly unfolded (denatured) states is substantial, which may guarantee the kinetic stability of the native states *in vivo*, besides their thermodynamic stability (44, 57, 58). The fact that thermal and chemical denaturation of many proteins is an irreversible process due to protein aggregation—especially multidomain and oligomeric-like PAH—precludes an in-depth analysis of protein stability in terms of unfolding thermodynamics (57). The unfolding enthalpy (ΔH) values can be obtained (59) but not other parameters such as the changes in heat capacity (ΔC_p), entropy (ΔS), or free energy (ΔG) upon unfolding. In situations where the irreversible alterations occur fast (in the scale of the experiment, min–h), kinetics must define the relevant stability of a protein which undergoes irreversible denaturation (57, 60). Simple and general models to study protein *kinetic stability* of proteins that unfold irreversibly are provided by the Lumry–Eyring scenarios (60, 61), and can be summarized as in Scheme 1, where the reversible N ⇌ U step is defined thermodynamically by the equilibrium constant k_u.

$$N \underset{k_f}{\overset{k_u}{\rightleftharpoons}} U \overset{k_0}{\longrightarrow} F$$

SCHEME 1

This equilibrium can be kinetically dissected into the unfolding and folding rate constants k_u and k_f, which are related to the unfolding/refolding activation energies (see Fig. 2A). The step U → F holds for the irreversible unfolding step, where F is a final state which cannot fold back to the U state. We may consider more complex scenarios, where more than one unfolded or partially unfolded states are populated (Fig. 2B) and therefore different states may undergo the irreversible step, requiring a more complicated mathematical description of the model (57). However, based on the assumption that native states are separated by a large kinetic barrier from any nonnative state, the presence of intermediates would potentially affect the kinetic behavior of the system quantitatively but not qualitatively (for a thorough discussion, see (57)). If the unfolded state U does not significantly accumulate during this reaction, the model can be simplified to N → F, which is characterized by a strongly temperature-dependent first-order rate constant k (60). Adherence of the first-order kinetic behavior to this model at different temperatures, strongly supports the applicability of this very simple model and suggests that the rate-limiting step in the irreversible denaturation of the protein is the unfolding rate (N → U) (57, 60). This approach allows to study the kinetic stability both from operational (as a half-life for the process) and energetic perspectives (the activation energy obtained from an Arrhenius analysis equals the activation enthalpy; see (56) for a recent example).

C. Folding and Stability of hPAH

The stability of hPAH is a matter of interest due to its link to PKU as a misfolding disease (30, 32, 37). Most of the PKU mutations expressed *in vitro* display various degrees of decreased stability, as seen by reduced thermal stability and resistance toward proteolytic degradation, as well as a high tendency to aggregate (62–66). However, little is known about the unfolding mechanisms and relevant stabilities of hPAH. The unfolding of PAH has been investigated by urea as well as guanidium hydrochloride denaturation (67, 68). It has been shown that hPAH populates an unfolding intermediate characterized by a high tendency to aggregate (68), separated from the native state by only ~4 kcal/mol *(68)*. This value is within the range found for partially unfolded conformations in other protein systems *(43)* and consistent with a very low population of the intermediate under native conditions (even though small changes in native state stability could lead to significant populations of the intermediate). The unfolding transitions could not be interpreted in terms of changes in the oligomeric structure since the hPAH tetramer seems to progressively and simultaneously unfold and dissociate by urea, showing a denaturant concentration dependence which did not fit with the simple three-state model proposed *(68)*. Attempts to renature the unfolded protein were not successful, reducing the validity of the thermodynamic parameters obtained from fitting to a three-state model *(68)*. Denaturation by denaturants thus may give information on the kinetic stability of wild-type (wt) and mutant hPAH tetramers, but are not expected to provide quantitative information on the unfolding thermodynamics.

The thermal stability and domain unfolding of hPAH have been studied by our group using a combination of spectroscopic and calorimetric methods *(59, 69, 70)*. Infrared spectroscopy was used to monitor hPAH thermal unfolding by following the decrease in α-helix content and the increase in a signal associated to protein aggregation *(69)*. Later, differential scanning calorimetry (DSC) combined with far-UV circular dichroism (CD) were applied to obtain more detailed and functional information of the thermal unfolding of hPAH *(59)*. DSC is particularly suited to study complex unfolding reactions since it may provide a full description of the unfolding process (ΔH, ΔS, ΔG, and ΔC_p). The thermal unfolding of hPAH was again characterized as an irreversible process due to sample aggregation at high temperatures *(59)*. However, careful analysis on the scan-rate dependence of the thermal transition showed that the process was not kinetically controlled in a significant temperature range, allowing the interpretation of at least ΔH in thermodynamic terms. DSC analyses showed that hPAH populates at least two thermal intermediates. By comparing the predicted ΔH from well-established structure–energetics relationships *(71, 72)* and the experimental calorimetric ΔH, it was shown that the low-temperature transition with midpoint melting temperature ($T_m \sim 45\,°C$) leads to the unfolding of four regulatory domains in the tetramer (Fig. 1, darker smaller domains), while the high-temperature transition ($T_m \sim 54\,°C$) represents the unfolding of only two catalytic domains *(59)*. The absence of protein concentration dependence ruled out tetramer dissociation within these two unfolding transitions. At higher temperatures, the "unfolded" state aggregates causing an endothermic distortion of the calorimetric profile. Moreover, both transitions are not described by simple two-state scenarios (based on the calorimetric to van't Hoff enthalpies ratios) indicating a remarkable complexity of the thermal unfolding process *(59)*. Regarding the effect of phosphorylation on the enzyme stability it was recently shown by a combination of CD, DSC, partial

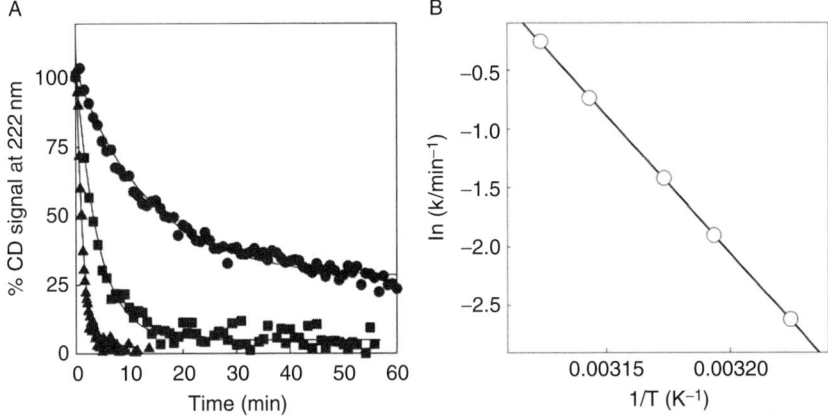

FIG. 3. Irreversible denaturation of hPAH at different temperatures monitored by far-UV CD spectroscopy. (A) Kinetic unfolding traces at different temperatures: 37 °C (circles), 42 °C (squares), and 47 °C (triangles). (B) Arrhenius plot for the irreversible denaturation; the first-order rate constants (k) were determined by fitting the kinetic CD data (A) to a single exponential. Experiments were performed in 20 mM K-Hepes, 200 mM KCl, pH 7, using 5 μM protein (in subunit), 10-fold molar excess of Fe(II) and 1-mm-path length quartz cuvettes.

proteolysis, and structure-derived electrostatic analysis that phosphorylation of the enzyme at Ser16 only locally affected the structure of the regulatory domain, while the overall stability of the enzyme was not affected *(70)*.

Lately, we have investigated the stability and aggregation tendency of hPAH at temperatures close to the T_m for the regulatory domain (Calvo, A., *et al.*, unpublished results). As seen in Fig. 3, kinetic CD measurements show that tetrameric hPAH significantly unfolds at relatively low—even physiological—temperature. At all tested temperatures, the kinetic data are well described by a single exponential decay function and the loss of CD signal in 1 h was ~70% at 37 °C and ~100% at temperatures ≥42 °C. Loss of secondary structure in the full-length hPAH was irreversible, as seen by incubation for 5 min at 37 °C and immediate cooling of the samples down to 10 °C (data not shown). The temperature dependence of the irreversible unfolding rates shows a linear dependence on an Arrhenius plot (Fig. 3B), yielding a value for the activation energy (E_a) of 45.8 ± 0.2 kcal/mol. The good adherence of the kinetic behavior to a first-order rate process is consistent with a two-state irreversible process. Actually, the value of E_a obtained is within the range of values obtained for other protein systems undergoing two-state irreversible denaturation (typically 50–100 kcal/mol) *(57, 73, 74)*. These results suggest that the native tetrameric hPAH is not stable at physiological temperature, at least in the absence of other stabilizing agents that might be found *in vivo* (see later on the effect of natural ligands such as BH_4, Section V).

III. Misfolding

A. Intracellular Control of Misfolding: The Quality Control System

To investigate misfolding diseases, it is important to consider protein folding inside the human cell, which is a much more complex and interactive process than protein folding *in vitro* (75). Proteins have to fold in an adverse environment characterized by high temperatures (37 °C), molecular crowding and abundant unfolded polypeptides (76), where folding intermediates would have a tendency to interact and eventually aggregate. To maintain the correct folding of thousands of proteins, the cell has developed specific mechanisms known as the QCS (77, 78). The QCS includes molecular chaperones and associated cochaperones, factors and cofactors that promote cotranslational protein folding and preserve the metastable conformation of the proteins. In addition, the molecular chaperones repair misfolded proteins, and upon failure of folding, they participate in the transfer of the misfolded proteins to the degradation systems, notably the UPP in the cytoplasm, to avoid accumulation of misfolded proteins and formation of aggregates.

Molecular chaperones recognize and bind partially unfolded states of polypeptides, normally via hydrophobic patches, and promote their correct folding. Chaperones can be both *generic* (public) and *specific* (private) (79). PAH is not known to have private chaperones and it is therefore considered to be attended by the public chaperones (Fig. 4A), which are highly conserved proteins, found in every organism and every cell where they work coordinately (7). Heat shock proteins (HSP) was the first denomination to these proteins because they were induced during stress situations (e.g., temperature). Molecular chaperones are divided according to their subunit molecular mass: small HSPs, HSP40, HSP60 (including the chaperonins GroEL in prokaryotes and TRiC in eukaryotes), HSP70, HSP90, and HSP100 (7, 81, 82). The chaperonins are double-ring-shaped chaperones, represented by the well-studied bacterial GroEL and its cochaperone GroES (83). GroEL is composed of two heptameric rings that form two cavities where the polypeptide is encapsulated and helped to fold in a sequestered environment. Eukaryotic TRiC is made of eight different subunits and it does not use a cochaperone, but instead it opens and closes the cage with α-helical extensions of its apical domains (84). TRiC works, as GroEL, through cycles of ATP hydrolysis. Its substrate size ranges from 30 to 120 kDa, thus being able to fold proteins larger than the expected size fitting inside the cage (85). The main function of the chaperones is to help proteins to correctly fold and refold but they are also able to sequester misfolded proteins and try to prevent toxic interactions. It is established that chaperone overexpression alleviates toxicity in neurodegenerative diseases (86). Chaperones are in addition mediators and

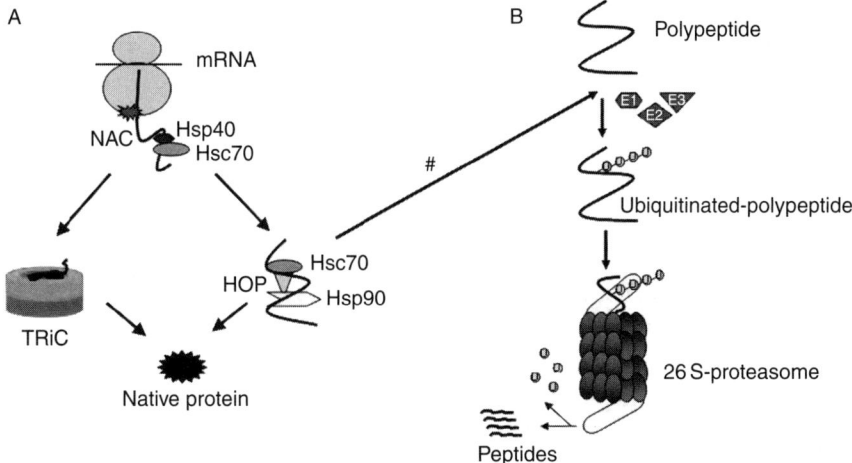

FIG. 4. Quality control system (QCS) in the eukaryotic cytosol. (A) Protein assisted folding for newly synthesized polypetides, involving the interactions with different molecular chaperones; NAC, nascent polypeptide-associated complex; HOP, Hsp organizing protein; for the names of other components see text (Section IIIA and IIIB). (B) Ubiquitin–proteasome pathway (UPP) for the degradation of defective proteins. U, ubiquitin; E1, ubiquitin activating enzyme; E2, ubiquitin-conjugating enzyme; E3, ubiquitin ligase. The Hsc70–HSP90 complex interacts with the UPP system through cochaperones (#), for example, CHIP (7, 80). (See Color Insert.)

controllers of the degradation of misfolded proteins (87, 88), and it has been established that some chaperones physically interact with members of the UPP (8, 88). This is the most important non-lysosomal system to degrade misfolded cellular proteins in eukaryotes (89).

Some events, like genetic mutations, elevated temperature, and oxidative stress result in increased number of misfolded peptides (see below). In those instances, the QCS is overwhelmed and unfolded proteins accumulate and aggregate. Conformational diseases highlight the importance of the QCS response, but also show the inability of chaperones and the UPP to cope with all misfolding, and, importantly, point to new therapeutic possibilities for misfolding diseases (see Section V).

B. Protein Folding and Degradation in the Cytosol

In eukaryotic cells, most of the proteins fold in the endoplasmic reticulum (ER) or in the cytosol. There appears to be more misfolding diseases associated to defects in folding and trafficking of proteins that fold in the secretory pathway, notably ER, than associated to cytosolic proteins (90). Although the general mechanisms are shared in both environments, specific differences can be found. The present understanding of the folding process and the QCS operating in the cytosol—relevant to the folding of PAH and PKU mutants—has increased during the last years *(7, 8, 80)* (schematized in Fig. 4). To avoid nonnative protein interactions due to macromolecular crowding in the cytosol *(91)*, the nascent polypeptide binds to ribosome-related factors, such as the nascent polypeptide-associated complex (NAC), which protects the emerging proteins from misfolding and aggregation. In an ATP independent manner, the polypeptide is then released from the ribosome and transferred to other chaperones, like TRiC or Hsp70. The eukaryotic cytosol contains two types of chaperones of the Hsp70 family, that is, the constitutively expressed forms (Hsc70) and the stress-inducible forms (Hsp70), both usually requiring cochaperones, such as Hsp40 *(82)*. Hsc70 is implicated in the folding of large and multidomain proteins that do not fit into the central cavity of the chaperonins *(92)*. Hsc70 interacts with HSP90 through the Hsp organizing protein (HOP), forming a multichaperone machinery implicated in ATPase-dependent folding of compact folding intermediates (Fig. 4A). The Hsc70–HSP90 complex has a central role in quality control, and can modulate its activity from folding to degradation directed (Fig. 4) depending on interacting cofactors and cochaperones in a highly regulated manner. The complex targets defective proteins for degradation in the UPP through certain connecting proteins such as CHIP *(7, 88, 93)*. Proteasome-mediated degradation is ubiquitin dependent. Ubiquitination consists on the attachment of ubiquitin (76 residues) to the protein to be degraded. Four ubiquitin units are usually needed for recognition by the proteasome and ubiquitination is an ordered enzymatic process involving ubiquitin activating and ubiquitin-conjugating enzymes, and ubiquitin ligase which transfers ubiquitin to a lysine in the target protein *(94)* (Fig. 4B). The proteasome complex consists of a 19S cap and the 20S core. The former component cleaves the ubiquitin moieties, unfolds the polypeptide and feeds it to the 20S core that cleaves the unfolded proteins into short peptides *(95)*.

Besides the chaperones and UPP the cell has an additional defense mechanism toward misfolding through the sequestration of aggregated proteins into a specialized structure called aggresome *(96)*. Aggresomes, which are in fact aggregates of aggregates, are usually enriched with a single aggregated protein species, but they also recruit machinery from the QCS. The function of the aggresome is unclear, but it has been linked with a "proteolysis center" or degrasome and with the autophagyc route. For excellent reviews on protein aggregation see *(5, 97, 98)*. It was believed that the ability to form amyloids was a property of just a handful of proteins, but it has been shown to be a generic feature under appropriate conditions *(99)*. Most conformational diseases are characterized by the accumulation of the misfolded proteins in amyloid fibrils and plaques. In the case of PAH, however, there is no indication that mutants aggregate intracellularly, suggesting that misfolded proteins are efficiently degraded by the UPP.

C. Physiological Processes Leading to Misfolding

1. MUTATIONS

About 90% of all mutations in human genes are point mutations or small deletions and insertions, which usually yield a protein with just one or few different amino acids. The rest are gross deletions, insertions, and rearrangements of the gene sequence *(30)*. Mutations often produce proteins with defective folding and result in misfolding diseases, which can be divided into two groups *(31)*: (i) the "gain-of-function" diseases, also known as "amyloid diseases," because the mutant protein is usually resistant to degradation and adopts a stable new conformation that aggregates and forms amyloidogenic deposits, toxic for the cell. Typical examples of this group of illnesses are neurodegenerative pathologies such as Parkinson's (PD), Alzheimer's (AD), or Huntington's disease *(31)*; and (ii) the "loss-of-function" diseases, where the misfolded protein is targeted for proteasome degradation and no function is found in the cell. Many metabolic diseases belong to this group, and PKU is a good example (see below), but cystic fibrosis, caused by mutations in the cystic fibrosis transmembrane regulator (CFTR) gene, is perhaps the disease that receives most attention because of its severity. The most frequent mutation, ΔF508-CFTR, is not able to fold correctly and it is retained at the ER and targeted for degradation *(100, 101)*. In the last years genotype–phenotype studies have been performed to correlate specific mutations with the degree of clinical symptoms in several conformation diseases *(37, 102)*.

Protein stability is a key determinant for protein function and regulation through links to flexibility and degradation, respectively. It has been proposed that during evolution mutations—both stabilizing and destabilizing, but without affecting protein function and catalysis—have been fixed on protein sequences to guarantee a minimum value for protein stability preserving flexibility. The frequency of a mutation on a certain position in a sequence may provide an estimation of its contribution (favorable or unfavorable) to protein stability. This is known as the "consensus concept" to protein stability *(103)*. In some cases, as thoroughly described in *(58)*, the minimum stability has a value remarkably close to the wt-stability and high enough to ensure thermodynamic stability, with high ΔG_{N-U} at optimal temperature growth conditions and therefore a high T_m for reversible denaturation. The consensus concept also implies that changes in ΔG_{N-U} are almost fully translated to changes in $\Delta G_{N-\ddagger}$; that is, to the unfolding kinetic barrier (Fig. 2) *(58, 104, 105)*. However, changes in $\Delta\Delta G_{N-U}$ are almost linearly reflected in changes in T_m values while changes in $\Delta\Delta G_{N-\ddagger}$ translate exponentially to unfolding rates *(58)*. Therefore, small changes in equilibrium stability might be translated to large changes in unfolding rates, compromising protein kinetic stability *in vivo* on mutation.

2. Posttranslational Modifications and Translational Misincorporation

Prior to adopting the final folded form the polypeptide released from the ribosome complex often undergoes a final step called *posttranslational modification (106)*. This is a complex process, not completely understood. Some of the modifications are enzymatically driven (proteolysis, acetylation, glycosylation, or phosphorylation) and others are spontaneous (deamidation, oxidation, or racemization). Although in most cases these modifications are beneficial and essential for the physiological function of the protein, there are pathological situations where we can find aberrant modifications playing an important role in the development of the diseases, and often related to misfolding and aggregation. In the case of type 2 diabetes mellitus, it is clear that the spontaneous deamidation of Asn or Gln residues of amylin (a polypeptide hormone implicated in this complex disease) could change the properties of the peptide and promote its aggregation *(107)*. Interestingly, it has been found that PAH is susceptible to nonenzymatic deamidation at asparagine residues *(108)*, a modification that is considered to function as a molecular clock for enzymes *(109)*. Deamidation might aggravate the misfolding at least for certain PKU mutants.

Amino acid substitutions largely result from mutations in genes, but substitutions can also be due to errors in the translation process. The aminoacyl-transfer RNA (tRNA) synthetases control that the correct amino acid is charged into the correct tRNA, discriminating in the pool of cellular amino acids. Large amino acids are usually sterically discriminated, but smaller ones could fit into the binding pocket and be incorporated in the nascent protein. Wrongly placed amino acids are usually cleared by the editing domain of the tRNA synthetase, but if a different residue is still incorporated the protein may misfold and probably aggregate. Translational misincorporation was recently described as a mechanism involved in generating misfolding mutants and neurodegeneration *(110, 111)*.

3. Other Processes Leading to Misfolding

a. Oxidative stress. The role of oxidative stress in misfolding and aggregation is actually well established, notably in PD and AD and other neurodegenerative diseases *(112)*. Neurons are more sensitive to oxidative stress and lipid peroxidation since the oxygen consumption and the amount of polyunsaturated fatty acids are higher in brain respect to other tissues. Concerning PAH it has been shown that its *in vivo* liver activity is protected from peroxide inactivation by selenium-dependent glutathione peroxidase activity *(113)*. We have also shown that wt and mutant PAH expressed in *in vitro* cell-free systems are rapidly inactivated via H_2O_2-mediated mechanisms which are retarded by the presence of BH_4 *(16)*. Oxidative stress has also been involved in the pathophysiology of the tissue damage found in PKU, which is probably not directly correlated to high Phe blood levels since it occurs both in treated and not treated patients *(114)*. The possibility should be considered that lipid peroxidation with consequent oxidative stress in PKU patients might worsen the misfolding in PKU mutants and be important in the modulation of PAH residual activity.

b. Temperature. There are numerous examples of disease-associated mutants with temperature-sensitive folding defects leading to different expression levels and residual activities depending on the expression temperature in eukaryotic cells *(115)*, including PKU mutants *(63, 65)*. It has then been proposed that some transient environmental factors like fever may temporarily affect metabolic outcome *(116)*. In the case of misfolding mutations in antithrombin linked to a risk of thrombosis, it was shown that the severe episodes of thrombosis occur in association with fever *(117)*. For PKU it appears that the efficacy on the control of plasma L-Phe by BH_4 supplementation is abolished in periods of infection *(118)*, likely due to an exacerbated folding defect caused by a raise in body temperature.

c. Proteins prone to aggregation. There are some proteins with a partial or total naturally disordered conformation inside the cell and that adopt a more ordered functional conformation via partner interaction. It is in fact estimated that more than 30% of the proteins in the eukaryote proteome are natively unfolded proteins *(119)*. These proteins are said to be "prone to aggregation," because their naturally unfolded states are predisposed to interact with other unfolded proteins and finally aggregate.

D. PKU as a Misfolding Disease Caused by Mutations

Since the PAH gene was mapped and cloned *(120)*, mutational analysis on PKU patients have been performed worldwide, leading to the striking finding that more than 500 different mutations in the PAH locus are disease related (http://www.pahdb.mcgill.ca/) *(33)*. Out of the 532 mutations compiled in February 2008 in the PAHdb, 327 (61.5%) are missense mutations leading to a single amino acid change in the primary sequence. The effects of about 100 of the mutations have been studied by in vitro expression analysis (also compiled in PAHdb). The picture emerging from these expression analyses is that the predominant molecular mechanism underlying PKU is a loss-of-function pathogenesis due to decreased stability and folding efficiency of the mutant proteins *(62–66)*. Interestingly, many mutants also show altered kinetic and/or regulatory properties compared to the wt-protein, specially affecting the regulation by L-Phe *(19, 121)*, while only a few can be considered as catalytic mutations with large reductions in specific activity or apparent affinities (see references and examples from (37)). Moreover, expression analysis under permissive conditions that alleviate folding defects such as subphysiological temperatures in eukaryotic cultures and overexpression of bacterial chaperonins in *Escherichia coli* led to remarkable increases in both the fraction of functional tetramers and PAH activity for many of the mutations (63, 65), highlighting the mutational effects on stability and folding efficiency. As a cytosolic protein, hPAH is expected to fold co- and/or posttranslationally upon interaction with different chaperone systems (see above and Fig. 4A). However, to our knowledge no studies have analyzed the chaperone requirement for PAH in eukaryotic cells. Misfolded or partially unfolded PAH might be targeted to degradation by the UPP (Fig. 4B) and Døskeland and coworkers have shown that hPAH is a substrate for the ubiquitin-conjugating enzyme (122). Nevertheless it has been reported that the degradation of hPAH in a cell-free expression system was not exerted by the ATP-dependent UPP (123) despite the finding that degradation of the homologous human tyrosine hydroxylase (hTH) indeed appears to be performed by this pathway in a similar system (124). Regarding studies on PAH degradation, consistent values have been obtained in independent measurements performed in hepatoma cells (125, 126) and cell-free systems (63, 123), providing a half-life of 8.6 ± 0.8 h for wt-PAH (mean value for these four independent reports).

The structural complexity of PAH and the absence of solid data on the folding mechanism of PAH *in vitro* or *in vivo* (see above) hinder the comprehensive structural-based analysis of mutational effects on PAH activity and stability. Still, results from expression analysis of PKU mutants have been interpreted in the light of the available crystal structures, essentially evaluating the possible alteration of native interactions and packing (36, 127). We have recently analyzed the energetic effects of PKU mutations on PAH stability using the semiempirical force field FoldX (37). The FoldX-based approach explicitly takes into account the energetic impact of mutations on the native state stability considering a suitable structural model, while the effects on the "unfolded" state are considered implicitly, by calculating different energy terms important for protein stability which have been weighted using experimental stability data (128). We used this approach to estimate the energetic impact of 80 mutations for which activity measurements had been performed on eukaryotic cells, and 45 mutations for which consistent genotypes had been reported as associated to a given mutation in patients (37). FoldX analysis provided two different energy parameters to evaluate the impact of mutations on stability: m values, which evaluated the energy impact due to structural rearrangements upon van der Waals clashes introduced by the mutation on the structure, and y_0 values, which provided estimations on the local effects of the mutation on the residue environment (Table I). High values of m and y_0 parameters correlated well with low residual activities *in vitro* and more importantly with severe phenotypes in patients, while low values of these parameters correlated with high *in vitro* activities and mild phenotypes (37) (Fig. 5). We then used the

TABLE I

Frequency Distribution and Mean Values of the m and y_0 Parameters (See Main Text) Determined from the Linear Fitting of $\Delta\Delta G$ versus Energetic Penalizations for Different Groups of PKU Mutations Classified by *in vitro* Activity or *in vivo*-Associated Phenotypes Calculated from 41 Mutations after Withdrawal of 5 Outliers. from (37)

Phenotype	m				y_0 (kcal/mol)			
	Frequency				Frequency			
	$m < 0.1$	$0.1 < m \leq 0.3$	$m > 0.3$	Mean ± S.D.	$y_0 \leq 3$	$3 < y_0 \leq 7$	$y_0 > 7$	Mean ± S.D.
Group (1) MHP	1	0	0	0	0.556	0.444	0	2.71 ± 1.44
Group (2) Mild PKU	0.177	0.411	0.411	0.15 ± 0.25	0.25	0.417	0.083	4.72 ± 2.71
Group (3) Severe PKU	0.333	0.056	0.611	1.19 ± 1.14	0.167	0.444	0.389	7.16 ± 5.04

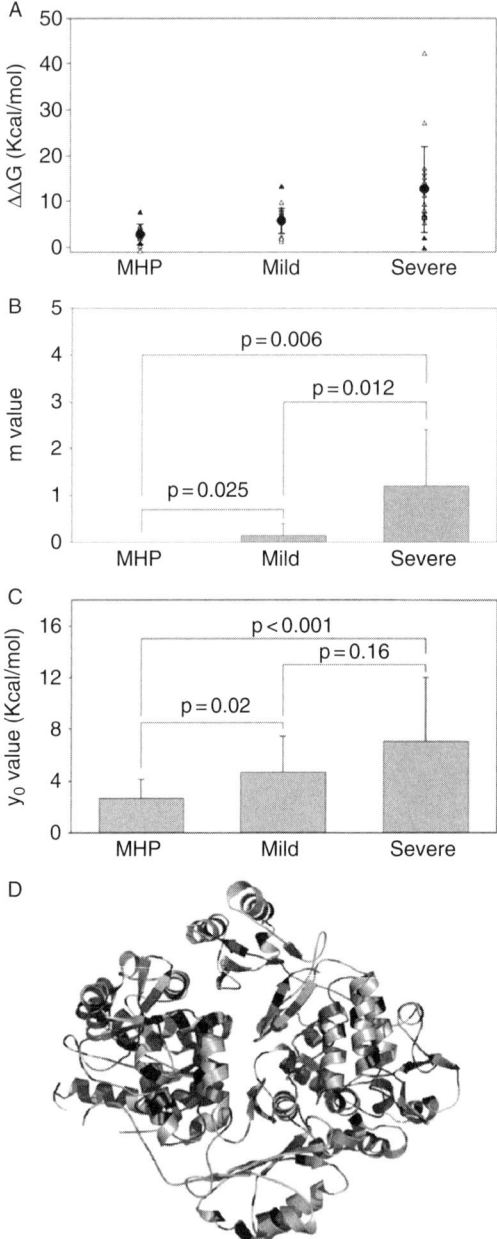

FIG. 5. Mutation-dependent destabilization and *in vivo* patient phenotype. (A) Calculated effect on $\Delta\Delta G$ (in kcal/mol dimer) for 46 PKU mutants classified by phenotypic groups at a 5-kcal/mol penalty. The mean ± SD $\Delta\Delta G$ values (*blackened circles*) for the three phenotypic groups,

guidelines provided by these analysis to perform predictions for the phenotype associated to 238 mutations with no clear phenotypes or very rare frequencies, presenting a structure–energetic framework that could be used to analyze new PKU mutations (37). This strategy can also be applied to mutations in other protein systems related to loss-of-function misfolding diseases.

E. Current Therapeutic Approaches to Treat PKU

Early diagnosis through newborn screening tests and restriction of L-Phe intake by using artificial dietary formulations have led to a remarkable success in preventing the major manifestations of the disease, including mental retardation (129). However, *diet therapy* has to be maintained "for life," it is relatively expensive and socially burdening, and if not continued it may affect fetus development during pregnancy. Currently, different strategies to partially or totally substitute low-Phe diet to treat PKU are available. BH_4 *supplementation* has been demonstrated to short- and long-term reduce L-Phe levels and increase L-Phe tolerance in mild and severe PKU phenotypes, increasing the L-Phe hydroxylation *in vivo* (17, 18, 130). The correction mechanisms are probably multifactorial involving the increase of subsaturating concentrations of cofactor intracellularly and stabilization of PAH mutant proteins against degradation/inactivation (15, 16, 19, 131) (see Section V). BH_4 supplementation thus faces PKU as a misfolding disease, overcoming the instability of PKU mutations through ligand binding. *Large neutral amino acids (LNAA) supplementation* is based on the reversal of the L-Phe-induced inhibition of the LNAA transport across the blood–brain barrier by the L-Type amino acid carrier (132). Both these supplementations may allow less (but still) restrictive L-Phe diets (133). Two alternative approaches are also currently being investigated to treat PKU, both envisioning complete substitution of the classic low-L-Phe diet, that is, *enzyme replacement therapy* using formulations of PEG-modified phenylalanine ammonia lyase (134) and (hepatic) *gene therapy* (135–137). However, both procedures are still in their experimental infancy and many questions need to be addressed for potential human application.

---

calculated using 41 mutations (*unblackened triangles*), were 2.8 ± 2.2, 5.7 ± 2.7, and 13.0 ± 9.5 kcal/mol for MHP (group 1), mild (group 2), and severe (group 3) phenotypes, respectively. Five outliers (*blackened triangles*) were removed for the calculation of the mean values. (B) and (C) Means ± SDs of m and y_0 values for the different phenotypic groups, calculated using individual fits for each mutation. P values are obtained from one-way ANOVA; $p < 0.05$ is considered statistically significant. (D) The predicted phenotype represented on the dimeric structure (PDB 2PHM). All misense PKU mutations (http://PAHdb, http://www.pahdb.mcgill.ca/) have been represented according to the predicted phenotype by FoldX analyses: MHP (blue), mild (yellow), and severe (*red*). Reproduced from Ref (37) with permission from Elsevier Limited. (See Color Insert.)

IV. Ligand Binding

A. Protein Stabilization by Specific Ligand Binding

Specific ligand binding to protein native states is usually accompanied by protein stabilization. This stabilizing effect can be ascribed to the effect of ligand binding on the folding/unfolding equilibrium and kinetics. As illustrated in Section II, protein unfolding can be generally described by a model based on the Lumry–Eyring scenarios (61) (Scheme 1). In this model, the rate of F formation (with rate constant k_0) follows first-order kinetics and is thus directly proportional to the concentration of U. Under these conditions, native state stabilization (higher value of ΔG_{N-U} in Fig. 2) may be achieved by any means that shift the equilibrium toward N, for instance, by destabilization of U by osmolyte binding (see later, Section V) or by stabilization of N by specific ligand binding. There are two limiting cases for the scenario depicted in Scheme 1 (57). The "upper limit" is reached when the U → F step is negligible (slow), and then ligand effects on protein stability are dictated by equilibrium thermodynamics, and therefore by the increase in ΔG_{N-U} (Fig. 2 and Scheme 2); that is, the stabilization exerted by the ligand on the native state free energy (9, 138). The lower limit is reached when the U → F step is fast, and therefore equilibrium thermodynamics are not applicable. In this instance, we thus have to consider the impact of ligand binding to the native state on the overall rate of denaturation, which in principle would be reduced due to an increase in $\Delta G_{N-\ddagger}$, that is, a decrease in unfolding rate (23, 139).

Based on Scheme 2, analytical expressions have been derived to evaluate the effect of ligand concentration on protein thermal stabilities (see for instance (9, 60)). In both upper and lower limits, protein stabilization will depend on the equilibrium-binding affinity ($K_{binding}$) and total ligand concentration (L), as well as on intrinsic properties of the unfolding reaction such as unfolding enthalpies or activation energies. An interesting prediction is that protein stability will be enhanced by the ligand even at ligand concentrations far beyond the saturation range, although these effects will be quantitatively different depending on the reversibility of the reaction (9). This prediction argues against the naïve concept of ligand-binding stabilization related to the

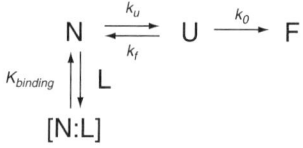

Scheme 2

saturation of protein-binding sites, and suggests that modulation of protein stability *in vivo* would not be limited by the ligand-binding affinity but also by the intracellular availability and solubility of a given ligand. This notion has large implications for recognizing the protein stabilization of both wt and misfolding mutants that can be obtained via supplementation of the natural chaperone ligands, and in particular to understand the stabilization of PAH by therapeutic BH_4 supplementation (see Section V).

The chaperoning role of specific ligands on protein folding and stability may be much more complex than just these "native state effects," since ligands can also bind to nonnative states and impact the folding reaction *(10)*. In some cases, the protein would not be fully folded (native) but in some sort of folding intermediate such as the common MG if the ligand is removed. In the absence of ligand, the protein may rely on this highly flexible state to display multitask function/interaction partnerships *(140)*. In fact, evidence supports that intermediates observed in protein folding *in vitro* may be closely related to protein conformational states that are important in various functional intracellular processes *(141, 142)*. We notice that MG states are often identified in proteins with a high functional mobility and flexibility and where cofactors and ligands are necessary to adopt the well-structured and -stabilized native 3D structure *(12, 140, 143)*. A MG state has however not been observed for the unfolding of PAH, and all together the results indicate that the stabilization of PAH by BH_4 is a native state effect, which occurs with a tradeoff in flexibility *(49)*.

B. Ligand-Binding Studies of PAH

PAH has been the subject of extensive research and experimental investigation for more than half a century. The elucidation of the crystal structure of the homologous rat tyrosine hydroxylase (TH) opened up a new era in the study of the aromatic amino acid hydroxylases *(144)*. The ability to access and study the information contained in the structure of TH made it possible to understand the vast amount of experimental data acquired throughout the preceding decades from a new perspective. Consequently, the structure of the catalytic domains of hPAH *(145)* and of human tryptophan hydroxylase 1 (TPH1) *(146)* were also solved by the Steven's group and collaborators. Several structures of truncated forms of mammalian PAH are nowadays available, including those of complexes with cofactor, catecholamine inhibitors and substrate analogs *(147, 148)*. The structures of the ligand-bound forms of the enzyme open for rational discussions on structure–function–energetics relationships.

1. Cofactor Binding

Tetrahydrobiopterin (BH_4), the natural cofactor for PAH, is a member of a large family of compounds known as pteridines, which are characterized by a two-ring structure containing a fused pyrimidine and pyrazine ring (rings to the

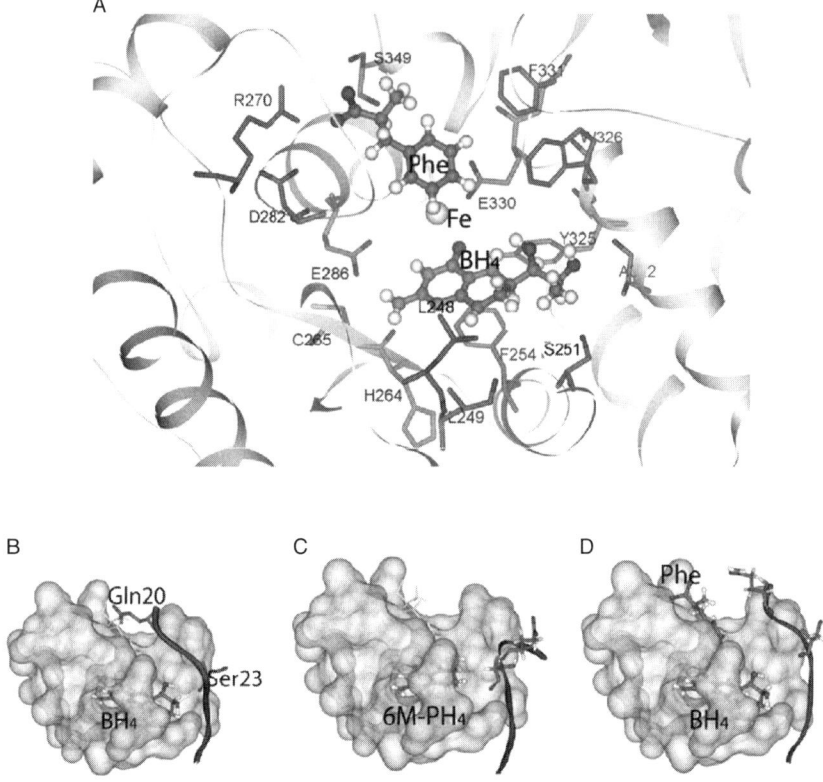

FIG. 6. Structural analyses. (A) Detailed structure of the active site region of hPAH with bound L-Phe and BH$_4$ based on NMR and docking analyses *(149)*. The iron is shown as a yellow sphere. (B) PAH·BH$_4$ and (C) PAH·6M·PH$_4$ complexes obtained by MD simulations *(150)*. The autoregulatory sequence is displayed as a red ribbon. (D) The PAH·L-Phe·BH$_4$ complex simulated at the same conditions as in *(150)*. In (B) but not in (C) and (D) the carbonyl O in Ser23 interacts with the dihydroxypropyl side chain of BH$_4$ and Gln20 occupies the L-Phe-binding site. (See Color Insert.)

left and to the right, respectively in (See BH$_4$ in Fig. 6A)). When there are an amino and an oxo groups in positions 2 and 4, respectively, of the pyrimidine ring (as in BH$_4$), the compound is referred to as a pterin. The term "biopterin" is reserved for pterins with a dihydroxypropyl group in position 6. Dihydrobiopterin (BH$_2$) is the oxidized form of BH$_4$, with a double bond between N5 and C6. The structure of BH$_2$ (and BH$_4$) in complex with PAH was first solved by NMR and docking *(149)* and subsequently confirmed by crystallographic studies *(151)* and molecular dynamics simulations *(150)*.

The cofactor-binding pocket in PAH is located at the bottom of the hydrophobic active site opening. The cofactor makes stacking interactions with an invariant phenylalanine (Phe254 in PAH) and the N3 and the amino group at C2 hydrogen bonds with the carboxylic group of a conserved glutamate residue (Glu286 in PAH) *(149)* (Fig. 6A). The cofactor-binding site is well conserved between the members of the aromatic amino acid hydroxylase family. Nevertheless, there are some specific substitutions, notably around the dihydroxypropyl-binding site, which appear to have implications for cofactor recognition and specificity *(152, 153)*. PAH and the other hydroxylases are dependent on a pterin cofactor to perform catalysis, but several analogs are able to substitute the natural cofactor BH_4 in the reaction. However, it is only BH_4 with the dihydroxypropyl side chain in 6R configuration that induces a negative regulatory effect on the enzyme and a large stabilization. This regulatory effect has been postulated to be induced by the binding to an allosteric-binding site, different from the active site *(154)*. But this hypothesis has not been confirmed, and several reports are in agreement with the regulatory effect being induced by the binding of BH_4 to the active site *(14, 149, 150)*. The thermodynamics of binding of BH_4 have been studied by isothermal titration calorimetry *(155)*. At neutral pH and 25 °C, BH_4 binding to PAH occurs with high affinity and 1:1 stoichiometry, further supporting that the cofactor uniquely binds to the active site. The binding is a strong exothermic process ($\Delta H = -11.8 \pm 0.4$ kcal/mol) accompanied by an entropic penalty ($T\Delta S = -3.4 \pm 0.4$ kcal/mol) *(155)*. Structure-based energetics calculations applied on the molecular dynamics simulated structures of the cofactor–enzyme complexes indicate that the large favorable enthalpic contribution is caused by the conformational rearrangement of the autoregulatory N-terminal sequence of PAH (up to residue 33 *(34)*) (Fig. 6B). The unfavorable entropic contribution is most probably associated to the reduction of conformational flexibility. The simulated structures of PAH with bound BH_4 show that the dihydroxypropyl side chain of the cofactor interacts with Ser23 in the N-terminal autoregulatory sequence and pulls the N-terminal into the active site resulting in Gln20 covering the substrate-binding site (Fig. 6B) *(150)*. When simulations are performed with 6-methyl-tetrahydropterin (6M-PH_4), where the side chain is a methyl group which is unable to interact with Ser23, there are no restrictions on the conformation of the N-terminal (Fig. 6C) and the substrate-binding site is more accessible. Hence, the binding of 6M-PH_4 is both enthalpically ($\Delta H = -3.3 \pm 0.3$ kcal/mol) and entropically ($T\Delta S = 3.2$ kcal/mol) driven *(155)*. The intrinsic thermodynamic parameters of BH_4 binding are in agreement with the negative modulatory effect of BH_4 that protects the resting forms of the enzyme in the absence of activating concentrations of substrate, as also seen by a decreased rate of limited proteolysis *(13, 14)* and increased T_m

(19, 131, 156). The calculated energetic parameters combined with the structural analysis thus aid to elucidate the regulatory effect and the large stabilization of PAH by BH_4 (see also discussion in *(15)* and Section V).

Several PKU-associated mutants present increased K_m in enzyme kinetic characterizations, and for some of them it has subsequently been shown that they are defective in BH_4 binding at equilibrium conditions *(16, 19)* (F39L, I65T, R68S, and to a lesser extent, A313T) (Table II). The energetic effects of

TABLE II
KINETIC AND EQUILIBRIUM-BINDING AFFINITY FOR BH_4 MEASURED FOR WT AND MUTANT TETRAMERIC PAH BY ENZYME KINETIC ANALYSES (DETERMINATION OF K_m) AND ISOTHERMAL TITRATION CALORIMETRY (DETERMINATION OF K_d)

PAH	K_m (µM)	K_d (µM)	References
Wt	26	2.7	*(19)*
F39L	44	8.4	*(19)*
I65T	40	3.9	*(19)*
R68S	30	9.0	*(19)*
D129G	37	n.d.	*(157)*
D143G	69	n.d.	*(158)*
N167I	14	n.d.	*(159)*
H170D	12	n.d.	*(19)*
E178G	29	n.d.	*(19)*
V190A	17	n.d.	*(19)*
N207D	8	n.d.	*(159)*
R252Q	33	n.d.	*(62)*
R252V	33	n.d.	*(62)*
R261Q	25	2.7	*(19)*
A300S	26	2.7	*(19)*
L308F	44	n.d.	*(19)*
A313T	24	3.4	*(19)*
A373T	22	n.d.	*(19)*
V388M	24	n.d.	*(19)*
E390G	29	n.d.	*(19)*
P407S	17	n.d.	*(19)*
R408Q	30	n.d.	Pey, A., unpublished results
R408W	35	n.d.	Pey, A., unpublished results
Y414C	22	n.d.	*(19)*

the mutations are comparatively small (<0.7 kcal/mol change in $\Delta G_{binding}$) *(19)*. Nevertheless, the binding of BH_4 to the mutants with decreased affinity shows a distinct thermodynamic profile characterized by increased entropic penalization with respect to wt-PAH, only partially counterbalanced by a favorable enthalpic contribution *(19)*. Taken together with the fact that these particular mutants also appear to be partially "preactivated," binding thermodynamics indicate that the mutations affect the regulatory conformational changes associated with BH_4 binding. The molecular mechanism for correction of low-affinity mutants has been associated to the activity increase and correction of the K_m (BH_4) effect *(17)* by "Michaelis–Menten kinetics" considerations, but binding energetics further complete this picture by illustrating the potential for conformational stabilization of the mutants by cofactor supplementation (Scheme 2).

2. SUBSTRATE BINDING

The structure of PAH in complex with L-Phe was first reported in 1999 *(149)* and has been confirmed by subsequent crystallographic studies of PAH in complex with two substrate analogs, that is, thienylalanine and norleucine *(148)*. The aromatic ring of the substrate is stacking with His285 while the NH3 and carboxyl group interact with Arg270 and Ser349, respectively (Fig. 6A). The substrate-binding site of PAH shows high sequence identity with the other hydroxylases and the substrate specificity seems to reside in the shape and electrostatics of the active site and, possibly, in the conformation of the amino acids coordinating the active site iron *(160, 161)*.

It is largely accepted that the regulatory domain of PAH mediates the access of substrates to the active site, mainly through the inhibiting autoregulatory N-terminal sequence, and that L-Phe activation leads to the removal of this inhibition *(34, 162)*. Molecular dynamics simulations performed in the presence of both BH_4 and L-Phe, result in bound structures where the N-terminal autoregulatory sequence does not occupy the substrate-binding site, which appears more open (Fig. 6D). Accordingly, the entropic penalty effected by BH_4 binding and associated to the conformational rearrangement of the N-terminal disappears for the catalytically competent L-Phe-activated enzyme ($\Delta H = -6.61 \pm 0.49$ kcal/mol and $T\Delta S = 0.99$ kcal/mol *(155)*). Finally, site directed mutagenesis in combination with molecular dynamics simulations have also aided to envision the details of the allosteric conformational changes in the tetramer, induced by L-Phe binding to the active site *(163)*.

V. Strategies to Correct Misfolding

A. The Chaperone Concept

As seen in Section III, the main purpose of the molecular chaperones in the QCS would be to entrap and isolate the folding intermediates from the environment, aiding to protein folding, and to target misfolded states to the degradative machinery, notably the UPP *(7, 8)*. But there are many cases in which the mutated proteins show some activity but are retained and further degraded by the QCS. For proteins folded in the ER, this is translated into a defect in the trafficking of the proteins which do not reach final destination. Though counterintuitive in a context of misfolding correction, inhibition of the chaperones—not in their role as folder-helpers but in their role as keepers (cages) as well as mediators to the proteasome (Fig. 4)—could aid to recover functionality in some cases. Inhibition of the chaperones by, for example, calcium pump inhibitors to deplete the calcium stores has been proposed as alternative intervention for cystic fibrosis in particular and misfolding diseases in general *(164)*. Nevertheless, several groups have failed to show correct trafficking of the ΔF508-CFTR mutant protein in human cell lines and animal studies *(165)*.

Investigating the molecular mechanism for the chaperoning function may also provide clues for therapeutic intervention. Structure–function studies of GroEL/GroES have shown that this system switches between two states in response to binding and/or hydrolysis of ATP; it binds the folding intermediates by solvent-exposed hydrophobic residues and changes to a hydrophilic contact surface prior to the release of the substrate *(166)*. These hydrophobic/hydrophilic oscillations have inspired the design of artificial chaperones based on polymeric thermoresponsive hydrogels with good functionality for biotechnological applications in folding reactors *(167)*. For therapeutic applications this approach appears, however, to be less attractive. On the other hand peptides and peptidomimetics could be of therapeutic potential *(168)* as shown for the cell-permeable peptide penetrating and its synthetic analog KLAL, which appear to act as chaperones and rescue a mutant of the vasopressin V2 receptor causing nephrogenic diabetes insipidus *(169)*. These peptides appear specific for the post-ER compartment and they were functional at very low micromolar concentrations, preserving the integrity of the membranes. However, the molecular mechanism for correct trafficking and recovery of functionality is not known and a direct interaction of the peptides with the mutants has not been demonstrated. Direct interaction of the chaperone-like peptides with the misfolded proteins has been shown for peptides derived from αA-crystallin (and αB-crystallin) *(170)*, and the C-terminal derived peptides of α-synuclein *(171)*. The application of peptides as synthetic chaperones opens the prospect of applying phage-display libraries to select specific or "private" chaperones for each misfolded protein. To our knowledge, however, this strategy has not been thoroughly investigated, and has not been considered for the treatment of PKU.

Finally, small molecules which bind to native or partially folded states may also present chaperoning function by affecting the partition of the protein between correctly folded and misfolded/aggregated/prone to aggregated states (see Section III). Small molecular weight chaperone-like compounds have a higher therapeutic potential than that of peptides and other chaperone-mimicking devices mentioned above. In this section, we discuss three classes of small molecules used and/or designed as chaperone-like compounds: chemical, pharmacological, and natural chaperones. The natural chaperone ligands may be considered as a particular case of the pharmacological chaperones when they are supplemented therapeutically to treat misfolding diseases, as is the case for BH_4.

B. Chemical Chaperones

The term *chemical chaperone* usually refers to small organic molecules such as osmolytes (glycerol, trimethylamine-N-oxide, proline, sucrose, etc.). These compounds have been found to stabilize the native protein structure in *in vitro* studies and to increase the yield of expressed proteins in prokaryote expression systems. Historically, glycerol and sucrose have been used to preserve proteins and enzymatic activity *(172)*. The principle behind osmolyte-induced stabilization is the preferential exclusion of the osmolyte from the peptide backbone, which is "osmophobic," with consequent unfavorable exposure of backbone and concomitant destabilization of the unfolded states (increase in ΔG_{N-U}; Fig. 2) in the presence of osmolyte *(173, 174)*. Moreover, protein mobility is also limited by elevating the solvent density, with consequent increase in stability. Other compounds as dimethyl sulfoxide (DMSO), 4-phenylbutyrate, and lipids/detergents are not osmolytes but can promote folding of mutant proteins. In so doing they are referred to as chemical chaperones. Direct effect of the chemical chaperones can also take place by binding and stabilizing either the native structure or late folding intermediates as suggested by Arakawa *et al. (115)* for the effect of glycerol on the mutant $\Delta F508$ of CFTR. Except for 4-phenylbutyrate which has been approved for use in urea-cycle disorders and for clinical trials for thalassemia and cystic fibrosis (for references see *(175)*), the use of chemical chaperones for therapeutic intervention in protein folding and aggregation diseases is not that clear *(115)*. The main reasons appear to be that they require high toxic concentrations for effective folding of mutant proteins. Moreover, chemical chaperones are not specific and stabilize also other nontarget proteins in the cell. It is nevertheless important to extract information on the effect of chemical chaperones to contribute and optimize the interventions using other approaches *(115, 175)*.

For PAH is has been shown that glycerol at a concentration of 0.4% increased 2–3-fold the yield and activity of wt and selected PKU-associated mutants (R270K and V388M) expressed in a prokaryote recombinant system *(176)*. It is however not clear if the chemical chaperones directly promote the folding of PAH in this prokaryote system or stabilization is an indirect effect via activation of the bacterial molecular chaperones *(177)*.

C. Pharmacological Chaperones

Pharmacological chaperones are low molecular weight compounds which, opposite to the chemical chaperones, are functional at low concentrations. The pharmacological chaperones must be cell permeable, and for those proteins which are translocated to the ER during folding, also ER permeable. There are a growing number of examples of their potential use for the correction of conformational diseases *(21, 22, 178, 179)*. Usually they are selected as specific stabilizers of the proteins associated to misfolding diseases (wt or mutants) either by screening strategies or by further developing the structure of natural ligands, for example, receptor antagonists. But a disadvantage might be that they bind very tightly at the functional binding sites, which can be a concern for the treatment of misfolding diseases involving receptors and enzymes if the compounds cannot easily be removed from the protein by the natural agonists and substrates. But in fact it has been shown that some of the most effective pharmacological chaperones are active site-directed specific molecules *(180)* which resemble active site inhibitors *(181)*. This may be a counterintuitive approach to treat diseases, but studies with lysosomal storage diseases (LSDs) indicate that the gain in function by misfolding correction counteracts and overcomes the inhibition by the chaperones *(180)*. This is probably due to the fact that the critical threshold of residual enzyme activity for disease manifestation is only 5–10% of normal level. In addition, the active site inhibitors usually inhibit competitively versus the natural substrates and agonists, which allows reaching a therapeutic activity threshold *in vivo* at locations and conditions with high concentration of substrates *(180)*. To increase specificity for the selected misfolded targets and to induce their rescue at low concentrations of pharmacological chaperones, it may be beneficial to increase the affinity of the compounds. A strategy to do so would be to use composite ligands containing fragments of binding units attached to a single backbone molecule. These ligands can bind to multiple sites in the same protein, which provides a higher affinity compared to the binding of the separated integrating units *(182)*.

In strict sense, the term *pharmacological chaperone* should indicate that the compounds accelerate folding or revert misfolding (by analogy with the molecular chaperone function). In fact it has been proposed that these compounds may bind to partially folded intermediates allowing to accelerate and complete the folding process *(90, 175)*. Pharmacological chaperones have also been defined as template molecules that "induce mutant proteins to adopt native-type-like conformations instead of improperly folded ones" *(178, 180)*. Based on these definitions pharmacological chaperones are folding-aid drugs that would modify folding kinetics for the target protein possibly by binding to on-pathway folding intermediates (Fig. 2B). But the term *pharmacological chaperone* has also often been associated to molecules that just bind to the native state of the wt and mutant proteins stabilizing their conformation and correct subunit assembly *(24)*. In this case and as discussed in Section IV, specific ligand binding to a native state is expected to increase the thermodynamic stability (ΔG_{N-U}) as well as to reduce the unfolding rates (increased $\Delta G_{N-\ddagger}$; Fig. 2) by lowering the free energy of the native state *(10)*, and the degree of stabilization will depend on ligand concentration (Scheme 2) *(9, 49, 60)*. Certainly, the proposed molecular mechanisms for the mode of action of different pharmacological chaperones may differ, but indeed there are not many cases in which the mechanisms have been studied in detail, mostly due to the fact that the proteins involved are complex, often multimeric. Like PAH (see Section II) these proteins do not fold reversibly, impeding detailed studies on the thermodynamic and kinetic effects of the chaperones on the wt and mutant proteins. In any case, the pharmacological chaperones have been found to rescue the mutants from proteasomal degradation leading to an elongation of the half-life of the protein and to an improved or normalized cellular localization and function *(90, 180)*.

D. Natural Chaperone Ligands

Natural substrates, cofactors, and inhibitors have effects on protein stability beyond their functional role in enzyme function by the same arguments as for other specific ligands and can be considered as natural chaperone ligands. Though the role of natural ligands such as vitamins is well documented for the treatment of metabolic diseases, the molecular explanation provided for the impact of ligand supplementation on enzyme activity has so far been focused on the correction of the kinetic (K_m) effect *(20)*.

In addition to the current case of stabilization of PAH by BH_4 (see below), we would like to bring in two interesting case studies on the potential role of natural chaperone ligands in human misfolding diseases. These refer to the biomedical relevant proteins transthyretin and Cu/Zn-superoxide dismutase (SOD), which on the other hand represent two different situations in terms of stability of the *apo-* (ligand-free) form. Transthyretin is a quite robust homotetrameric protein which is denatured by urea only at low protein concentrations where dissociation into monomers is favored *(183)*. It displays a significant kinetic stability which protects the enzyme against dissociation into monomeric species competent to amyloid fibril formation. Mutations associated to amyloidosis are known to reduce the kinetic barrier for unfolding *(183)*, while binding of the natural thyroxine and thyroxine-like ligands increase significantly the unfolding barrier *(24)*. On the other hand, SOD is a dimeric protein that folds first into native-like monomers and then dimerizes and binds catalytic Cu and structural Zn ions *(184)*. In this case, the ion-free apo-protein is marginally stable under physiological conditions and, therefore, ion binding seems to be necessary to reach high stability *(184)*. Mutations in SOD associated with amyotrophic lateral sclerosis are known to destabilize the protein by different mechanisms, including destabilization of the folded monomer (leading to fibril formation) or decreased metal-binding capacity to form the *holo*-enzyme *(184, 185)*. In fact even the wt-protein requires natural ligand binding to be stable, highlighting the biological role of the ligands enhancing protein stability under physiological conditions.

For PAH the role of BH_4 as a natural chaperone ligand was proven with studies with knockout mice lacking the 6-pyruvoyl-tetrahydropterin synthase (*PTPS*) gene (Fig. 1), and which present a complete deficiency in the endogenous cofactor biosynthesis *(15)*. Interestingly, the knockout mice showed normal *PAH* gene expression, as shown by mRNA quantification by quantitative reverse-transcription PCR analysis of liver tissue, but almost total lack of PAH protein and activity with respect to the wt-animals, underlining the chaperoning role of BH_4.

1. Tetrahydrobiopterin-Responsive PKU

BH_4 supplementation has been classically used to treat genetic deficiencies in the metabolic steps involved in the synthesis or regeneration of BH_4 (Fig. 1) *(17, 186)*. However, more than 20 years ago there appeared some reports on positive responses to BH_4 supplementation in patients with mutations in *PAH* *(187)*. Kure *et al.* (1999) paid specially attention to this phenomenon when they reported four PKU patients with a positive response to BH_4 supplementation, and, in accordance to what is inferred from other vitamin-responsive diseases *(20)*, they proposed that PKU mutants would display reduced affinity for the cofactor. Since then multiple reports have confirmed the efficiency in short- and long-term treatment with BH_4 in PKU patients, specially with mild forms of PKU *(17, 18, 130)*. A recent analysis has indicated that up to 80% of the mild PKU patients and 40% of the total PKU patients are potentially responsive *(188)*. Most of the mutations associated to BH_4 responsiveness show significant residual activities (more than 30% of the wt-specific activity) consistent with the mild phenotypes found in most responsive patients, while only a small fraction of the responsive alleles display significantly reduced affinity for the cofactor (16, 19, 157). Expression analyses on in vitro transcription–translation systems have shown that some of these PKU mutants are stabilized by BH_4 against proteolytic degradation, while the cofactor also exerts a general protective effect versus oxidative inactivation (16). From all these studies, it has been proposed that BH_4 responsiveness occurs through a multifactorial mechanism, involving a stabilization by BH_4 (natural chaperone effect), overcoming low-binding affinity in some mutants and increasing the saturation fraction of the enzyme under physiological turnover conditions (16, 19, 131). Experimental evidence for a substoichiometric binding of BH_4 to PAH in hepatocytes and mice liver has in fact been presented (15, 189). An increase in intracellular BH_4 due to pharmacological treatment will therefore raise the levels of free BH_4 intracellularly. The following expression (adapted from (49)) establishes a relationship between binding affinity, free BH_4 levels, and the stabilization of the PAH native state under equilibrium conditions, assuming that the "unfolded" state does not bind BH_4, and that the native state is fully saturated with BH_4:

$$\Delta\Delta G_{N-U} = -RT \ln\left(\frac{1}{1 + K_{binding}[BH_4]}\right), \quad (1)$$

where $\Delta\Delta G_{N-U}$ is the free-energy difference between the native state and an "unfolded" state not competent for BH_4 binding, $K_{binding}$ is the binding affinity constant for BH_4 (at 25 °C), $[BH_4]$ is the concentration of free BH_4, R is 1.987 cal/mol K and $T = 310.15$ K (37 °C). An increase of 5–10 μM in (free) BH_4 concentration will stabilize the wt-PAH protein by 0.65–0.95 kcal/mol, while this stabilization will be lower for a mutant with a decreased binding affinity for BH_4 [e.g., for R68S (K_d shown in Table II) the stabilization will be 0.27–0.46 kcal/mol, also for 5–10 μM BH_4 increase]. In the cases where the K_d values for BH_4 binding are not known, kinetic unfolding studies may also provide information on the stabilization by ligand supplementation. We have performed CD-monitored kinetic analyses of wt and certain PKU mutants (Fig. 7), in order to illustrate the concept and calculate mutation effects on native state stability, as follows:

$$\Delta\Delta G_{N-TS(mut-wt)} = RT \ln\left(\frac{k_{wt}}{k_{mut}}\right), \quad (2)$$

where $\Delta\Delta G_{N-TS(mut-wt)}$ is the activation free-energy difference between mutant and wt-PAH and k_{wt} and k_{mut} are the corresponding first-order rate constant for irreversible unfolding obtained for wt and mutant PAH proteins, respectively, at 37 °C (Fig. 7). For the mutant R68S, the kinetic stability is

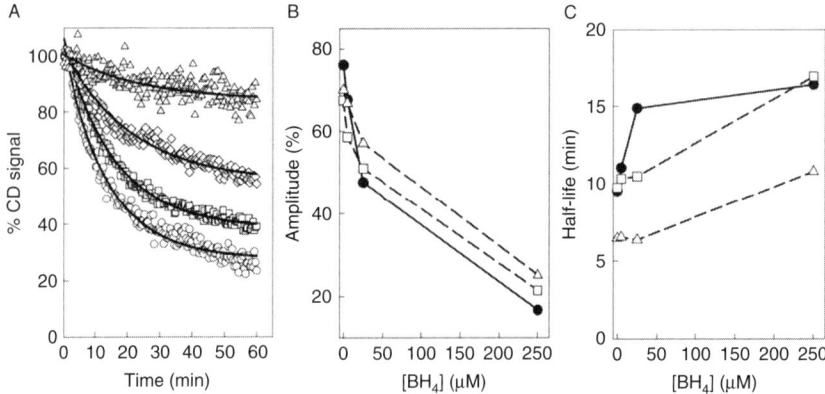

FIG. 7. Effect of BH$_4$ on the irreversible unfolding of wt and mutant hPAH tetrameric proteins at 37 °C. (A) Kinetic traces of wt-PAH in the absence (circles) or the presence of 5 μM (squares), 25 μM (diamonds), and 250 μM (triangles) of BH$_4$. Lines are fits to a first-order decay function. (B) Amplitudes and (C) half-lives for the irreversible unfolding of wt-PAH (circles) and of the mild-PKU-associated mutants R68S-PAH (squares) and D143G-PAH (triangles) in the presence of 0–250 μM BH$_4$.

similar to that measured for wt-PAH in the absence of ligand, but in the case of the D143G-PAH mutant, $\Delta\Delta G_{N-TS(mut-wt)}$ is −0.24 kcal/mol, and even higher in other mutations (e.g., −0.58 kcal/mol for R408W; data not shown). Therefore, it is obvious that a relatively small increase in the free concentration of BH$_4$ may overcome the kinetic destabilization introduced on the native state by mutations (see also Section IV). These general predictions would be general for most PKU missense mutations with decreased stability (37) (Fig. 5) and which are therefore susceptible to ligand-binding stabilization of the native state.

An important insight that arises from the CD-monitored experiments shown in Fig. 7 is that at physiological conditions (pH 7.0 and 37 °C), wt-PAH has a low kinetic stability and increasing concentrations of BH$_4$ from physiological (5 μM) to supraphysiological concentrations (250 μM), reduce the extent of the irreversible denaturation and unfolding rates (Fig. 7B and C). Thus, PAH seems to require BH$_4$ to reach the half-life that corresponds to physiological relevant half-lives (Fig. 7C). This *in vitro* result is in agreement with the *in vivo* stabilizing effect of BH$_4$ (15, 126, 190). In summary, these results reinforce the concept of BH$_4$ as natural chaperone ligand of PAH both in health and disease. BH$_4$ keeps the enzyme in a low-activity and high-stability state which is shifted to the active state when the L-Phe concentration raises (131, 189).

Data on BH_4-responsive PKU can be found at: BIODEF (International database of tetrahydrobiopterin deficiencies), BIOMDB (Database of mutations causing tetrahydrobiopterin deficiencies) and BIOPKU (International database of BH_4-responsive HPA/PKU), all at http://www.bh4.org/.

2. Potential Treatment of PKU by Pharmacological Chaperones

With respect to the total number of patients, PKU is at the limit of being a rare disease. But with 62% of the PKU mutations being missense mutations that result in different misfolding degrees of PAH (Fig. 5), the phenotype of the patients is very variable, ranging from mild to classical PKU. Actually, each of these graded phenotypes might be considered in itself as a rare disorder, pointing to PKU as a model disease for patient-tailored intervention and optimal for screening of pharmacological chaperones. The positive experience on the rescue of mild PKU mutations by BH_4 supplementation supports attempting the rescue of a larger number of patients. For the most severe classical forms of PKU, at least one could aim to ameliorate the current therapy with for-life L-Phe-free diet. In some way, the action of the pharmacological chaperones must overlap with that of BH_4, since they would be expected to bind to the native state (stabilizing it) and/or promote efficient folding in the cytosol. However, the possibility of drug-optimization on these likely inert leads seems more plausible than optimization of the natural cofactor, since improving BH_4 binding affinity may affect its catalytic efficiency and reaction coupling, leading to a suboptimal cofactor or even to a pterin compound exhibiting inhibitory properties on PAH or the other aromatic amino acid hydroxylases *(161)*.

VI. Concluding Remarks

To date, there are described six to seven thousand rare diseases and new ones are reported every week in the medical literature (http://www.orpha.net/). The heterogeneous nature of these diseases and the small number of patients in each group affect all aspects of their recognition and management, making proper diagnosis, registration of patients, understanding of disease mechanisms, treatment, and follow up very difficult and casual. In spite that rare diseases are serious chronic diseases, and often life-threatening, the academic research community and the pharmaceutical industry have traditionally given very low dedication to their understanding and therapeutic correction. This is mostly due to the low assumed social and economic impact associated to the small number of patients for each disease. But many of these diseases, and notably those caused by genetic defects, are associated to protein misfolding, both those where the misfolded proteins form amyloid structures or other aggregates, and those in which the proteins appear to be quickly degraded, as is the case for PKU. Integrating the patient-specific cases in a generic effective screening and therapeutic schemes appears as a synergetic and cost effective effort. Moreover, mutation-induced unfolding shares common elements with misfolding induced by processes such as oxidative stress, overstimulation of signal transduction pathways, and overexpression of proteins which operate at the onset of several diseases both rare and major (serpinopathies, cancer, neurodegeneration, and cardiovascular diseases, among other) (5). These insights may increase the interest of the pharmaceutical and medical community and ameliorate the "orphaned" nature of these diseases and patients and of the drugs to treat them.

The early view of protein structures as relatively rigid molecules with one unique active conformation has been replaced by the understanding that they are highly flexible and malleable entities. Computational approaches (such as FoldX analysis and MD simulations) can be used to evaluate the impact of mutations and ligand binding on protein structure, energetics, and dynamics. All together, the integration of experimental and computational biophysical methods provides structural and functional information on complexes of the proteins with, for example, natural chaperone ligands. Concretely, atomic-level information would certainly contribute to understand the molecular mechanism underlying the further development of stabilizing pharmacological chaperones with improved affinity and/or selectivity for the protein target. Structure-based drug design has matured into a discipline with impact on the pharmaceutical industry, moving it from a trial-and-error-based search to more focused and rapid search for promising drug candidates.

Acknowledgments

The authors want to thank Professor Arturo Muga for critical reading of the manuscript. Work described in this review from the authors' laboratory was supported in part by grants from the Research Council of Norway and Helse-Vest.

Since the submission of this manuscript, the stabilization of phenylalanine hydroxylase by pharmacological chaperones discovered through high throughout screening has been published (Pey, A.L., Ying, M., Cremades, N., Velazquez-Campoy, A., Scherer, T., Thony, B., Sancho, J. and Martinez, A. (2008) Identification of pharmacological chaperones as potential therapeutic agents to treat phenylketonuria. *J Clin Invest* **118**, 2858-2867). The selected compounds also significantly increased activity and steady-state PAH protein levels in cells transiently transfected with either wild-type PAH or PKU mutants and in liver of mice treated with small doses of the compounds.

References

1. Garcia-Mira, M. M., Sadqi, M., Fischer, N., Sanchez-Ruiz, J. M., and Munoz, V. (2002). Experimental identification of downhill protein folding. *Science* **298**, 2191–2195.

2. Munoz, V. (2007). Conformational dynamics and ensembles in protein folding. *Annu. Rev. Biophys. Biomol. Struct.* **36**, 395–412.
3. Uversky, V. N., Gillespie, J. R., and Fink, A. L. (2000). Why are "natively unfolded" proteins unstructured under physiologic conditions? *Proteins* **41**, 415–427.
4. Sickmeier, M., Hamilton, J. A., LeGall, T., Vacic, V., Cortese, M. S., Tantos, A., Szabo, B., Tompa, J., Chen, J., Uversky, V. N., Obradovic, Z., and Dunker, A. K. (2007). DisProt: The database of disordered proteins. *Nucleic Acids Res.* **35**, D786–D793.
5. Dobson, C. M. (2003). Protein folding and misfolding. *Nature* **426**, 884–890.
6. Foguel, D., and Silva, J. L. (2004). New insights into the mechanisms of protein misfolding and aggregation in amyloidogenic diseases derived from pressure studies. *Biochemistry* **43**, 11361–11370.
7. Young, J. C., Agashe, V. R., Siegers, K., and Hartl, F. U. (2004). Pathways of chaperone-mediated protein folding in the cytosol. *Nat. Rev. Mol. Cell Biol.* **5**, 781–791.
8. Dai, Q., Qian, S. B., Li, H. H., McDonough, H., Borchers, C., Huang, D., Takayama, S., Younger, H. Y., Ren, H. Y., Cyr, D. M., and Patterson, C. (2005). Regulation of the cytoplasmic quality control protein degradation pathway by BAG2. *J. Biol. Chem.* **280**, 38673–38681.
9. Sanchez-Ruiz, J. M. (2007). Ligand effects on protein thermodynamic stability. *Biophys. Chem.* **126**, 43–49.
10. Wittung-Stafshede, P. (2002). Role of cofactors in protein folding. *Acc. Chem. Res.* **35**, 201–208.
11. Sancho, J. (2006). Flavodoxins: Sequence, folding, binding, function and beyond. *Cell. Mol. Life Sci.* **63**, 855–864.
12. Landfried, D. A., Vuletich, D. A., Pond, M. P., and Lecomte, J. T. (2007). Structural and thermodynamic consequences of b heme binding for monomeric apoglobins and other apoproteins. *Gene* **398**, 12–28.
13. Iwaki, M., Phillips, R. S., and Kaufman, S. (1986). Proteolytic modification of the amino-terminal and carboxyl-terminal regions of rat hepatic phenylalanine hydroxylase. *J. Biol. Chem.* **261**, 2051–2056.
14. Solstad, T., Stokka, A. J., Andersen, O. A., and Flatmark, T. (2003). Studies on the regulatory properties of the pterin cofactor and dopamine bound at the active site of human phenylalanine hydroxylase. *Eur. J. Biochem.* **270**, 981–990.
15. Thony, B., Ding, Z., and Martinez, A. (2004). Tetrahydrobiopterin protects phenylalanine hydroxylase activity *in vivo*: Implications for tetrahydrobiopterin-responsive hyperphenylalaninemia. *FEBS Lett.* **577**, 507–511.
16. Pey, A. L., Perez, B., Desviat, L. R., Martinez, M. A., Aguado, C., Erlandsen, H., Gamez, A., Stevens, M., Thorolfsson, M., Ugarte, M., and Martínez, A. (2004). Mechanisms underlying responsiveness to tetrahydrobiopterin in mild phenylketonuria mutations. *Hum. Mutat.* **24**, 388–399.
17. Blau, N., and Erlandsen, H. (2004). The metabolic and molecular bases of tetrahydrobiopterin-responsive phenylalanine hydroxylase deficiency. *Mol. Genet. Metab.* **82**, 101–111.
18. Levy, H. L., Milanowski, A., Chakrapani, A., Cleary, M., Lee, P., Trefz, F. K., Whitley, C. B., Feillet, A. S., Feigenbaum, A. S., Bebchuk, J. D., Christ-Schmidt, H., and Dorenbaum, A. (2007). Efficacy of sapropterin dihydrochloride (tetrahydrobiopterin, 6R-BH4) for reduction of phenylalanine concentration in patients with phenylketonuria: A phase III randomised placebo-controlled study. *Lancet* **370**, 504–510.
19. Erlandsen, H., Pey, A. L., Gamez, A., Perez, B., Desviat, L. R., Aguado, C., Koch, R., Surendran, S., Tyring, S., Matalon, R., Scriver, C. R., Ugarte, M. *et al.* (2004). Correction of kinetic and stability defects by tetrahydrobiopterin in phenylketonuria patients with certain phenylalanine hydroxylase mutations. *Proc. Natl. Acad. Sci. USA* **101**, 16903–16908.

20. Ames, B. N., Elson-Schwab, I., and Silver, E. A. (2002). High-dose vitamin therapy stimulates variant enzymes with decreased coenzyme binding affinity (increased K(m)): Relevance to genetic disease and polymorphisms. *Am. J. Clin. Nutr.* **75**, 616–658.
21. Bernier, V., Lagace, M., Bichet, D. G., and Bouvier, M. (2004). Pharmacological chaperones: Potential treatment for conformational diseases. *Trends Endocrinol. Metab.* **15**, 222–228.
22. Conn, P. M., Ulloa-Aguirre, A., Ito, J., and Janovick, J. A. (2007). G protein-coupled receptor trafficking in health and disease: Lessons learned to prepare for therapeutic mutant rescue in vivo. *Pharmacol. Rev.* **59**, 225–250.
23. Hammarstrom, P., Wiseman, R. L., Powers, E. T., and Kelly, J. W. (2003). Prevention of transthyretin amyloid disease by changing protein misfolding energetics. *Science* **299**, 713–716.
24. Johnson, S. M., Wiseman, R. L., Sekijima, Y., Green, N. S., Adamski-Werner, S. L., and Kelly, J. W. (2005). Native state kinetic stabilization as a strategy to ameliorate protein misfolding diseases: A focus on the transthyretin amyloidoses. *Acc. Chem. Res.* **38**, 911–921.
25. Pey, A. L., and Martínez, A. (2006). The phenylalanine hydroxylase system. *In* "PKU and BH4. Advances in phenylketonuria and tetrahydrobiopterin" (N. Blau, Ed.), pp. 67–91. SPS Verlagsgesellschaft MBH, Heilbronn, Germany.
26. Scriver, C. R., and Kaufman, S. (2001). Hyperphenylalaninemia:phenylalanine hydroxylase deficiency. *In* "The Metabolic and Molecular bases of Inherited Disease" (C. R. Scriver, A. L. Beaudet, D. ValleW. S. Sly, Eds.), 8th edn., pp. 1667–1724. McGraw-Hill, New York.
27. Scriver, C. R., Hurtubise, M., Konecki, D., Phommarinh, M., Prevost, L., Erlandsen, H., Stevens, P. J., Waters, P. J., Ryan, S., McDonald, D., and Sarkissian, C. (2003). PAHdb 2003: What a locus-specific knowledgebase can do. *Hum. Mutat.* **21**, 333–344.
28. Følling, A. (1934). Über Ausscheidung von Phenylbrenztraubensäure in den Harn als Stoffwechselanomalie in Verbindung mit Imbezilität. *Hoppe-Seylers Zeitschrift für physiologische Chemie* **227**, 169–176.
29. Penrose, L. S. (1946). Phenylketonuria—A problem in eugenics. *Lancet* **1**, 949–953.
30. Gregersen, N., Bross, P., Andrese, B. S., Pedersen, C. B., Corydon, T. J., and Bolund, L. (2001). The role of chaperone-assisted folding and quality control in inborn errors of metabolism: Protein folding disorders. *J. Inherit. Metab. Dis.* **24**, 189–212.
31. Gregersen, N. (2006). Protein misfolding disorders: Pathogenesis and intervention. *J. Inherit. Metab. Dis.* **29**, 456–470.
32. Dobson, C. M. (2004). Principles of protein folding, misfolding and aggregation. *Semin. Cell Dev. Biol.* **15**, 3–16.
33. Scriver, C. R. (2007). The PAH gene, phenylketonuria, and a paradigm shift. *Hum. Mutat.* **28**, 831–845.
34. Kobe, B., Jennings, I. G., House, C. M., Michell, B. J., Goodwill, K. E., Santarsiero, B. D., Stevens, R. G., Cotton, R. G., and Kemp, B. E. (1999). Structural basis of autoregulation of phenylalanine hydroxylase. *Nat. Struct. Biol.* **6**, 442–448.
35. Flatmark, T., and Stevens, R. C. (1999). Structural insight into the aromatic amino acid hydroxylases and their disease-related mutant forms. *Chem. Rev.* **99**, 2137–2160.
36. Erlandsen, H., and Stevens, R. C. (1999). The structural basis of phenylketonuria. *Mol. Genet. Metab.* **68**, 103–125.
37. Pey, A. L., Stricher, F., Serrano, L., and Martinez, A. (2007). Predicted effects of missense mutations on native-state stability account for phenotypic outcome in phenylketonuria, a paradigm of misfolding diseases. *Am. J. Hum. Genet.* **81**, 1006–1024.
38. Fusetti, F., Erlandsen, H., Flatmark, T., and Stevens, R. C. (1998). Structure of tetrameric human phenylalanine hydroxylase and its implications for phenylketonuria. *J. Biol. Chem.* **273**, 16962–16967.

39. Jahn, T. R., and Radford, S. E. (2008). Folding versus aggregation: Polypeptide conformations on competing pathways. *Arch. Biochem. Biophys.* **469**, 100–117.
40. Dill, K., and Chan, H. S. (1997). From levinthal to pathways to funnels. *Nat. Struct. Mol. Biol.* **4**, 10–19.
41. Brockwell, D. J., and Radford, S. E. (2007). Intermediates: Ubiquitous species on folding energy landscapes? *Curr. Opin. Struct. Biol.* **17**, 30–37.
42. Campos, L. A., Bueno, M., Lopez-Llano, J., Jimenez, M. A., and Sancho, J. (2004). Structure of stable protein folding intermediates by equilibrium phi-analysis: The apoflavodoxin thermal intermediate. *J. Mol. Biol.* **344**, 239–255.
43. Cremades, N., Sancho, J., and Freire, E. (2006). The native-state ensemble of proteins provides clues for folding, misfolding and function. *Trends Biochem. Sci.* **31**, 494–496.
44. Han, J. H., Batey, S., Nickson, A. A., Teichmann, S. A., and Clarke, J. (2007). The folding and evolution of multidomain proteins. *Nat. Rev. Mol. Cell. Biol.* **8**, 319–330.
45. Cooper, A. (1999). Thermodynamic analysis of biomolecular interactions. *Curr. Opin. Chem. Biol.* **3**, 557–563.
46. Thomson, J. A., Shirley, B. A., Grimsley, G. R., and Pace, C. N. (1989). Conformational stability and mechanism of folding of ribonuclease T1. *J. Biol. Chem.* **264**, 11614–11620.
47. Genzor, C. G., Beldarrain, A., Gomez-Moreno, C., Lopez-Lacomba, J. L., Cortijo, M., and Sancho, J. (1996). Conformational stability of apoflavodoxin. *Protein Sci.* **5**, 1376–1388.
48. Ibarra-Molero, B., and Sanchez-Ruiz, J. M. (1997). Are there equilibrium intermediate states in the urea-induced unfolding of hen egg-white lysozyme. *Biochemistry* **36**, 9616–9624.
49. Luque, I., Leavitt, S. A., and Freire, E. (2002). The linkage between protein folding and functional cooperativity: Two sides of the same coin? *Annu. Rev. Biophys. Biomol. Struct.* **31**, 235–256.
50. Tsai, C. J., Ma, B., Sham, Y. Y., Kumar, S., and Nussinov, R. (2001). Structured disorder and conformational selection. *Proteins* **44**, 418–427.
51. Pervushin, K., Vamvaca, K., Vogeli, B., and Hilvert, D. (2007). Structure and dynamics of a molten globular enzyme. *Nat. Struct. Mol. Biol.* **14**, 1202–1206.
52. Dunker, A. K., Lawson, J. D., Brown, C. J., Williams, R. M., Romero, P., Oh, J. S., Oldfield, A. M., Campen, A. M., Ratliff, C. M., Hipps, K. W., Ausio, J., Nissen, M. S. et al. (2001). Intrinsically disordered protein. *J. Mol. Graph. Model.* **19**, 26–59.
53. Gianni, S., Ivarsson, Y., Jemth, P., Brunori, M., and Travaglini-Allocatelli, C. (2007). Identification and characterization of protein folding intermediates. *Biophys. Chem.* **128**, 105–113.
54. Scalley, M. L., and Baker, D. (1997). Protein folding kinetics exhibit an Arrhenius temperature dependence when corrected for the temperature dependence of protein stability. *Proc. Natl. Acad. Sci. USA* **94**, 10636–10640.
55. Naganathan, A. N., Doshi, U., Fung, A., Sadqi, M., and Munoz, V. (2006). Dynamics, energetics, and structure in protein folding. *Biochemistry* **45**, 8466–8475.
56. Rodriguez-Larrea, D., Minning, S., Borchert, T. V., and Sanchez-Ruiz, J. M. (2006). Role of solvation barriers in protein kinetic stability. *J. Mol. Biol.* **360**, 715–724.
57. Plaza del Pino, I. M., Ibarra-Molero, B., and Sanchez-Ruiz, J. M. (2000). Lower kinetic limit to protein thermal stability: A proposal regarding protein stability *in vivo* and its relation with misfolding diseases. *Proteins* **40**, 58–70.
58. Godoy-Ruiz, R., Ariza, F., Rodriguez-Larrea, D., Perez-Jimenez, R., Ibarra-Molero, B., and Sanchez-Ruiz, J. M. (2006). Natural selection for kinetic stability is a likely origin of correlations between mutational effects on protein energetics and frequencies of amino acid occurrences in sequence alignments. *J. Mol. Biol.* **362**, 966–978.
59. Thórólfsson, M., Ibarra-Molero, B., Fojan, P., Petersen, S. B., Sanchez-Ruiz, J. M., and Martínez, A. (2002). L-phenylalanine binding and domain organization in human phenylalanine hydroxylase: A differential scanning calorimetry study. *Biochemistry* **41**, 7573–7585.

60. Sanchez-Ruiz, J. M. (1992). Theoretical analysis of Lumry–Eyring models in differential scanning calorimetry. *Biophys. J.* **61**, 921–935.
61. Lumry, R., and Eyring, H. (1954). Conformation changes of proteins. *J. Phys. Chem.* **58**, 110–120.
62. Bjørgo, E., Knappskog, P. M., Martínez, A., Stevens, R. C., and Flatmark, T. (1998). Partial characterization and three-dimensional-structural localization of eight mutations in exon 7 of the human phenylalanine hydroxylase gene associated with phenylketonuria. *Eur. J. Biochem.* **257**, 1–10.
63. Gámez, A., Pérez, B., Ugarte, M., and Desviat, L. R. (2000). Expression analysis of phenylketonuria mutations. Effect on folding and stability of the phenylalanine hydroxylase protein. *J. Biol. Chem.* **275**, 29737–29742.
64. Gjetting, T., Petersen, M., Guldberg, P., and Guttler, F. (2001). In vitro expression of 34 naturally occurring mutant variants of phenylalanine hydroxylase: Correlation with metabolic phenotypes and susceptibility toward protein aggregation. *Mol. Genet. Metab.* **72**, 132–143.
65. Pey, A. L., Desviat, L. R., Gamez, A., Ugarte, M., and Perez, B. (2003). Phenylketonuria: Genotype-phenotype correlations based on expression analysis of structural and functional mutations in PAH. *Hum. Mutat.* **21**, 370–378.
66. Waters, P. J. (2003). How PAH gene mutations cause hyper-phenylalaninemia and why mechanism matters: Insights from *in vitro* expression. *Hum. Mutat.* **21**, 357–369.
67. Parniak, M. A. (1989). Limited denaturation stimulates tetrahydrobiopterin dependent activity of rat liver phenylalanine hydroxylase. *In* Chemistry and Biology of Pteridines" (H.-C. Curtius, S. Ghisla, and N. Blau, Eds.), pp. 656–659. de Gruyter, Berlin.
68. Kleppe, R., Uhlemann, K., Knappskog, P. M., and Haavik, J. (1999). Urea-induced denaturation of human phenylalanine hydroxylase. *J. Biol. Chem.* **274**, 33251–33258.
69. Chehin, R., Thorolfsson, M., Knappskog, P. M., Martínez, A., Flatmark, T., Arrondo, J. L., and Muga, A. (1998). Domain structure and stability of human phenylalanine hydroxylase inferred from infrared spectroscopy. *FEBS Lett.* **422**, 225–230.
70. Miranda, F. F., Thórólfsson, M., Teigen, K., Sanchez-Ruiz, J. M., and Martínez, A. (2004). Structural and stability effects of phosphorylation: Localized structural changes in phenylalanine hydroxylase. *Protein Sci.* **13**, 1219–1226.
71. Luque, I., Gómez, J., and Freire, E. (1998). Structure-based thermodynamic design of peptide ligands: Application to peptide inhibitors of the aspartic protease endothiapepsin. *Proteins* **30**, 74–85.
72. Hilser, V. J., Gómez, J., and Freire, E. (1996). The enthalpy change in protein folding and binding: Refinement of parameters for structure-based correlations. *Proteins* **26**, 123–133.
73. Kreimer, D. I., Shnyrov, V. L., Villar, E., Silman, I., and Weiner, L. (1995). Irreversible thermal denaturation of Torpedo californica acetylcholinesterase. *Protein Sci.* **4**, 2349–2357.
74. Duy, C., and Fitter, J. (2005). Thermostability of irreversible unfolding alpha-amylases analyzed by unfolding kinetics. *J. Biol. Chem.* **280**, 37360–37365.
75. Kelly, J. W., and Balch, W. E. (2006). The integration of cell and chemical biology in protein folding. *Nat. Chem. Biol.* **2**, 224–227.
76. Ellis, R. J. (2001). Macromolecular crowding: An important but neglected aspect of the intracellular environment. *Curr. Opin. Struct. Biol.* **11**, 114–119.
77. Outeiro, T. F., and Tetzlaff, J. (2007). Mechanisms of disease II: Cellular protein quality control. *Semin. Pediatr. Neurol.* **14**, 15–25.
78. McClellan, A. J., Tam, S., Kaganovich, D., and Frydman, J. (2005). Protein quality control: Chaperones culling corrupt conformations. *Nat. Cell Biol.* **7**, 736–741.
79. Anelli, T., and Sitia, R. (2008). Protein quality control in the early secretory pathway. *EMBO J.* **27**, 315–327.
80. Hartl, F. U., and Hayer-Hartl, M. (2002). Molecular chaperones in the cytosol: From nascent chain to folded protein. *Science* **295**, 1852–1858.

81. Fink, A. L. (1999). Chaperone-mediated protein folding. *Physiol. Rev.* **79**, 425–449.
82. Bukau, B., Weissman, J., and Horwich, A. (2006). Molecular chaperones and protein quality control. *Cell* **125**, 443–451.
83. Horwich, A. L., Low, K. B., Fenton, W. A., Hirshfield, I. N., and Furtak, K. (1993). Folding *in vivo* of bacterial cytoplasmic proteins: Role of GroEL. *Cell* **74**, 909–917.
84. Valpuesta, J. M., Martin-Benito, J., Gomez-Puertas, P., Carrascosa, J. L., and Willison, K. R. (2002). Structure and function of a protein folding machine: The eukaryotic cytosolic chaperonin CCT. *FEBS Lett.* **529**, 11–16.
85. Frydman, J., Nimmesgern, E., Ohtsuka, K., and Hartl, F. U. (1994). Folding of nascent polypeptide chains in a high molecular mass assembly with molecular chaperones. *Nature* **370**, 111–117.
86. Muchowski, P. J., and Wacker, J. L. (2005). Modulation of neurodegeneration by molecular chaperones. *Nat. Rev. Neurosci.* **6**, 11–22.
87. Bercovich, B., Stancovski, I., Mayer, A., Blumenfeld, N., Laszlo, A., Schwartz, A. L., and Ciechanover, A. (1997). Ubiquitin-dependent degradation of certain protein substrates *in vitro* requires the molecular chaperone Hsc70. *J. Biol. Chem.* **272**, 9002–9010.
88. Meacham, G. C., Patterson, C., Zhang, W., Younger, J. M., and Cyr, D. M. (2001). The Hsc70 co-chaperone CHIP targets immature CFTR for proteasomal degradation. *Nat. Cell Biol.* **3**, 100–115.
89. Glickman, M. H., and Ciechanover, A. (2002). The ubiquitin–proteasome proteolytic pathway: Destruction for the sake of construction. *Physiol. Rev.* **82**, 373–428.
90. Cohen, F. E., and Kelly, J. W. (2003). Therapeutic approaches to protein-misfolding diseases. *Nature* **426**, 905–909.
91. van den Berg, B., Ellis, R. J., and Dobson, C. M. (1999). Effects of macromolecular crowding on protein folding and aggregation. *EMBO J.* **18**, 6927–6933.
92. Thulasiraman, V., Yang, C. F., and Frydman, J. (1999). *In vivo* newly translated polypeptides are sequestered in a protected folding environment. *EMBO J.* **18**, 85–95.
93. Connell, P., Ballinger, C. A., Jiang, J., Wu, Y., Thompson, L. J., Hohfeld, J., and Patterson, C. (2001). The co-chaperone CHIP regulates protein triage decisions mediated by heat-shock proteins. *Nat. Cell Biol.* **3**, 93–96.
94. Berke, S. J., Schmied, F. A., Brunt, E. R., Ellerby, L. M., and Paulson, H. L. (2004). Caspase-mediated proteolysis of the polyglutamine disease protein ataxin-3. *J. Neurochem.* **89**, 908–918.
95. Yao, T., and Cohen, R. E. (2002). A cryptic protease couples deubiquitination and degradation by the proteasome. *Nature* **419**, 403–407.
96. Kopito, R. R. (2000). Aggresomes, inclusion bodies and protein aggregation. *Trends Cell Biol.* **10**, 524–530.
97. Dobson, C. M. (2004). Experimental investigation of protein folding and misfolding. *Methods* **34**, 4–14.
98. Cavagnero, S., and Jungbauer, L. M. (2005). Painting protein misfolding in the cell in real time with an atomic-scale brush. *Trends Biotechnol.* **23**, 157–162.
99. Chiti, F., Webster, P., Taddei, N., Clark, A., Stefani, M., Ramponi, G., and Dobson, C. M. (1999). Designing conditions for *in vitro* formation of amyloid protofilaments and fibrils. *Proc. Natl. Acad. Sci. USA* **96**, 3590–3594.
100. Farinha, C. M., and Amaral, M. D. (2005). Most F508del-CFTR is targeted to degradation at an early folding checkpoint and independently of calnexin. *Mol. Cell. Biol.* **25**, 5242–5252.
101. Jorgensen, M. M., Bross, P., and Gregersen, N. (2003). Protein quality control in the endoplasmic reticulum. *APMIS Suppl.* 86–91.

102. Andresen, B. S., Bross, P., Udvari, S., Kirk, J., Gray, G., Kmoch, S., Chamoles, N., Knudsen, I., Winter, B., Wilcken, B., Yokota, I., Hart, K. et al. (1997). The molecular basis of medium-chain acyl-CoA dehydrogenase (MCAD) deficiency in compound heterozygous patients: Is there correlation between genotype and phenotype? Hum. Mol. Genet. **6**, 695–707.
103. Steipe, B. (2004). Consensus-based engineering of protein stability: From intrabodies to thermostable enzymes. Methods Enzymol. **388**, 176–186.
104. Rath, A., and Davidson, A. R. (2000). The design of a hyperstable mutant of the Abp1p SH3 domain by sequence alignment analysis. Protein Sci. **9**, 2457–2469.
105. Pey, A. L., Rodriguez-Larrea, D., Bomke, S., Dammers, S., Godoy-Ruiz, R., Garcia-Mira, M. M., and Sanchez-Ruiz, J. M. (2008). Engineering proteins with tunable thermodynamic and kinetic stabilities. Proteins **71**, 165–174.
106. Uy, R., and Wold, F. (1977). Posttranslational covalent modification of proteins. Science **198**, 890–896.
107. Nilsson, M. R., Driscoll, M., and Raleigh, D. P. (2002). Low levels of asparagine deamidation can have a dramatic effect on aggregation of amyloidogenic peptides: Implications for the study of amyloid formation. Protein Sci. **11**, 342–349.
108. Solstad, T., and Flatmark, T. (2000). Microheterogeneity of recombinant human phenylalanine hydroxylase as a result of nonenzymatic deamidations of labile amide containing amino acids effects on catalytic and stability properties. Eur. J. Biochem. **267**, 6302–6310.
109. Robinson, N. E. (2002). Protein deamidation. Proc. Natl. Acad. Sci. USA **99**, 5283–5288.
110. Rochet, J. C. (2006). Errors in translation cause selective neurodegeneration. ACS Chem. Biol. **1**, 562–566.
111. Lee, J. W., Beebe, K., Nangle, L. A., Jang, J., Longo-Guess, C. M., Cook, S. A., Davisson, J. P., Sundberg, J. P., Schimmel, P., and Ackerman, S. L. (2006). Editing-defective tRNA synthetase causes protein misfolding and neurodegeneration. Nature **443**, 50–55.
112. Bieschke, J., Zhang, Q., Bosco, D. A., Lerner, R. A., Powers, E. T., Wentworth, P., Jr., and Kelly, J. W. (2006). Small molecule oxidation products trigger disease-associated protein misfolding. Acc. Chem. Res. **39**, 611–619.
113. Milstien, S., Dorche, C., and Kaufman, S. (1990). Studies on the interaction of a thiol-dependent hydrogen peroxide scavenging enzyme and phenylalanine hydroxylase. Arch. Biochem. Biophys. **282**, 346–351.
114. Sitta, A., Barschak, A. G., Deon, M., Terroso, T., Pires, R., Giugliani, R., Dutra-Filho, C. S., Wajner, M., and Vargas, C. R. (2006). Investigation of oxidative stress parameters in treated phenylketonuric patients. Metab. Brain Dis. **21**, 287–296.
115. Arakawa, T., Ejima, D., Kita, Y., and Tsumoto, K. (2006). Small molecule pharmacological chaperones: From thermodynamic stabilization to pharmaceutical drugs. Biochim. Biophys. Acta **1764**, 1677–1687.
116. Bross, P., Pedersen, P., Winter, V., Nyholm, M., Johansen, B. N., Olsen, R. K., Corydon, M. J., Andresen, H., Eiberg, H., Kolvraa, S., and Gregersen, N. (1999). A polymorphic variant in the human electron transfer flavoprotein alpha-chain (alpha-T171) displays decreased thermal stability and is overrepresented in very-long-chain acyl-CoA dehydrogenase-deficient patients with mild childhood presentation. Mol. Genet. Metab. **67**, 138–147.
117. Bruce, D., Perry, D. J., Borg, J. Y., Carrell, R. W., and Wardell, M. R. (1994). Thromboembolic disease due to thermolabile conformational changes of antithrombin Rouen-VI (187 Asn– > Asp). J. Clin. Invest. **94**, 2265–2274.
118. Trefz, F. K., Scheible, D., Frauendienst-Egger, G., Korall, H., and Blau, N. (2005). Long-term treatment of patients with mild and classical phenylketonuria by tetrahydrobiopterin. Mol. Genet. Metab. **86**(Suppl 1), S75–S80.
119. Fink, A. L. (2005). Natively unfolded proteins. Curr. Opin. Struct. Biol. **15**, 35–41.

120. Woo, S. L., Lidsky, A. S., Guttler, F., Chandra, T., and Robson, K. J. (1983). Cloned human phenylalanine hydroxylase gene allows prenatal diagnosis and carrier detection of classical phenylketonuria. *Nature* **306**, 151–155.
121. Perez, B., Desviat, L. R., Gomez-Puertas, P., Martinez, A., Stevens, R. C., and Ugarte, M. (2005). Kinetic and stability analysis of PKU mutations identified in BH(4)-responsive patients. *Mol. Genet. Metab.* **86**, 11–16.
122. Døskeland, A. P., and Flatmark, T. (1996). Recombinant human phenylalanine hydroxylase is a substrate for the ubiquitin-conjugating enzyme system. *Biochem. J.* **319**, 941–945.
123. Waters, P. J., Parniak, M. A., Akerman, B. R., Jones, A. O., and Scriver, C. R. (1999). Missense mutations in the phenylalanine hydroxylase gene (PAH) can cause accelerated proteolytic turnover of PAH enzyme: A mechanism underlying phenylketonuria. *J. Inherit. Metab. Dis.* **22**, 208–212.
124. Døskeland, A. P., and Flatmark, T. (2002). Ubiquitination of soluble and membrane-bound tyrosine hydroxylase and degradation of the soluble form. *Eur. J. Biochem.* **269**, 1561–1569.
125. Baker, R. E., and Shiman, R. (1979). Measurement of phenylalanine hydroxylase turnover in cultured hepatoma cells. *J. Biol. Chem.* **254**, 9633–9639.
126. Aguado, C., Perez, B., Ugarte, M., and Desviat, L. R. (2006). Analysis of the effect of tetrahydrobiopterin on PAH gene expression in hepatoma cells. *FEBS Lett.* **580**, 1697–1701.
127. Jennings, I. G., Cotton, R. G., and Kobe, B. (2000). Structural interpretation of mutations in phenylalanine hydroxylase protein aids in identifying genotype-phenotype correlations in phenylketonuria. *Eur. J. Hum. Genet.* **8**, 683–696.
128. Guerois, R., Nielsen, J. E., and Serrano, L. (2002). Predicting changes in the stability of proteins and protein complexes: A study of more than 1000 mutations. *J. Mol. Biol.* **320**, 369–387.
129. Matalon, K. M. (2006). Dietary recommendations in the USA. *In* "PKU and BH4. Advances in phenylketonuria and tetrahydrobiopterin" (N. Blau, Ed.), pp. 220–231. SPS Verlagsgesellschaft, Heilbronn.
130. Muntau, A. C., Roschinger, W., Habich, M., Demmelmair, H., Hoffmann, B., Sommerhoff, C. P., and Roscher, A. A. (2002). Tetrahydrobiopterin as an alternative treatment for mild phenylketonuria. *N. Engl. J. Med.* **347**, 2122–2132.
131. Pey, A. L., and Martinez, A. (2007). Tetrahydrobiopterin for patients with phenylketonuria. *Lancet* **370**, 462–463.
132. Matalon, R., Surendran, S., Matalon, K. M., Tyring, S., Quast, M., Jinga, W., Ezell, E., and Szucs, S. (2003). Future role of large neutral amino acids in transport of phenylalanine into the brain. *Pediatrics* **112**, 1570–1574.
133. Santos, L. L., Magalhaes Mde, C., Januario, J. N., Aguiar, M. J., and Carvalho, M. R. (2006). The time has come: A new scene for PKU treatment. *Genet. Mol. Res.* **5**, 33–44.
134. Gamez, A., Wang, L., Sarkissian, C. N., Wendt, D., Fitzpatrick, P., Lemontt, J. F., Scriver, C. R., and Stevens, R. C. (2007). Structure-based epitope and PEGylation sites mapping of phenylalanine ammonia-lyase for enzyme substitution treatment of phenylketonuria. *Mol. Genet. Metab.* **91**, 325–334.
135. Chen, L., and Woo, S. L. (2005). Complete and persistent phenotypic correction of phenylketonuria in mice by site-specific genome integration of murine phenylalanine hydroxylase cDNA. *Proc. Natl. Acad. Sci. USA* **102**, 15581–15586.
136. Harding, C. O., Gillingham, M. B., Hamman, K., Clark, H., Goebel-Daghighi, E., Bird, A., and Koeberl, D. D. (2006). Complete correction of hyperphenylalaninemia following liver-directed, recombinant AAV2/8 vector-mediated gene therapy in murine phenylketonuria. *Gene Ther.* **13**, 457–462.

137. Ding, Z., Georgiev, P., and Thony, B. (2006). Administration-route and gender-independent long-term therapeutic correction of phenylketonuria (PKU) in a mouse model by recombinant adeno-associated virus 8 pseudotyped vector-mediated gene transfer. *Gene Ther.* **13**, 587–593.
138. Luque, I., and Freire, E. (2002). Structural parameterization of the binding enthalpy of small ligands. *Proteins* **49**, 181–190.
139. Friedler, A., Veprintsev, D. B., Hansson, L. O., and Fersht, A. R. (2003). Kinetic instability of p53 core domain mutants: Implications for rescue by small molecules. *J. Biol. Chem.* **278**, 24108–24112.
140. Leal, S. S., and Gomes, C. M. (2007). Studies of the molten globule state of ferredoxin: Structural characterization and implications on protein folding and iron-sulfur center assembly. *Proteins* **68**, 606–616.
141. Kuwajima, K. (1992). Protein folding *in vitro*. *Curr. Opin. Biotechnol.* **3**, 462–467.
142. Ptitsyn, O. B. (1995). Molten globule and protein folding. *Adv. Protein Chem.* **47**, 83–229.
143. Kuwajima, K. (1996). The molten globule state of alpha-lactalbumin. *Faseb. J.* **10**, 102–109.
144. Goodwill, K. E., Sabatier, C., Marks, C., Raag, R., Fitzpatrick, P. F., and Stevens, R. C. (1997). Crystal structure of tyrosine hydroxylase at 2.3 A and its implications for inherited neurodegenerative diseases. *Nat. Struct. Biol.* **4**, 578–585.
145. Erlandsen, H., Fusetti, F., Martínez, A., Hough, E., Flatmark, T., and Stevens, R. C. (1997). Crystal structure of the catalytic domain of human phenylalanine hydroxylase reveals the structural basis for phenylketonuria. *Nat. Struct. Biol.* **4**, 995–1000.
146. Wang, L., Erlandsen, H., Haavik, J., Knappskog, P. M., and Stevens, R. C. (2002). Three-dimensional structure of human tryptophan hydroxylase and its implications for the biosynthesis of the neurotransmitters serotonin and melatonin. *Biochemistry* **41**, 12569–12574.
147. Erlandsen, H., Flatmark, T., Stevens, R. C., and Hough, E. (1998). Crystallographic analysis of the human phenylalanine hydroxylase catalytic domain with bound catechol inhibitors at 2.0 A resolution. *Biochemistry* **37**, 15638–15646.
148. Andersen, O. A., Stokka, A. J., Flatmark, T., and Hough, E. (2003). 2.0A resolution crystal structures of the ternary complexes of human phenylalanine hydroxylase catalytic domain with tetrahydrobiopterin and 3-(2-thienyl)-L-alanine or L-norleucine: Substrate specificity and molecular motions related to substrate binding. *J. Mol. Biol.* **333**, 747–757.
149. Teigen, K., Frøystein, N.Å., and Martínez, A. (1999). The structural basis the recognition of phenylalanine and pterin cofactors by phenylalanine hydroxylase: Implications for the catalytic mechanism. *J. Mol. Biol.* **294**, 807–823.
150. Teigen, K., and Martínez, A. (2003). Probing cofactor specificity in phenylalanine hydroxylase by molecular dynamics simulations. *J. Biomol. Struct. Dyn.* **20**, 733–740.
151. Erlandsen, H., Bjørgo, E., Flatmark, T., and Stevens, R. C. (2000). Crystal structure and site-specific mutagenesis of pterin-bound human phenylalanine hydroxylase. *Biochemistry* **39**, 2208–2217.
152. Teigen, K., Dao, K. K., McKinney, J. A., Gorren, A. C., Mayer, B., Froystein, N. A., Haavik, J., and Martinez, A. (2004). Tetrahydrobiopterin binding to aromatic amino acid hydroxylases. Ligand recognition and specificity. *J. Med. Chem.* **47**, 5962–5971.
153. Hodneland, E., and Teigen, K. (2007). A simple method to calculate the accessible volume of protein-bound ligands: Application for ligand selectivity. *J. Mol. Graph. Model.* **26**, 429–433.
154. Shiman, R., Xia, T., Hill, M. A., and Gray, D. W. (1994). Regulation of rat liver phenylalanine hydroxylase. II. Substrate binding and the role of activation in the control of enzymatic activity. *J. Biol. Chem.* **269**, 24647–24656.
155. Pey, A. L., Thórólfsson, M., Teigen, K., Ugarte, M., and Martínez, A. (2004). Thermodynamic characterization of the binding of tetrahydropterins to phenylalanine hydroxylase. *J. Am. Chem. Soc.* **126**, 13670–13678.

156. Pey, A. L., Martinez, A., Charubala, R., Maitland, D. J., Teigen, K., Calvo, A., Pfleiderer, W., Wood, J. M., and Schallreuter, K. U. (2006). Specific interaction of the diastereomers 7(R)- and 7(S)-tetrahydrobiopterin with phenylalanine hydroxylase: Implications for understanding primapterinuria and vitiligo. *Faseb. J.* **20,** 2130–2132.
157. Aguado, C., Perez, B., Garcia, M. J., Belanger-Quintana, A., Martinez-Pardo, M., Ugarte, M., and Desviat, L. R. (2007). BH4 responsiveness associated to a PKU mutation with decreased binding affinity for the cofactor. *Clin. Chim. Acta* **380,** 8–12.
158. Knappskog, P. M., Eiken, H. G., Martinez, A., Bruland, O., Apold, J., and Flatmark, T. (1996). PKU mutation (D143G) associated with an apparent high residual enzyme activity: Expression of a kinetic variant form of phenylalanine hydroxylase in three different systems. *Hum. Mutat.* **8,** 236–246.
159. Carvalho, R. N., Solstand, T., Bjorgo, E., Barroso, J. F., and Flatmark, T. (2003). Deamidations in recombinant human phenylalanine hydroxylase. Identification of labile asparagine residues and functional characterization of Asn– > Asp mutant forms. *J. Biol. Chem.* **278,** 15142–15152.
160. Daubner, S. C., Melendez, J., and Fitzpatrick, P. F. (2000). Reversing the substrate specificities of phenylalanine and tyrosine hydroxylase: Aspartate 425 of tyrosine hydroxylase is essential for L-DOPA formation. *Biochemistry* **39,** 9652–9661.
161. Teigen, K., McKinney, J. A., Haavik, J., and Martinez, A. (2007). Selectivity and affinity determinants for ligand binding to the aromatic amino acid hydroxylases. *Curr. Med. Chem.* **14,** 455–467.
162. Wang, G.-A., Gu, P., and Kaufman, S. (2001). Mutagensis of the regulatory domain of phenylalanine hydroxylase. *Proc. Natl. Acad. Sci.* **98,** 1537–1542.
163. Thórólfsson, M., Teigen, K., and Martínez, A. (2003). Activation of phenylalanine hydroxylase: Effect of substitutions at Arg68 and Cys237. *Biochemistry* **42,** 3419–3428.
164. Egan, M. E., Pearson, M., Weiner, S. A., Rajendran, V., Rubin, D., Glockner-Pagel, J., Canny, K., Du, K., Lukacs, G. L., and Caplan, M. J. (2004). Curcumin, a major constituent of turmeric, corrects cystic fibrosis defects. *Science* **304,** 600–602.
165. Grubb, B. R., Gabriel, S. E., Mengos, A., Gentzsch, M., Randell, S. H., Van Heeckeren, A. M., Knowles, M. L., Drumm, M. L., Riordan, J. R., and Boucher, R. C. (2006). SERCA pump inhibitors do not correct biosynthetic arrest of deltaF508 CFTR in cystic fibrosis. *Am. J. Respir. Cell. Mol. Biol.* **34,** 355–363.
166. Muga, A. and Moro, F. (2008). Thermal adaptation of heat shock proteins. *2008 Curr. Prot. Pept. Sci. In press.*
167. Jones, H., Dalmaris, J., Wright, M., Steinke, J. H., and Miller, A. D. (2006). Hydrogel polymer appears to mimic the performance of the GroEL/GroES molecular chaperone machine. *Org. Biomol. Chem.* **4,** 2568–2574.
168. Janovick, J. A., Goulet, M., Bush, E., Greer, J., Wettlaufer, D. G., and Conn, P. M. (2003). Structure–activity relations of successful pharmacologic chaperones for rescue of naturally occurring and manufactured mutants of the gonadotropin-releasing hormone receptor. *J. Pharmacol. Exp. Ther.* **305,** 608–614.
169. Oueslati, M., Hermosilla, R., Schonenberger, E., Oorschot, V., Beyermann, M., Wiesner, B., Schmidt, J., Klumperman, J., Rosenthal, W., and Schulein, R. (2007). Rescue of a nephrogenic diabetes insipidus-causing vasopressin V2 receptor mutant by cell-penetrating peptides. *J. Biol. Chem.* **282,** 20676–20685.
170. Tanaka, N., Tanaka, R., Tokuhara, M., Kunugi, S., Lee, Y. F., and Hamada, D. (2008). Amyloid fibril formation and chaperone-like activity of peptides from alphaA-crystallin. *Biochemistry.*
171. Kim, T. D., Paik, S. R., and Yang, C. H. (2002). Structural and functional implications of C-terminal regions of alpha-synuclein. *Biochemistry* **41,** 13782–13790.
172. Frigon, R. P., and Lee, J. C. (1972). The stabilization of calf-brain microtubule protein by sucrose. *Arch. Biochem. Biophys.* **153,** 587–589.

173. Bolen, D. W., and Baskakov, I. V. (2001). The osmophobic effect: Natural selection of a thermodynamic force in protein folding. *J. Mol. Biol.* **310,** 955–963.
174. Auton, M., and Bolen, D. W. (2005). Predicting the energetics of osmolyte-induced protein folding/unfolding. *Proc. Natl. Acad. Sci. USA* **102,** 15065–15068.
175. Loo, T. W., and Clarke, D. M. (2007). Chemical and pharmacological chaperones as new therapeutic agents. *Expert Rev. Mol. Med.* **9,** 1–18.
176. Leandro, P., Lechner, M. C., Tavares de Almeida, I., and Konecki, D. (2001). Glycerol increases the yield and activity of human phenylalanine hydroxylase mutant enzymes produced in a prokaryotic expression system. *Mol. Genet. Metab.* **73,** 173–178.
177. Diamant, S., Eliahu, N., Rosenthal, D., and Goloubinoff, P. (2001). Chemical chaperones regulate molecular chaperones *in vitro* and in cells under combined salt and heat stresses. *J. Biol. Chem.* **276,** 39586–39591.
178. Ulloa-Aguirre, A., Janovick, J. A., Brothers, S. P., and Conn, P. M. (2004). Pharmacologic rescue of conformationally-defective proteins: Implications for the treatment of human disease. *Traffic* **5,** 821–837.
179. Chaudhuri, T. K., and Paul, S. (2006). Protein-misfolding diseases and chaperone-based therapeutic approaches. *Febs. J.* **273,** 1331–1349.
180. Fan, J. Q. (2008). A counterintuitive approach to treat enzyme deficiencies: Use of enzyme inhibitors for restoring mutant enzyme activity. *Biol. Chem.* **389,** 1–11.
181. Wiseman, R. L., and Balch, W. E. (2005). A new pharmacology–drugging stressed folding pathways. *Trends Mol. Med.* **11,** 347–350.
182. Handl, H. L., Vagner, J., Han, H., Mash, E., Hruby, V. J., and Gillies, R. J. (2004). Hitting multiple targets with multimeric ligands. *Expert Opin. Ther. Targets* **8,** 565–586.
183. Hammarstrom, P., Jiang, X., Hurshman, A. R., Powers, E. T., and Kelly, J. W. (2002). Sequence-dependent denaturation energetics: A major determinant in amyloid disease diversity. *Proc. Natl. Acad. Sci. USA* **99**(Suppl. 4), 16427–16432.
184. Lindberg, M. J., Tibell, L., and Oliveberg, M. (2002). Common denominator of Cu/Zn superoxide dismutase mutants associated with amyotrophic lateral sclerosis: Decreased stability of the apo state. *Proc. Natl. Acad. Sci. USA* **99,** 16607–16612.
185. Hayward, L. J., Rodriguez, J. A., Kim, J. W., Tiwari, A., Goto, J. J., Cabelli, D. E., Valentine, J. S., and Brown, R. H., Jr. (2002). Decreased metallation and activity in subsets of mutant superoxide dismutases associated with familial amyotrophic lateral sclerosis. *J. Biol. Chem.* **277,** 15923–15931.
186. Bernegger, C., and Blau, N. (2002). High frequency of tetrahydrobiopterin-responsiveness among hyperphenylalaninemias: A study of 1,919 patients observed from 1988 to 2002. *Mol. Genet. Metab.* **77,** 304–313.
187. Niederwieser, H. C., and Curtius, H. C. (1985). Tetrahydrobiopterin deficiencies in hyperphenylalaninemia. In "Inherited Diseases of Amino Acid Metabolism" (H. Bickel and U. Wachtel, Eds.), pp. 104–121. Georg Thieme, Stuttgart.
188. Zurfluh, M. R., Zschocke, J., Lindner, M., Feillet, F., Chery, C., Burlina, A., Stevens, R. C., Thony, B., and Blau, N. (2008). Molecular genetics of tetrahydrobiopterin-responsive phenylalanine hydroxylase deficiency. *Hum. Mutat.* **29,** 167–175.
189. Mitnaul, L. J., and Shiman, R. (1995). Coordinate regulation of tetrahydrobiopterin turnover and phenylalanine hydroxylase activity in rat liver cells. *Proc. Natl. Acad. Sci. USA* **92,** 885–889.
190. Scavelli, R., Ding, Z., Blau, N., Haavik, J., Martinez, A., and Thony, B. (2005). Stimulation of hepatic phenylalanine hydroxylase activity but not Pah-mRNA expression upon oral loading of tetrahydrobiopterin in normal mice. *Mol. Genet. Metab.* **86,** 153–155.

The Endoplasmic Reticulum: Crossroads for Newly Synthesized Polypeptide Chains

TITO CALÌ[*], OMAR VANONI[*], AND MAURIZIO MOLINARI

Institute for Research in Biomedicine,
CH-6500 Bellinzona, Switzerland

[*]Equal contribution

I. Protein Translocation and Maturation in the
  Mammalian ER 136
  A. Arriving at the ER Membrane 136
  B. Polypeptide Translocation into the ER 136
  C. Emerging in the ER Lumen 138
  D. Addition of Preassembled Oligosaccharides onto Nascent Chains 140
  E. Recruiting Calnexin and Calreticulin 142
  F. Cycling in the Calnexin Chaperone System 143
  G. Leaving the ER 145
  H. Deletion of Individual Members of the Calnexin Chaperone System 146
  I. Deletion of Other Chaperones 150
  J. The Fate of Newly Synthesized, Folding-Competent
    Polypeptides: A Summary 151
II. Substrate Recognition and Dislocation into the Cytosol for ERAD 152
  A. Terminally Misfolded Polypeptides, but also Folding Intermediates and
    Native Proteins can Become ERAD Substrates 152
  B. Understanding ERAD 153
  C. A Lag Phase Before Destruction 154
  D. Extensive De-Mannosylation to Deviate Misfolded Proteins for ERAD ... 155
  E. Disposal of Nonglycosylated Proteins 157
  F. Directing ERAD Substrates to the Retro-Translocation Site 158
  G. The Mammalian Orthologs of Yos9p 160
  H. Macroautophagy 161
  I. Do Autophagy-Like Processes Regulate ERAD Activity? The Concept of
    ERAD Tuning 161
  J. The Fate of Folding-Defective Polypeptides: A Summary 162
  References 163

About one third of the proteins synthesized in mammalian cells are cotranslationally translocated in the endoplasmic reticulum (ER). Newly synthesized polypeptides emerging in the ER lumen are exposed to machineries that assist conformational maturation and to machineries designed to interrupt futile folding attempts and to destroy terminally misfolded proteins. Cell and organism homeostasis depends on the regulated balance between these two activities. In this review, we summarize recent progresses made in the characterization of protein folding, quality control, and degradation in the mammalian ER; we survey the literature on the several knockouts of ER-resident proteins made available in recent years; we compare conflicting models explaining how these processes are regulated in yeast and in mammalian cells; we eventually propose the existence of regulatory mechanisms named *ERAD tuning* that operate at steady state in mammalian cells to optimize protein biogenesis and rely on the selective segregation from the folding environment and destruction of ER-resident regulators of protein disposal.

I. Protein Translocation and Maturation in the Mammalian ER

A. Arriving at the ER Membrane

The ER lumen is site of maturation for all secretory proteins, for proteins destined for the plasma membrane and for the endocytic and exocytic compartments. Most of the proteins that start their journey throughout the cell in the ER are characterized by the presence of a short N-terminal hydrophobic address tag, the signal sequence *(1)*. When this short sequence of about 20 residues emerges from the ribosome, it associates with an RNA/protein complex, the signal recognition particle *(2)*. Polypeptide elongation is substantially slowed until the ribosome engages a proteinaceous channel in the ER membrane, the Sec61 complex *(3)*. Only then synthesis is resumed and nascent chains are vectorially discharged across the ER membrane (Fig. 1) *(4–6)*. The highly hydrophobic signal sequence is integrated in the membrane and is normally removed from the nascent chains by a pentameric complex in the ER membrane, the signal peptidase *(7)*. Signal peptide cleavage often occurs cotranslationally *(8)*. However, examples are known of proteins whose folding is facilitated by persistent anchoring of the N-terminus to the membrane resulting from a delayed cleavage of the signal peptide (e.g., the human immunodeficiency virus (HIV) glycoprotein gp160 *(9)*).

B. Polypeptide Translocation into the ER

The translocation channel consists of a heterotrimeric complex (Sec61 $\alpha\beta\gamma$) with the α subunit forming the actual channel *(10)*. Several accessory proteins are also associated with or are in close proximity to the translocon, that is, the

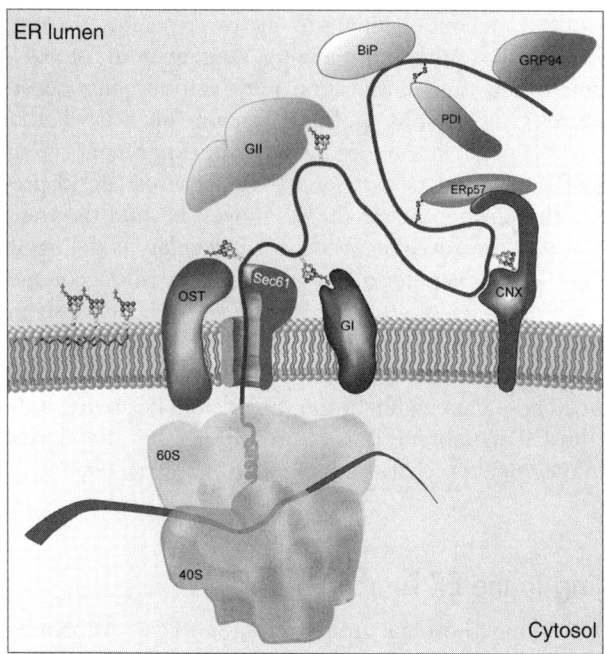

Fig. 1. Cotranslational protein translocation into the ER lumen. Nascent chains are cotranslationally injected into the ER lumen through the Sec61 translocon. The large and small ribosomal subunits at the cytosolic face of the ER membrane are labeled with 60S and 40S, respectively. Asparagine residues in appropriate sequons are covalently modified with the addition of preassembled oligosaccharides (the 2 N-acetylglucosamine residues are in yellow, the 4 α1,2-bonded mannose residues in gray, and the three terminal glucose residues in red. The shape and color code of the saccharide units in the protein-bound glycan are the same in Figs. 1–5). Nascent chains associate with a variety of ER-resident molecular chaperones and folding enzymes. OST is oligosaccharyl transferase; GI and GII are glucosidase I and II, respectively; CNX is calnexin, PDI is protein disulfide isomerase. (See Color Insert.)

translocating chain-associating membrane protein (TRAM), the translocon-associated complex (TRAP), the signal peptidase complex, the oligosaccharyl transferase (OST), and several ER-resident molecular chaperones such as BiP, Grp94, calnexin, and calreticulin that associate with incoming nascent chains at the luminal side of the translocation pore (Fig. 1). The driving force for cotranslational protein translocation is given by chain elongation and the hydrolysis of GTP during translation (11). Polypeptide translocation may also occur posttranslationally. In this case, the translocation channel also comprises a tetrameric Sec62/Sec63 complex (12, 13). Posttranslational translocation requires ATP hydrolysis because it occurs by a ratcheting mechanism in which BiP/Grp78, an ATPase of the Hsp70 family inside the ER lumen, inhibits backward movement of the translocating chain (14).

The size of the aqueous channel serving for protein translocation is matter of some debate, with a proposed diameter ranging from 8 to 60 Å *(15)*. The ribosomal tunnel and the translocation pore can certainly accommodate α-helical structures *(16, 17)* (Fig. 1), but it remains unclear whether folding of larger structures is possible. An elegant series of experiments recently showed that folding of a fragment of the Semliki forest virus capsid protein is only possible when the polypeptide chain has moved beyond the translocon pore *(18)*. These results are revealing because molecular modeling shows that a cylindrical pore with a diameter of 40–60 Å should easily accommodate capsid folding. Thus, a diameter of 40–60 Å for the active channel seems too large. On the other hand, it has been shown that several membrane segments do assemble within the translocation pore before integration into the ER membrane *(19–24)*. These data highlight the remarkable flexibility of the translocation site in the ER membrane possibly resulting from the capacity of Sec61 complexes to coordinate their lateral openings in a large channel.

C. Emerging in the ER Lumen

After a 100 Å long ribosomal tunnel *(25)*, a short gap of about 20 Å between the ribosome and the translocon and the 45 Å long translocation pore, the nascent polypeptide chain emerges in the ER lumen (Fig. 1). Unstructured nascent chains that emerge from the translocation pore expose hydrophobic patches, unpaired cysteines and other aggregation-prone determinants that during the folding process will be buried inside the native conformers. The intervention of ER-resident molecular chaperones such as BiP/Grp78 (with several cofactors) *(26–28)* or Grp94 *(29–31)* (Table I) is crucial to maintain nascent chains in a folding-competent state. Similarly, the engagement of exposed unpaired cysteines in mixed disulfides with members of the protein

TABLE I
BiP, BiP Cofactors, and Grp94

Family	Protein	Function	References
Hsp70	Grp78/BiP	Conventional chaperone	*(26)*
Hsp40	ERdj1/Mtj1 ERdj2/hSec63 ERdj3/HEDJ/ERj3/ ABBP-2 Erdj4/Mdj1 ERdj5/JPDI	Cofactors for BiP Substrate delivery; enhancement of ATP hydrolysis	*(274, 275) (13, 276)* *(277–279)* *(280–282)* *(283, 284)*
GrpE-like	BAP/Sil1 Grp170	Cofactors for BiP Nucleotide exchange factors	*(285–287)*
Hsp90	Grp94/endoplasmin/ERp99	Conventional chaperone	*(31)*

disulfide insomerase (PDI) superfamily such as PDI, ERp72, and ERp57 *(32–34)* (Table II) hampers formation of nonnative intermolecular disulfides and facilitates the oxidative phases of polypeptide maturation.

Altogether, chaperone association inhibits aggregation of unfolded polypeptides and facilitates the activity of folding enzymes that catalyze rate-limiting reactions of the polypeptide folding (e.g., formation of inter and intramolecular disulfide bonds *(35)* and *cis/trans* isomerization of peptidyl–prolyl bonds *(36)*). Chaperone association also retains nonnative polypeptides in the ER lumen because chaperones display ER-retention or ER-retrieval sequences such as C-terminal KDEL-like signals for luminal chaperones and cytosolic KKXX signals for ER membrane proteins of type I *(37)*. ER-resident chaperones are often organized in multiprotein complexes *(38, 39)* that possibly form cages or local environments in which newly synthesized polypeptides explore conformations that eventually lead to the native state, which is the one with the lowest Gibbs free energy *(40)*.

Formation of inter- and intramolecular disulfide bonds that covalently link cysteines is a common modification that starts immediately when nascent polypeptide chains enter into the ER lumen *(34, 41)*. Several members of the PDI superfamily, most of which play unknown roles in protein biogenesis and

TABLE II
HUMAN PDI FAMILY MEMBERS WITH AT LEAST ONE THIOREDOXIN-LIKE DOMAIN

Protein	Active site motif	Postulated activity
ERdj5	CXHC, CXPC, CXPC, CXPC	Reductase
PDIr	CXHC, CXHC, CXXC	Insufficient catalyst, lacks essential Glu
ERp72	CXHC, CXHC, CXHC	Oxidase/isomerase
ERp46	CXHC, CXHC, CXHC	Oxidase
PDI	CXHC, CXHC	Oxidase/isomerase
PDIp	CXHC, CXHC	Oxidase/isomerase
PDILT	SXXS, SXXC	Oxidase/isomerase
ERp57	CXHC, CXHC	Reductase/isomerase
P5	CXHC, CXHC	Oxidase
ERp44	CXXS	Retention of Ero 1
ERp18	CXHC	Inefficient catalyst, lacks essential Glu
TMX	CXXC	?
TMX2	SXXC	?
TMX3	CXHC	Oxidase
TMX4	CXXC	Inefficient catalyst, lacks essential Glu

quality control, catalyze protein oxidation and reshuffling of nonnative disulfides. Family members contain Cys-X-X-Cys (CXXC) active site motifs in thioredoxin domains (Table II). For example, PDI has two catalytic domains (a and a') divided by two inactive thioredoxin-like, substrate-binding domains (b and b') *(35)*. In family members acting as oxidases, thus promoting disulfides formation, the cysteines in the CXXC motif are disulfide-bonded. They act as an electron acceptor and leave the substrate-oxidation-reaction in a reduced state. Reductases disassemble disulfide bonds, for example, when protein unfolding is required for translocation across the ER membrane into the cytosol of toxin subunits or ERAD substrates *(42–45)*. In this case, the CXXC motif is initially reduced and leaves the reaction in an oxidized state. Finally, isomerases, which play a crucial role in adjusting the unique set of native intramolecular disulfides, enter and leave the reaction with a reduced active site.

Much less is known about one other rate-limiting reaction occurring during protein folding in the ER lumen, that is, the isomerization of prolyl–peptidyl bonds. In particular, despite several ER-resident members of the peptidyl-prolyl *cis/trans* isomerases (PPI) superfamily have been described (Table III), their involvement in polypeptide maturation *in vivo* is poorly understood.

D. Addition of Preassembled Oligosaccharides onto Nascent Chains

Most of the polypeptides emerging in the ER lumen receive N-glycans *(46)*, but important exceptions do exist. For example, albumin, the most abundant secretory protein produced in the liver, is not glycosylated.

The OST, a dimer of nine different subunits *(47)* (Table IV) is strategically positioned at the exit of the translocation pore (Fig. 1). The OST scans the sequence of nascent polypeptide chains emerging from the translocation pore in search for asparagine-any amino acid but proline-serine/threonine (Asn-Xxx-Ser/Thr) consensus sequences *(48)*. The hydroxyl group of the serine or threonine residue in the consensus sequence is brought in close proximity to

TABLE III
MEMBERS OF THE HUMAN PPI FAMILY

Protein	Notes
CyclophilinB/CypB	Part of large multichaperone complexes in the ER *(38, 39, 288)*
FKBP2/FKBP13	ER-stress induced *(289)*
FKBP7/FKBP23	Modulation of BiPs ATPase *(290, 291)*
FKBP10/FKBP65	Association with BiP-bound substrates *(292)*

TABLE IV
GLYCOSYLATING, SUGAR PROCESSING, AND SUGAR BINDING PROTEINS IN THE ER

Protein	Family	Function
Oligosaccharyl transferase		Addition of $Glc_3Man_9GlcNAc_2-$
Glucosidase I	GH Family 63	Removal of glucose-1
Glucosidase II	GH Family 31	Removal of glucose-2 and -3
Calnexin	Lectin	Binding to $Glc_1Man_{9-5}GlcNAc_2$
Calreticulin	Lectin	Binding to $Glc_1Man_{9-5}GlcNAc_2$
GT1	GT Family 24	Readdition of glucose-3 on Mannose A
ERGIC53, VIPL, VIP36	L-type lectin	High-mannose lectins
ER α1,2-mannosidase I	GH Family 47	Preferential removal of Mannose B
EDEM1	GH Family 47	Binding and processing of $Man_xGlcNAc_2$
EDEM2	GH Family 47	Binding and processing of $Man_xGlcNAc_2$
EDEM3	GH Family 47	Binding and processing of $Man_xGlcNAc_2$
OS-9.1, OS-9.2, erlectin	Lectins?	ER-retention/ERAD

the amide group of the asparagine, which is activated *(49)* to accept the covalent addition of a preassembled oligosaccharide whose composition (three glucose, nine mannose and two N-acetylglucosamine residues, $Glc_3Man_9GlcNAc_2$, Fig. 2) is conserved in plants, fungi, and mammals *(50)*. The OST transfers this preassembled oligosaccharide from a lipid donor in the ER membrane (dolichol-pyrophosphate) onto the nascent chain *(51)* as soon as the acceptor asparagine has emerged for 40–45 Å, corresponding to about 12–14 residues, from the translocation pore *(52)* (Fig. 1). The addition of oligosaccharides, bulky hydrophilic appendices that extend for about 30 Å from the polypeptide backbone, dramatically changes the biophysical properties of unstructured nascent chains substantially increasing their solubility. As thoroughly discussed in the next sections, N-glycosylation will also determine the fate of the associated polypeptide chain in many different ways. For example, if a glycosylation site is located in the first 50 residues or so, then addition of the oligosaccharide sterically hinders binding of chaperones such as BiP to the polypeptide backbone. Instead, lectin-like chaperones will associate and assist glycoprotein maturation *(53)*. If the first N-glycan is more downstream in the sequence, then the polypeptide will associate with BiP or other peptide backbone-binding chaperones. In this case, the folding polypeptides may only subsequently be handed off to the lectin-chaperone system *(53–57)*. Finally, if the polypeptide is folding-defective, the slow removal of individual mannose residues from the N-glycan will tag it for extraction from the ER folding environment and degradation *(58)* (Sect. II.D).

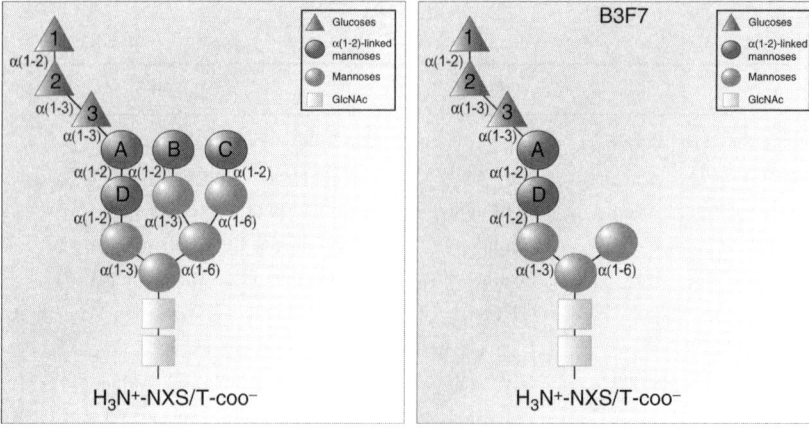

FIG. 2. Structure of core oligosaccharides. The panel on the left shows the three-antennary oligosaccharide covalently attached to the side chain of an asparagine in the N-X-S/T consensus sequence for N-linked glycosylation. Branches A, B, and C are those that display the terminal mannose residues A, B, and C, respectively. The panel on the right shows the aberrant oligosaccharide used in cell lines with defective synthesis of mannosylphosphoryldolichol (e.g., B3F7). (See Color Insert.)

E. Recruiting Calnexin and Calreticulin

Calnexin is a type I protein and calreticulin is its luminal paralog (Table IV). Both proteins are lectin-chaperones, but calreticulin is also involved in calcium storage in the ER lumen because it contains a highly acidic C-terminal domain that binds 18 calcium ions with low affinity (Kd of about 2 mM (59)). Calnexin and calreticulin are retained in the ER lumen by a cytosolic KKXX-like and a luminal KDEL-motif, respectively. They both contain a single carbohydrate-binding domain adopting a leguminous lectin-like β-sandwich fold (60) and a long hairpin of 140 Å for calnexin and of approximately 110 Å for calreticulin, the proline-rich P-domain. The tips of the calnexin and calreticulin P-domains interact with the b' domain of the oxidoreductase ERp57 (61–64). ERp57 is therefore in the best position to catalyze maturation of newly synthesized glycoproteins for which, disulfide bond formation is a rate-limiting reaction (34, 65, 66).

Few seconds after addition onto nascent chains, the N-linked oligosaccharide becomes accessible to the glucosidase I, a type II membrane glycoprotein of the glycosyl hydrolase (GH) family 63 (Table IV). The glucosidase I removes the outermost glucose-1 when the N-glycan is located 72 residues from the ribosome P-site, which is also the distance required for addition of the N-glycan (Fig. 1). This reaction is therefore immediate and shows that the OST and the glucosidase I active sites are in close proximity to each other (67). The second glucose is also rapidly removed by another α-glucosidase, the glucosidase II (Fig. 1). This is a soluble heterodimeric glycanase of the GH family 31 (Table IV) composed of a catalytic α subunit and a regulatory β subunit (68). The glucosidase II also removes glucose-3. However, being glucose-3 differently oriented in space than glucose-2 (69), this last cleavage needs transient separation and repositioning of the glucosidase II active site. This time window is possibly exploited by calnexin and calreticulin to associate with the mono-glucosylated trimming intermediate of the N-linked glycan processing (67).

The globular sugar-binding domain of calnexin and calreticulin accommodates the entire mono-glucosylated branch A of a protein-bound oligosaccharide (Fig. 2) within a concave β-sheet in which the glucose ring sits on the side chain of methionine-189 and is hydrogen bonded to tyrosine-165, lysine-167, tyrosine-186, and glutamic acid-217 in calnexin (60).

Calnexin and calreticulin association slows protein folding and renders it more efficient (70), for example by inhibiting formation of nonnative disulfide bonds. In some cases in fact, N-glycans are appropriately positioned so that calnexin/calreticulin association will protect cysteines from premature oxidation. The influenza virus hemagglutinin (HA) with its peculiar head-to-tail structure represents a paradigmatic example of such a situation. The N-terminal cysteine-14 of the HA must covalently pair with the C-terminal cysteine-466 (71). With an average synthesis rate of 4 residues/s, oxidation of cysteine-14 has therefore to be delayed by about 100 s for efficient maturation of the newly synthesized protein. N-glycans at position 8, 22, and 38 insure cotranslational calnexin binding (41, 53, 72) that prevents immediate oxidation of cysteine-14 with one of the eight cysteines preceding the appropriate partner in the primary sequence.

F. Cycling in the Calnexin Chaperone System

Removal of the two outermost glucose residues occurs cotranslationally and generates the mono-glucosylated intermediate of the N-glycan trimming that associates with calnexin/calreticulin and ERp57 (Fig. 3, step 1) (34, 65, 66, 73–78). Association can persist for several minutes and certainly continues after chain termination. The finding that the innermost glucose-3 of a misfolded glycoprotein retained in the ER lumen was removed and readded (79) and the identification of a large luminal protein member of the glycosyl transferase family 24, the UDP-glucose:glycoprotein glucosyltransferase (GT1, Table IV) that reglucosylates nonnative glycoproteins (80) led Ari Helenius and coworkers to propose an elegant cycle of de/reglucosylation determining substrate dissociation/reassociation with calnexin and calreticulin: the calnexin cycle (78). In this model, which has been better defined in the last 15 years but is still valid today, proteins eventually released from calnexin/calreticulin are deglucosylated by glucosidase II. This prevents their immediate reassociation with the lectin-like chaperones (Fig. 3, step 2, or step 3a). If the polypeptide has attained the native structure, it is

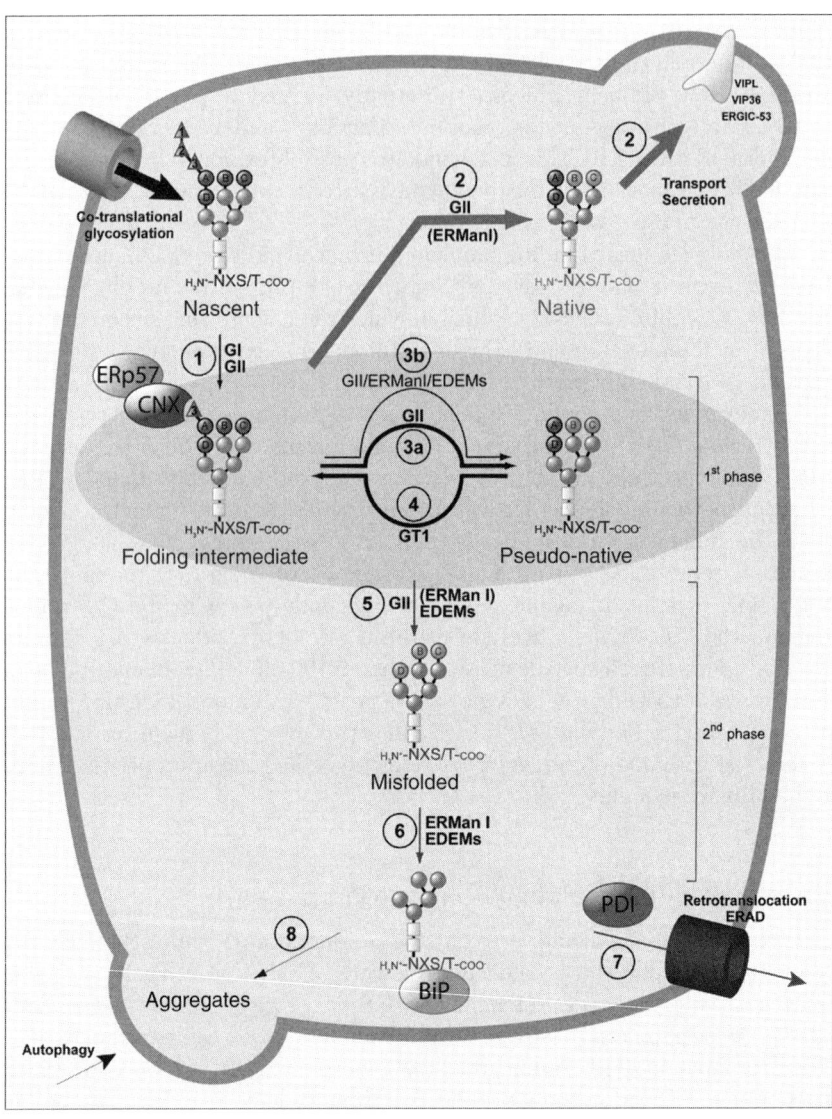

FIG. 3. The fate of folding-competent and folding-defective glycoproteins. The model shows the fate of newly synthesized, folding competent glycopolypeptides that are eventually transported at their site of activity through the secretory pathway (green arrows). Folding-defective polypeptides are trapped in a first phase of retention-based ER quality control (the calnexin cycle) and, eventually in a second phase of retention-based ER quality control (the BiP/PDI system). Terminally misfolded polypeptides are subjected to extensive de-mannosylation in the mammalian ER. ERManI is ER α1,2-mannosidase I; EDEMs stays for EDEM1, EDEM2, and EDEM3. (See Color Insert.)

released from the ER and is transported at the final intra- or extracellular destination through the secretory pathway (step 2). If the polypeptide is terminally misfolded and extensively demannosylated (Sect. II.D), it is handed off to BiP/PDI to be prepared for translocation into the cytosol and degradation (steps 5–7) *(42, 81)*. However, if the polypeptide released from calnexin has pseudo-native conformation, GT1 will selectively reglucosylate the terminal mannose on the oligosaccharide branch A to promote reassociation with calnexin and/or calreticulin and prolong retention in the folding environment (Fig. 3, step 4) *(82)*.

Several mechanistic features of glycoprotein folding are conserved between mammals and yeast. The core oligosaccharide added onto nascent chains has the same composition and in both organisms is added onto Asn-Xxx-Ser/Thr glycosylation sequons. Several proteins involved in the mammalian calnexin/calreticulin chaperone system have functional orthologs in *Saccharomyces cerevisiae*. GlsI, GlsII, and Cne1p are the glucosidase I, glucosidase II and calnexin orthologs, respectively. Moreover, the yeast oxidoreductase Mpd1p interacts with Cne1p to form a functional complex that enhances the catalytic activity of the oxidoreductase *(83)*, similarly to the functional calnexin:ERp57 complex *(34, 65, 66)*. However, other components are missing in *S. cerevisiae*, for example, calreticulin and, most importantly, GT1. Thus, polypeptide cycling on/off calnexin does only occur in the mammals and in more complex yeast strains such as *S. pombe* that possesses a GT ortholog *(84)*.

G. Leaving the ER

Properly folded and completely assembled proteins are eventually released from the ER in transport vesicles coated with Cytosolic Coat Protein II (COPII) that bud at ER exit sites *(85)*. In yeast, COPII-coated cargo vesicles are directly delivered to the Golgi compartment *(86)*. In mammalian cells, it has been proposed that they undergo homotypic fusion to generate a stationary ER-Golgi intermediate compartment (ERGIC) from which cargo proteins reach the *cis*-Golgi in COPI-coated vesicles *(87–89)*. Transmembrane proteins can directly interact with the cytosolic COPII coat, while soluble cargo proteins may require specific receptors for recruitment in COPII-coated vesicles *(90)*. It has recently been shown that ER export of some glycosylated proteins is facilitated by several leguminous-type lectins located in the ER (VIPL), cycling between ER and ERGIC (ERGIC53) or between ERGIC and *cis*-Golgi (VIP36) *(91)* (Table IV). ERGIC53 certainly is the best-characterized one. It is involved in export of several soluble cargo proteins such as pro-cathepsin Z and pro-cathepsin C *(92, 93)*, coagulation factors V and VIII *(94)* and α 1-antitrypsin *(95)*.

H. Deletion of Individual Members of the Calnexin Chaperone System

Several general (BiP, PDI, Grp94, CypB,...), substrate-specific (Hsp47 for collagen, tapasin for MHC class I loading complex, egasyn for β-glucuronidase, RAP for LDL receptors,...), and tissue-specific (PDILT, PDIp, calmegin,...) chaperones, enzymes, and escorting factors contribute to polypeptide maturation in the mammalian ER *(96)*. However, the mechanisms regulating maturation and quality control of *N*-glycosylated polypeptides in the calnexin chaperone system are the best studied and understood. For example, individual knockouts for each member of the glycoprotein-dedicated folding machinery have been generated. Cell lines lacking glucosidase I *(97)*, glucosidase II *(98)*, calnexin *(99)*, calreticulin *(100)*, ERp57 *(101)*, and GT1 *(102)* are available. Most of them have been derived from mouse embryos and their characterization led to better understand the role of each factor in the process of glycoprotein maturation.

1. GLUCOSIDASES-DEFICIENCY

Glucosidase I and glucosidase II sequentially remove the outermost and the middle glucose immediately after transfer of the preassembled oligosaccharides from the dolichol lipid donor in the ER membrane onto nascent polypeptide chains. Their intervention is required to generate the monoglucosylated trimming intermediate that binds to calnexin and calreticulin.

Lec23 is a ricin-resistant CHO cell line *(97)* expressing inactive glucosidase I characterized by a serine to phenylalanine amino acid substitution in the active site *(103)*. PhaR2.7 is a mouse lymphoma cell line selected for resistance to the cytotoxic effects of Phaseolus vulgaris leukoagglutinating lectin *(98)* characterized by the absence of glucosidase II activity. In these two cell lines, generation of the mono-glucosylated intermediate of *N*-glycan trimming does not occur and newly synthesized glycoproteins remain tri(in Lec23) or diglucosylated (in PhaR2.7). A similar phenotype is obtained upon cell exposure to castanospermine or deoxynojirimycin derivatives, which are specific inhibitors of the α-glucosidases *(104)*.

Analysis of protein association with calnexin or calreticulin revealed that it was abolished or substantially perturbed in these cells and that glycoprotein folding was generally accelerated and less efficient *(70, 105–109)*. Importantly, the viability of these cell lines revealed the dispensability of the calnexin chaperone system in cultured cells. However, the inherited glucosidase I deficiency named congenital disorder of glycosylation type IIb has lethal outcome early after birth. Moreover, the embryonic lethality in mice of several knockouts revels a critical role of members of the calnexin chaperone system possibly restricted to specific organs or developmental phases.

2. Calreticulin- and Calnexin-Deficiency

Calreticulin and calnexin associate with most, if not all glycoproteins that are synthesized in the mammalian ER. The calreticulin knockout is embryonic lethal at E14.5. Lethality results from impaired cardiac development and defective execution of calcium-dependent signaling events and can fully be ascribed to the loss of the calcium-binding activity of calreticulin *(100, 111)*. Instead, calnexin knockout mice are carried to full term. About 50% of the newborns die within 48 h and the survivors show growth defects and motor disorders associated with a dramatic loss of large myelinated nerve fibers *(99)*. Similarly to inactivation of α-glucosidases, deletions of calreticulin, calnexin (and calmegin, the testis-specific homologue of calnexin *(112, 113)*) are well tolerated in cultured cells. Despite binding to a large number of nascent glycoproteins, their deletion only prevents maturation of few of them, thus proving the existence of surrogate folding machineries that can intervene when the chaperones of choice are busy or inactive. Analysis of knockout cells revealed the crucial role of calreticulin in assembly and loading of class I MHC complexes with appropriate immunogenic peptides to be presented at the cell surface *(114)* (Fig. 4 and Sect. I.H.3). It also revealed that maturation of influenza virus HA obligatorily requires calnexin assistance *(72, 115)*.

Calnexin and calreticulin associate with a distinct set of substrates *(55, 72, 116–121)* even though they share the same specificity for mono-glucosylated protein-bound oligosaccharides. A possible reason of substrate-selectivity is their different topology *(116)*. Surprisingly, maturation of calnexin substrates was not characterized by enhanced association with calreticulin upon calnexin-deletion *(72)*. Relevant exceptions were viral calnexin substrates expressed in infected cells. For example, the G-protein of the vesicular stomatitis virus (VSV) did transiently associate with calreticulin upon calnexin-deletion in VSV-infected cells, but not when ectopically expressed in transfected cells. These data led to suggest that viral infections may subvert the normal glycoprotein recognition by the ER lectins *(72)* and may explain why inactivation of the calnexin/calreticulin cycle affects viral replication and infectivity but not cell viability *(122–127)*.

3. ERp57-Deficiency

ERp57 is a member of the PDI superfamily of ER-resident oxidoreductases (Table II). It is associated with calnexin and calreticulin and facilitates oxidative maturation of newly synthesized N-glycosylated polypeptides *(35)*. ERp57-deletion is embryonic lethal but it does not impair cell viability and it does not elicit an acute unfolded protein response *(101, 128)*. This confirms that, even though most glycopolypeptides make use of the calnexin chaperone system, their maturation can proceed under the assistance of surrogate folding machineries.

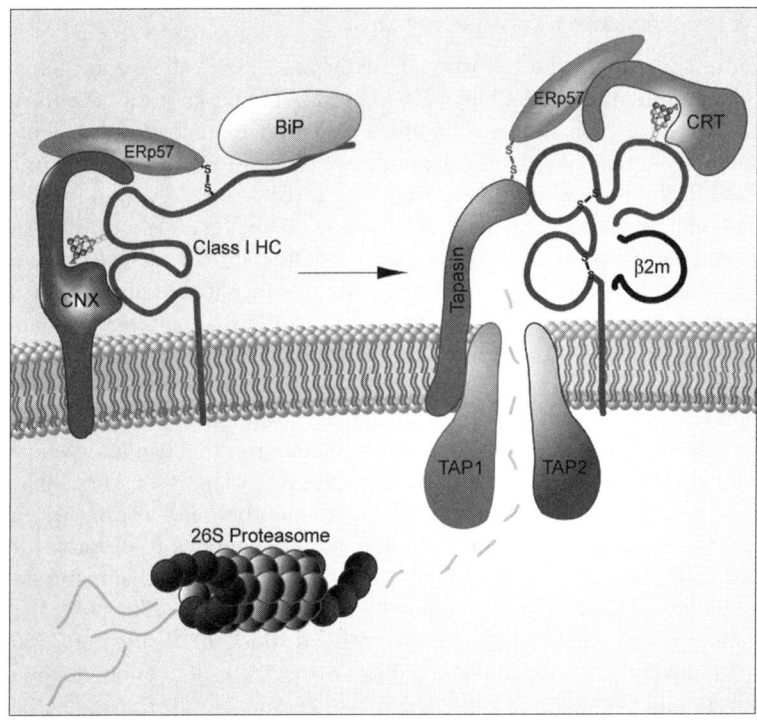

FIG. 4. The antigen-loading complex. The model shows biogenesis and loading with immunogenic peptides of the MHC class I complex. Antigenic peptides are generated by cytosolic proteasomes and are imported in the ER lumen through the TAP complex. Note that in the loading complex ERp57 forms a stable, covalent complex with tapasin. (See Color Insert.)

ERp57 is also part of the MHC class I loading complex *(129, 130)* where it is engaged in a stable disulfide bond with tapasin. Tapasin is an ER-resident protein that recruits MHC class I molecules to be loaded with antigenic peptides to the TAP peptide transporter located in the ER membrane (Fig. 4) *(131)*. It has been reported that the peptide-loading complex contains between 20% (steady state) and 85% (upon interferon-γ treatment) of the cellular ERp57 *(101, 131)*.

Analysis of primary B-cells and fibroblasts lacking ERp57 showed a strong reduction of recruitment of MHC class I in the peptide loading complex *(130)*. This was surprising because it had been assumed that tapasin was required and sufficient for this function. Comparison of wild type versus knockout cells also revealed that the covalent association of ERp57 induces a conformational change in tapasin that enhances affinity for MHC class I molecules, prolongs their retention in the loading complex and possibly allows a better selection of the antigenic peptide *(101)*. A role of ERp57 in facilitating the conformational breathing required for high-affinity loading of peptides in the MHC class I antigen-binding cleft has also been proposed *(132, 133)*.

Back-transfection of mouse fibroblasts lacking ERp57 with substrate-trapping ERp57 mutants in which one or both CXXC active sites were mutated to CXXA, led to the identification of several cellular ERp57 substrates *(134)*. The two mutated catalytic sites were shown to be equally efficient in trapping substrate proteins, with the exception of tapasin and few others that only formed mixed disulfides with the active site of sub domain *a (134)*.

Analysis of protein maturation in cells lacking ERp57 revealed that ERp72 (Table II) can act as a surrogate oxidoreductase in assisting maturation of glycoproteins *(128)*. ERp72 is the only member of the PDI superfamily that shares the residues that in ERp57 are involved in association with the calnexin and calreticulin P-domains *(35)*. In contrast to ERp57, however, substrate association with calnexin and calreticulin was dispensable for ERp72 intervention *(128)*. On the mechanistic site, analysis of an obligate calnexin/ERp57 substrate, the influenza virus HA, revealed that deletion of ERp57 exclusively hampered the posttranslational phase of HA maturation consisting in reshuffling of intramolecular disulfide bonds to the native set. The cotranslational phase consisting in the oxidation of cysteines emerging in the ER lumen progressed very efficiently even without ERp57 *(128)*. These findings and data showing that ERp57 is in the reduced state in the ER lumen at steady state *(135–137)* suggest that ERp57 acts as an isomerase/reductase, rather than as an oxidase.

4. GT1-Deficiency

GT1 prolongs substrate retention in the calnexin chaperone system by reglucosylating oligosaccharides on nonnative polypeptides prematurely released from calnexin *(78)*. GT1 deletion is embryonic lethal for most homozygous mice (E13), but few of them survive until birth *(102)*. Viability of cultured cells lacking GT1 is apparently normal. Unbiased comparison of protein biogenesis in cells with and without GT1 revealed the existence of cellular and viral glycoproteins that attain the native structure in a single round of association with calnexin. Kinetics of release from calnexin and folding efficiency of these proteins were the same in the two cell lines. It also showed that other proteins are normally subjected to multiple binding events to complete maturation. In fact, in cells lacking GT1 they were prematurely released from calnexin and their folding efficiency dropped. The finding that GT1-deletion substantially delayed release from calnexin of few cellular proteins and of the influenza virus HA was unpredicted. These data showed that GT1 and/or a GT1-associated folding enzyme (e.g., the seleno-cysteine-containing oxidoreductase Sep15 that forms a 1:1 complex with GT1 *(138))* regulate the conformational maturation that might be required for release from calnexin of select substrates *(139)*. The data also implied that calnexin can act as a bona fide molecular chaperone that retains nonnative glycoprotein conformers (even in cells lacking GT1), a function that can be inferred from *in vitro* experiments (reviewed in *(140)*).

The consequences of GT1-deletion on ER retention of misfolded conformers have been studied in cells expressing the tsO45 G protein *(141)*. At the permissive temperature (32°), the G protein rapidly attains the native structure in the calnexin chaperone system ($T_{1/2}$ is about 15 min) *(54)*. At the nonpermissive temperature (39°), it remains associated with calnexin for at least 60 min *(78)*. This was ascribed to a rapid turnover ($T_{1/2}$ of 5–10 min *(142)*) of the terminal glucose displayed by nonnative G protein, which is repeatedly removed by glucosidase II to be then rapidly readded by GT1 *(78, 79)*. Surprisingly, deletion of GT1 did not accelerate release of the misfolded tsO45 G protein from calnexin for at least 60 min, showing that the misfolded protein was not cycling on/off calnexin. Beyond the 60 min, a second phase of retention-based ER quality control was characterized by a slow release of the misfolded G protein that entered in BiP-associated disulfide-bonded aggregates. Deletion of GT1 substantially accelerated entry of terminally misfolded polypeptides in this second phase *(102)*. Evidently, cycling of misfolded glycopolypeptides in the calnexin chaperone system is activated very late as if repeated releases from dynamic constrictions caused by calnexin binding would only be exploited as a very last attempt to eventually fold defective polypeptides. Our data revealed that cycling in the calnexin chaperone system represents the *first phase* of retention-based ER quality control (Fig. 3). In cells with and without GT1, the terminally misfolded tsO45 G *(102)* or folding-defective ERAD candidates *(42)* are eventually released from the calnexin cycle to be trapped in the *second phase* of retention-based quality control relying on BiP association. This phase precedes substrate dislocation into the cytosol for disposal (Fig. 3).

I. Deletion of Other Chaperones

The individual knockouts of calreticulin, GT1, ERp57 are embryonic lethal at E13–14 while mice lacking calnexin die early after birth. The reason for lethality has been established only for calreticulin and is the loss of the calcium-binding function of this protein, rather than the loss of its chaperone activity *(100, 111)*. In all cases, however, cultured cells show surprisingly mild phenotypes and do not show symptoms of acute unfolded protein response. This shows that for most (but not all) cargo proteins translocated in the ER alternative folding machineries can be activated that efficiently support maturation.

Deletions of *private* chaperones allowed assessment of the function of the chaperone-of-interest in very specific processes. For example deletion of tapasin, a component of the antigen-loading complex, resulted in obvious phenotypes related to impaired antigen presentation *(143)*. Deletion of Hsp47, a collagen-specific chaperone, resulted in a clear defect in collagen biosynthesis *(144)*.

Deletion of *tissue-specific* chaperones has also been reported. An interesting case is deletion of calmegin, the testis-specific isoform of calnexin. Calmegin binds transiently to a large number of mono-glucosylated proteins synthesized in sperm cells, but its deletion only affects maturation of very few of them, namely fertilins *(112, 113)*. Consistently with lack of consequences on other glycoproteins maturation, sperm number, viability, or motility are unaffected in calmegin-deficient mice, while the fertilins regulated sperm's capacity to bind to the egg's zona pellucida is lost. This causes male sterility *(112, 145)*.

Deletion of conventional chaperones such as Grp94 *(146)* and BiP *(147)* has also been reported. Both deletions are embryonic lethal (E7.5 for Grp94 and E3.5 for BiP). Few or no data are available on phenotypes linked to the loss of their chaperone function. The studies are hampered, at least in the case of BiP, by the pleiotropic roles of this abundant ER protein that seals the translocon, acts as a molecular ratchet to facilitate posttranslational protein import in the ER, contributes to calcium homeostasis and serves as a chaperone *(28)*. The essential role of BiP is underscored by the very early lethality of the homozygous embryos and by the identification of BiP as the target substrate of SubAB, one of the most potent bacterial toxins responsible for hemolytic uremic syndrome outbreaks *(148)*.

J. The Fate of Newly Synthesized, Folding-Competent Polypeptides: A Summary

To summarize, the ER is the first station of the secretory pathway. Ribosomes attached at the cytosolic face of the ER membrane cotranslationally insert two classes of proteins in the ER lumen: firstly, the ER-resident proteins that operate in the ER lumen where they are retained by KDEL-like or KKXX-like sequences or through association with retained proteins; secondly, the transiting-cargo that will normally leave the ER lumen only upon acquisition of the correct tertiary and quaternary architecture.

Polypeptide maturation in the ER lumen is assisted by canonical chaperones that utilize ATP such as Grp94 and BiP with its several cofactors ERdj 1–5, BAP/Sil1 and Grp170 (Table I), by lectin-chaperones (calnexin and calreticulin, Table IV) and by a variety of enzymes that facilitate protein folding. Some of them (one OST, two α-glucosidases, one glucosyltransferase, and several $\alpha1,2$-

mannosidases, Table IV), add onto nascent chains and process oligosaccharides. Others, catalyze rate-limiting reactions such as covalent cross-linking of cysteines (about 15 members of the PDI superfamily, Table II) and *cis/trans* isomerization of peptidyl–prolyl bonds (at least four different members of the PPI superfamily, Table III). The ER lumen also contains a series of substrate- and tissue-specific chaperones that take care of individual, or of a restricted population of clients *(96)*. Also, specialized chaperones, lectins, and enzymes localized in the ER lumen or in the ER-Golgi intermediate compartment such as ERGIC-53, ERp44, Ero1 may intervene to facilitate polymerization of oligomeric proteins (reviewed in *(149)*).

Exposure of hydrophobic patches, unpaired reactive cysteines, or other less-well defined nonnative structures may elicit association of a series of ER-resident proteins, whose binding inhibits exit of nonnative cargo from the folding environment. Tightness of retention-based ER quality control is not absolute and secretion of unstable protein conformers that pass ER quality control but are subsequently misfolding outside the cell may occur (e.g., transthyretin mutants *(150, 151)*).

Upon successful completion of the folding program, the vast majority of native proteins are incorporated into COPII-coated vesicles to be released from the folding compartment *(85)*. Concentration of certain glycoproteins in transport vesicles requires intervention of specialized lectins (Table IV). VIP36 and VIPL preferentially associate with native glycoproteins displaying three mannose residues, but no glucose on the oligosaccharide branch A; ERGIC-53 seems to have low affinity but broader specificity as it also binds mono-glucosylated *N*-glycans *(91)*.

II. Substrate Recognition and Dislocation into the Cytosol for ERAD

A. Terminally Misfolded Polypeptides, but also Folding Intermediates and Native Proteins can Become ERAD Substrates

Genome replication is an accurate process with error-rates of less than 1 every 10^{10} nucleotides warranted by DNA editing and repair mechanisms *(152, 153)*. Transcription and translation, the latter with 1 error every 1000–10,000 bases, are less efficient processes *(96)*. It may therefore happen that, despite genetic integrity, individual nascent chains emerging in the ER lumen carry mutations, deletions, or truncations that may prevent polypeptide folding. It may also happen that polypeptides with correct primary structure enter off-pathways eventually leading to irreversible misfolding. Moreover, certain polypeptides have an intrinsic low capacity to attain a transport-competent structure, the most evoked example being the cystic fibrosis transmembrane conductance regulator (CFTR) with a folding efficiency below 30% *(154)*. It is unclear how many of the newly synthesized polypeptide chains will not acquire a native structure. Values ranging from an amazing 30% *(155)* to much less *(156)* have been reported. In any case, efficient removal from the folding environment of these physiologic by-products of protein biogenesis is required to maintain ER homeostasis *(157)*.

Under pathologic conditions such as ER stress or viral infection, the fraction of newly synthesized polypeptides successfully completing maturation may substantially drop. For example, acute ER stress may enhance the activity of the ER degradation machinery to such an extent that polypeptide disposal may start in advance of termination of polypeptide folding programs causing co-translational or immediate post-translational degradation (reviewed in *(158, 159)*). As a second example, viral gene products may associate and selectively target cellular proteins for destruction (e.g., the MHC class I heavy chain and the viral receptor CD4 are rapidly degraded in cells infected with the cytomegalovirus (HCMV) *(160)* and the HIV *(161)*, respectively). As a third example, negative feedback mechanisms do exist to adapt the intracellular level and activity of rate-limiting enzymes of specific metabolic pathways to the cellular demand for the pathway's final product (e.g., the turnover of the rate-limiting enzyme for cholesterol biosynthesis, the 3-hydroxy-3-methylglutaryl coenzyme A reductase *(162)*, is selectively enhanced under conditions of high sterol levels *(163, 164)*).

B. Understanding ERAD

Native proteins are rapidly released from the ER into the secretory pathway. Therefore, the vast majority of the cargo present in the ER lumen is either unfolded or misfolded. Unfolded chains are intermediates of a productive folding program that will eventually attain the native structure if retained long enough in the folding environment. Terminally misfolded chains, on the other hand, have irreparably failed their folding attempts and must rapidly be removed from the ER lumen and degraded, otherwise they will accumulate and will inhibit the compartmental capacity to deal with nascent chains incessantly emerging in the ER lumen. The efficient execution of these complex tasks is crucial for cell and organism survival and their manipulation may offer therapeutic approaches to cure or alleviate the symptoms of conformational diseases caused by defective protein folding *(159, 165–167)*.

The mechanisms regulating protein disposal from the mammalian ER have been established in some detail by careful analysis of the fate of several model substrates expressed in cultured cells. Amongst them, proteins that do not attain the native structure because they carry mutations (e.g., α 1-antitrypsin

Z *(168)*), deletions (e.g., the CFTR ΔF508 *(169)*, or β-secretase splice variants *(42)*), or truncations (e.g., the Null$_{\text{hong kong}}$ variant of α 1-antitrypsin *(170)*). Other examples are offered by orphan subunits of oligomeric complexes ectopically expressed in cultured cells (e.g., the asialoglycoprotein receptor subunit H2a *(171)*, the T cell receptor α-subunit *(172)* or orphan immunoglobulin chains *(173, 174)*) or cellular targets of viral gene products (e.g., the MHC class I molecules in cells expressing the HCMV immunoevasins US2 and US11 *(160)*). Moreover, yeast genetics in *S. cerevisiae* *(175)* paved the way for the identification of several ER-resident, transmembrane, and cytosolic proteins that regulate protein quality control. The mammalian system is much more complex, but several folding, quality control and ERAD regulators operating in mammalian cells have functional orthologs in yeast. In Sect. II.D we will highlight few differences between the yeast and the mammalian systems.

Altogether, it is clear that the existing models on function of folding, quality control, and degradation machineries have been generated upon analysis of a limited set of model proteins synthesized in select model systems. It still remains unclear how far we can go with generalizations of data collected from these studies.

C. A Lag Phase Before Destruction

Initially, there is no difference between a folding-competent and a folding-defective polypeptide. They are both inserted cotranslationally within the ER lumen in an unfolded state and are retained for some time in the folding machinery. Folding-competent polypeptides eventually attain the native structure and escape chaperone-mediated retention in the ER lumen. Folding-defective ones are initially subjected to folding attempts, and are therefore normally not degraded immediately after synthesis. The phase of *futile folding attempts* can be visualized as a *lag* phase that precedes degradation onset *(42, 172, 176–180)* and reviewed in *(159)*. Folding-defective glycoproteins spend most of this lag phase in the calnexin chaperone system. As long as folding-competent and folding-defective polypeptides are in the calnexin chaperone system, they are protected from premature degradation *(42)*.

How exactly folding intermediates necessitating longer retention in the folding machinery are distinguished from terminally misfolded conformers that must be extracted from the ER and degraded, is still a matter of intensive study. For glycoproteins, N-glycan processing plays a crucial role in these decisions *(159)*. Nascent chains are decorated with a preassembled 14-saccharides glycan ($Glc_3Man_9GlcNAc_2$–, Fig. 3, nascent). The initial removal of the two terminal glucose residues is an irreversible process that generates the mono-glucosylated trimming intermediate that recruits calnexin and calreticulin (Fig. 3, step 1). The removal of glucose-3 that follows substrate release from calnexin/calreticulin (Fig. 3, step 2, or step 3a) is on the other hand a reversible reaction. In fact, if the fully deglucosylated oligosaccharide is displayed on a polypeptide chain that elicits GT1-recognition, glucose-3 can be readded to prolong retention of the folding polypeptide in the calnexin chaperone system (and delay deviation of nonnative polypeptides into the ERAD pathway) (Fig. 3, step 4). During the off-phase, N-glycans may eventually become accessible to ER mannosidase of the glycosyl hydrolase 47 (GH47) family (these are α1,2-mannosidases *(158)*, Table IV). α1,2-mannosidases can potentially remove four mannoses from protein-bound oligosaccharide (Fig. 3, step 5). N-glycans with reduced number of mannoses displayed on a polypeptide exposing nonnative determinants that elicit persistent association with ER-resident factors, for example BiP (Fig. 3, step 6) represents a potent signal for polypeptide disposal (Fig. 3, step 7).

D. Extensive De-Mannosylation to Deviate Misfolded Proteins for ERAD

Native proteins are packaged into COPII-coated transport vesicles and are delivered through the secretory pathway at their final intra or extracellular destination. Immature polypeptides must be retained in the ER lumen until achievement of their native structure. Terminally misfolded polypeptides must be translocated into the cytosol to be degraded by 26S proteasomes. Efficient execution of these three tasks, that is, secretion of native structures, retention of folding intermediates and disposal of terminally misfolded polypeptides maintains ER homeostasis.

To channel terminally misfolded glycoproteins for degradation, the *futile* cycles of release and reassociation with calnexin/calreticulin regulated by the counteracting activities of the glucosidase II and GT1 must eventually be interrupted. Current models claim that while temporary detached from calnexin, nonnative glycopolypeptides may become accessible to one *(181, 182)* or more *(183)* ER-resident α1,2-mannosidases. These will remove one (mannose B *(181)*) or, sequentially, up to 4 α1,2-bonded mannose residues *(182, 183)* (Fig. 3, step 3b). Substrate de-mannosylation makes the associated polypeptide a weaker ligand for calnexin and calreticulin *(184)*, a better substrate for glucosidase II *(185)* (conflicting results have been published on this, though *(186)*) and a worst substrate for GT1 *(187)*. When a polypeptide with low mannose content is eventually released from calnexin and calreticulin the reglucosylation necessary to reassociate becomes less and less efficient. Consistently, selective inhibition of ER α1,2-mannosidases with the alkaloid kifunensine retards release of folding-defective polypeptides from the calnexin cycle *(42)* and delays ERAD *(188–190)*.

S. cerevisiae lacks a GT1 ortholog that delay onset of degradation by prolonging retention of nonnative polypeptides in the calnexin chaperone system *(191)*. Removal of a single mannose residue from the N-glycan branch B (Fig. 2) by the MnsI is apparently sufficient to tag terminally misfolded polypeptides for disposal *(192)*. Cumulating evidences indicate that requirements for mammalian ERAD may be different. Firstly, already in the early 1990s, it has been shown that extensive de-mannosylation precedes protein disposal from the mammalian ER *(171, 190)*. This was recently confirmed by several studies showing that N-glycans on folding-defective polypeptides are extensively processed to Man_{5-6} in the mammalian ER *(171, 182, 190, 193–197)*. Secondly, removal of a single mannose residue also characterizes the fate of native proteins that are released from the ER into the secretory pathway *(198)*. It is therefore unlikely that this single mannose-processing event represents a strong signal for disposal from the mammalian ER. Furthermore, several studies have shown that mannose removal is still required for glycoprotein degradation in mannosyl-phosphoryldolichol-deficient cell lines *(199)*. These cells are characterized by addition onto nascent chains of incomplete oligosaccharides that only display removable α1,2-bonded terminal mannoses on branch A *(200, 201)* (Fig. 2, B3F7 mutant CHO line). Certainly, in the mammalian system, removal of mannose A (Fig. 2) has irreversible consequences because this saccharide is the only residue that can be reglucosylated by GT1 and its cleavage irreversibly extracts folding-defective polypeptides from the calnexin chaperone system (Fig. 3, steps 5–6).

Who is operating the extensive de-mannosylation of folding-defective polypeptides occurring in the mammalian ER? The mammalian ortholog of MnsI is the ER α1,2-mannosidase I (ERManI), a member of the GH47 family of α1,2-mannosidases. Similarly to the yeast protein, ERManI specifically removes mannose B from protein-bound oligosaccharides, unless its activity is tested *in vitro* at *unphysiologically* high concentrations. In this case, recombinant ERManI removes mannose B as well as other α1,2-bonded mannoses *(202)*. It is possible that this enzyme is enriched in specialized subregions of the mammalian ER and extensively de-mannosylates terminally misfolded polypeptides *(182, 203)*. However, the specificity of α1,2-mannosidases is conferred by the dimension of the carbohydrate binding site where an arginine residue at the bottom of the catalytic site (Arg273 in the yeast MnsIp, Arg461 in human ERManI) plays a critical role in reducing the degree of freedom of the oligosaccharide entering the active site *(204)*. α1,2-mannosidases with broader specificity (e.g., Golgi-resident α1,2-mannosidases of *Pennicillium citrinum*) that can trim $Man_9GlcNAc_2-$ to $Man_5GlcNAc_2-$ are characterized by the presence of a smaller, uncharged residue at this position *(205)*. The conversion of MnsIp and ERManI in mannosidases that efficiently generate the $Man_5GlcNAc_2-$ final product requires the replacement of Arg273 (or 461) with a leucine or a glycine *(204)*.

An additional possibility is that other α1,2-mannosidases contribute to the generation of the ERAD signal in the mammalian ER. Likely candidates are three recently characterized ER-resident members of the GH47 family, namely EDEM1, EDEM2, and EDEM3 *(183)*. Notably, the three EDEM proteins are characterized by the presence of a glycine residue at the bottom of their putative active sites. This would confer to them the capacity to generate $Man_5GlcNAc_2$-oligosaccharides. There is an ongoing debate whether the EDEM proteins are active mannosidases. Certainly, they conserve the structure of the catalytic site as well as all the catalytic residues present in the other mannosidases of the GH47 family *(158, 206)*. Moreover, it has been shown that up-regulation of EDEM1 and EDEM3 enhances de-mannosylation of ERAD substrates *(201, 207)* and that this does not occur if the ectopically expressed EDEM1 or EDEM3 carry a mutation of a single residue in their *putative* catalytic sites *(201, 207)*. It must also be mentioned that recombinant *T. cruzi* EDEM is an active α1,2-mannosidase that extensively processes Man_9–$GlcNAc_2$– to Man_5–$GlcNAc_2$– *(208)*.

E. Disposal of Nonglycosylated Proteins

The fate of nonglycosylated proteins that enter the ER is much less clear compared to the fate of *N*-glycosylated ones. Certainly, folding-defective glycosylated and nonglycosylated polypeptides are both translocated into the cytosol to eventually be degraded by 26S proteasome. It can be postulated that folding-defective glycoproteins are subjected to two phases of retention-based ER quality control, the first in the calnexin chaperone system and the second in the BiP/PDI chaperone system (Fig. 3), while nonglycosylated polypeptides must bypass the first phase and can only rely on the second *(102)*. Analysis of few folding defective polypeptides and of bacterial toxins that exploit the same machineries to invade the host cell cytosol revealed that recruitment of BiP and/or PDI immediately precedes translocation of terminally misfolded proteins into the cytosol *(42, 209–212)*. Recent work showed that dislocation into the cytosol of nonglycosylated κ light chain requires derlin-1 and HERP *(174)*, while the glycosylated ERAD substrate $Null_{hong\ kong}$ requires derlin-2 and -3 (both forming a complex with EDEM1), but neither derlin-1 *(213)* nor HERP *(174)*. Since HERP is located in the ER membrane with most of its volume exposed to the cytosol *(214)*, its interaction with nonglycosylated ERAD substrates was proposed to be indirect or to require a partial dislocation of the misfolded protein into the cytosol *(174)*. If HERP plays the same role as its yeast ortholog Usa1p, then an implication of these data would be that in mammals, the soluble, misfolded NHK is dislocated into the cytosol and degraded without intervention of the Hrd1p/synoviolin complex (see Sect. II.F) because formation and stability of the complex should rely on the presence of HERP.

F. Directing ERAD Substrates to the Retro-Translocation Site

Both in yeast and in mammalian cells, the processes that regulate recognition and translocation of terminally misfolded polypeptides into the cytosol are ill defined. Substrate ubiquitylation facilitates proteasomal degradation and requires an activating, a conjugating and a ligating enzyme (E1, E2, and E3, respectively). Recently, many aspects and several factors involved in these processes have been characterized in *S. cerevisiae* where distinct machineries are involved in recognition, retro-translocation, and polyubiquitylation of polypeptides displaying folding defects in their luminal, transmembrane, or cytosolic domains *(215)*. These complex machineries are organized around two RING finger E3 ubiquitin ligases, Doa 10p (TEB4/MARCH VI in mammals) and Hrd 1p/Der3p (synoviolin and gp78 in mammals) *(216, 217))*.

Membrane polypeptides with cytosolic lesions are targeted to the ERAD-C machinery (Doa10p complex). The Doa10p complex comprises the E2 (ubiquitin-conjugating) enzymes Ubc6p (Ube2j1 and Ube2j2 in mammals) and Ubc7p (Cue1p and Ube2g2) with their membrane-connector Cue 1p as well as a substrate extractor complex containing the AAA-ATPase Cdc48p (p97 in mammals) with the Ufd1 and Np14 cofactors and the connector Ubx2p *(215, 218–220)*.

Transmembrane and soluble proteins with luminal defects are targeted to the ERAD-M and ERAD-L pathways (Hrd1p/Der3p complex, Fig. 5 shows one possible arrangement of the corresponding mammalian synoviolin complex). The Hrd1p/Der3p (synoviolin) complex contains the cytosolic factors also found in the Doa10p complex but several additional transmembrane (Der 1p (Usa1p), Usa1p (HERP), Hrd3p (Sel1L)), and luminal proteins (e.g., Yos9p (OS-9.1, OS-9.2, erlectin in mammals) and Kar2p (BiP) *(219, 221, 222))*.

The mammalian system is much more complex and has a greater number of E3 enzymes involved in ERAD. Some of them are spanning the ER membrane, for example synoviolin *(223)*, gp78 *(224)*, and TEB4 *(225)*. Synoviolin and gp78 are involved in disposal of several substrates from the ER lumen *(217, 226)* while for TEB4 no cellular substrate has been described, yet. Other mammalian E3 ubiquitin ligases are cytosolic proteins (e.g., parkin, CHIP, and the SCF-Fbs1/Fbs2 complex involved in degradation of Pael-R *(227)*, CFTRΔF508 *(228)* and select glycoproteins *(229)*, respectively).

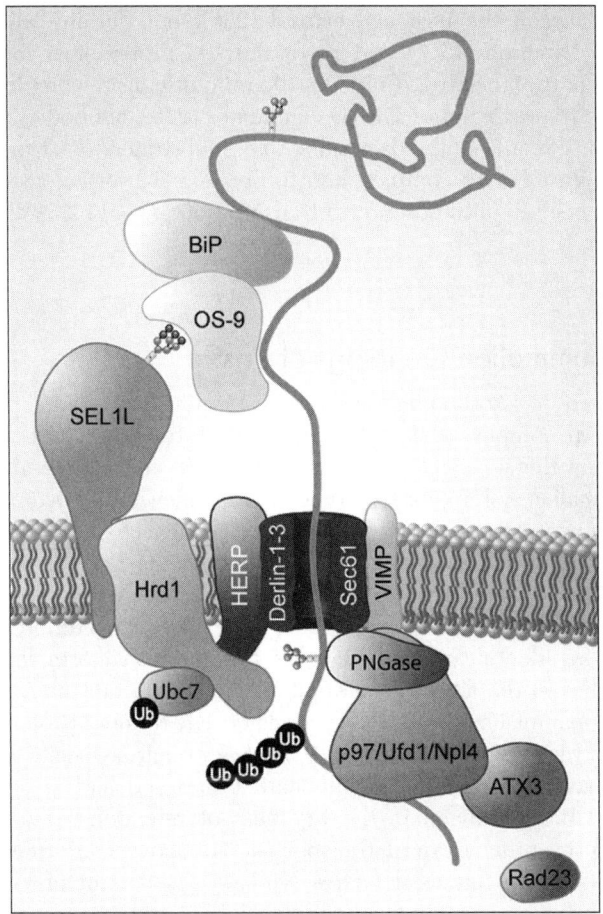

FIG. 5. A putative model of a complex regulating translocation of terminally misfolded polypeptides from the ER lumen into the cytosol. (See Color Insert.)

Despite several highly cited publications seemed to have solved the issue, it remains unclear how misfolded proteins cross the ER membrane during their translocation into the cytosol. Several candidates such as Sec61 and derlin proteins as well as protein-independent mechanisms such as lipid droplets formation (230) could be involved in the process. For example, dislocating MHC class I heavy chains in cells expressing immunoevasins were cross-linked to Sec61 (231); yeast Sec61 mutants exhibit severe defects in degradation of soluble, but not of membrane-bound folding-defective polypeptides (232–235); Sec61 is used for retro-translocation of toxin subunits that subvert the ERAD pathway to invade the host cell cytosol *(236)*, but derlin-1 also offers an essential contribution *(45)*; Sec61 binds the 19S proteasomal subunit *(237)*; Sec61 is not part of the Hrd1p and Doa10p ubiquitin ligase complexes regulating cytosolic translocation of ERAD candidates *(216)*; antibodies to Derlin-1, but not against Sec61, inhibit disposal of select substrates *(43)*. Der1p, which is part of the yeast Hrd1p complex, and the mammalian orthologs derlin-1, -2, and -3 have been implicated in retro-translocation *(43, 213, 238–241)*.

G. The Mammalian Orthologs of Yos9p

In *S. cerevisiae*, Yos9p binds and targets terminally misfolded polypeptides to the Hrd1p complex *(219, 221, 242–245)*. Intriguingly, although several examples show that most of the components of yeast machineries are conserved in the mammalian system, for long time, it was believed that mammalian OS-9 was a cytosolic protein, associated with the cytosolic face of the ER membrane *(246)*. This study was followed by several other publications, in which experimental design and interpretation of the results were based on the assumption that OS-9 was a cytosolic protein *(190, 247, 248)*. The different topology hampered the identification of OS-9 as the functional ortholog of Yos9p. Recent studies in the lab of Ron Kopito *(249)* and in our lab *(294)* demonstrated that mammalian OS-9 is a glycosylated ER-resident protein expressed in two splice variants, OS-9.1 and OS-9.2. OS-9 variants associate with folding-defective glycoproteins, but not with native ones and have a dual function within the ER: they maintain the tightness of retention-based ER quality control by preventing forward transport of nonnative conformers (this has been shown for the folding-defective $Null_{hong\ kong}$ variant of α1-antitrypsin *(294)*) and they participate in disposal of misfolded proteins from the mammalian ER (*(249)* and *(294)*).

Our studies show that transcription of both OS-9 variants is enhanced upon activation of the Ire1/Xbp1 pathway in cells exposed to acute ER stress. Analysis of transcriptional regulation of the mammalian orthologs of the components of the yeast Hrd1p complex reveals that all components are inducible upon ER-stress, but only OS-9 variants and synoviolin require activation of the Ire1/Xbp1 pathway, whereas Sel1L is regulated by ATF6 *(250)* and *(294)*. We postulate that *retention of misfolded* versus *facilitation of disposal* functions of mammalian OS-9 variants might depend on formation of multiprotein complexes as reported for the yeast proteins, and that activation of individual stress-response pathways (Ire1-regulated versus ATF6-regulated) in specific tissues or under specific stress conditions could enhance one or the other function of OS-9.

H. Macroautophagy

Macroautophagy is a unique intracellular process in which membrane-bound compartments engulf organelles and macromolecules and deliver them to lysosomes for destruction. Mammalian cells display a low level of constitutive autophagy (baseline autophagy) that regulates normal turnover of cytosolic components. Autophagic activity can strongly be induced upon nutrient deprivation to play pleiotropic roles in a variety of cytoprotective functions *(251–254)*. Autophagy can also be induced upon ER stress to insure cell survival and to counterbalance ER proliferation *(255–257)*. Macroautophagy is abolished upon deletion of the *atg5* gene, which is essential for autophagosome formation *(258–260)*. In cells subjected to several hours of nutrient deprivation *(261)*, or in which acute accumulation of insoluble protein aggregates triggers ER stress *(262, 263)*, deletion of ATG5 results in accumulation of aberrant polypeptides in the ER lumen (reviewed in *(264)*). Thus, autophagy certainly contributes with proteasomal degradation to the clearance of otherwise indigestible protein aggregates accumulating in the cytosol or in the ER. Autophagy may also be activated under acute stress conditions *(265)*. However, an involvement of autophagy in degradation of misfolded proteins from the ER at steady state must be an exception, rather than a rule. Several reports show in fact that lysosome and autophagy inhibitors do not normally affect ERAD *(266–269)*, which is, on the other hand, profoundly affected by interference with the activity of the ubiquitylating or proteasomal machineries.

I. Do Autophagy-Like Processes Regulate ERAD Activity? The Concept of ERAD Tuning

ER-resident molecular chaperones and folding enzymes are long-living proteins. Exceptions, however, do exist. For example, it has been reported that the ER α1,2-mannosidase I, a crucial regulator of ERAD, is characterized by rapid turnover *(270)*. Our studies reveal that EDEM1 as well has unconventional short half-life for an ER-resident protein *(293)*.

An interesting electron microscopy and confocal immunofluorescence analysis recently revealed the presence of EDEM1 in small ER-derived vesicles that lack conventional ER markers and a recognizable cytosolic coat such as COPII. The destination or function of these vesicles has not been established, but the occasional presence of ERAD substrates led to propose a possible role of this vesicular transport out of the ER in removal of misfolded proteins from the ER lumen *(271)*. The unpublished studies performed in our lab led us to challenge this hypothesis and to postulate that a vesicular transport out of the ER regulates the rapid turnover of EDEM1, and possibly of other short-living ERAD regulators. The selective segregation of factors regulating protein disposal from the mammalian ER would reduce the competition between folding and degradation machineries operating in the ER lumen at the advantage of the protein folding process *(293)*. Our data show that in cells with defective EDEM1 turnover (e.g., cells lacking basal autophagy), the intralumenal level of this mannosidase is aberrantly elevated. This enhances disposal from the ER lumen of misfolded proteins and, most significantly, substantially reduces efficiency in the maturation of folding-competent glycopolypeptides caused by premature interruption of ongoing folding programs *(293)*. Our postulate is that at steady state, in unstressed cells, the ER protein folding machinery must be offered a kinetic advantage over the protein degradation machinery operating in the same compartment to deal with unstructured nascent chains. We propose that mechanisms that we define as *ERAD tuning* contribute to maintenance of ER homeostasis, which is required for optimal function of the ER protein factory. These posttranslational mechanisms operate in mammalian cells to selectively remove from the ER folding compartment ERAD operators that, if present in excessive concentration, could prematurely interrupt productive polypeptide maturation *(293)*.

J. The Fate of Folding-Defective Polypeptides: A Summary

In the ER lumen, there is an intrinsic difficulty to distinguish terminally misfolded polypeptides to be selected for degradation from unstructured intermediates of the folding programs that will eventually complete their maturation. N-glycans appended to polypeptide chains may facilitate this task. Polypeptides entering a folding program expose oligosaccharides with nine mannose residues. This oligosaccharide structure is best suited for retention and/or cycling in the calnexin chaperone system that offers appropriate folding conditions and prevents unwanted, premature degradation. Instead, folding-defective polypeptides that have spent already some time in unproductive folding attempts are eventually exposed to ER-resident $\alpha1,2$-mannosidases that can sequentially remove up to four mannose residues. Extensive demannosylation inhibits and, upon removal of the mannose residue A, eventually fully prevents substrate retention in the calnexin chaperone system. It certainly therefore represents a strong signal for disposal from the mammalian ER.

Cumulating evidences show that the percentage of newly synthesized protein that is folded as opposed to degraded is strongly affected by the kinetic competition between conformational maturation and recognition for disposal in the ER lumen. As such, any change in protein folding versus degradation rate will determine the percentage of the newly synthesized proteins that will eventually acquire native structure (reviewed in *(96, 159, 272, 273)*).

Several disease-linked polypeptide mutations do not compromise the function, but delay polypeptide maturation resulting in polypeptide disposal in advance of maturation. In all these cases, pharmacologic intervention can be hypothesized to accelerate maturation of the mutated polypeptide or to delay the onset of its disposal to such an extent that the folding program can successfully be completed ((*159, 165, 166*) and references therein).

REFERENCES

1. Blobel, G., and Dobberstein, B. (1975). Transfer of proteins across membranes. I. Presence of proteolytically processed and unprocessed nascent immunoglobulin light chains on membrane-bound ribosomes of murine myeloma. *J. Cell. Biol.* **67**, 835–851.
2. Walter, P., and Blobel, G. (1982). Signal recognition particle contains a 7S RNA essential for protein translocation across the endoplasmic reticulum. *Nature* **299**, 691–698.
3. Deshaies, R. J., and Schekman, R. (1987). A yeast mutant defective at an early stage in import of secretory protein precursors into the endoplasmic reticulum. *J. Cell Biol.* **105**, 633–645.
4. Redman, C. M., and Sabatini, D. D. (1966). Vectorial discharge of peptides released by puromycin from attached ribosomes. *Proc. Natl. Acad. Sci. USA* **56**, 608–615.
5. Redman, C. M., Siekevitz, P., and Palade, G. E. (1966). Synthesis and transfer of amylase in pigeon pancreatic micromosomes. *J. Biol. Chem.* **241**, 1150–1158.
6. Sabatini, D. D., and Blobel, G. (1970). Controlled proteolysis of nascent polypeptides in rat liver cell fractions. II. Location of the polypeptides in rough microsomes. *J. Cell. Biol.* **45**, 146–157.
7. Hegde, R. S., and Bernstein, H. D. (2006). The surprising complexity of signal sequences. *Trends Biochem. Sci.* **31**, 563–571.
8. Daniels, R., Kurowski, B., Johnson, A. E., and Hebert, D. N. (2003). N-linked glycans direct the cotranslational folding pathway of influenza hemagglutinin. *Mol. Cell* **11**, 79–90.
9. Li, Y., Luo, L., Thomas, D. Y., and Kang, C. Y. (2000). The HIV-1 Env protein signal sequence retards its cleavage and down-regulates the glycoprotein folding. *Virology* **272**, 417–428.
10. Van den Berg, B., Clemons, W. M., Jr., Collinson, I., Modis, Y., Hartmann, E., Harrison, S. C., and Rapoport, T. A. (2004). X-ray structure of a protein-conducting channel. *Nature* **427**, 36–44.
11. Connolly, T., and Gilmore, R. (1986). Formation of a functional ribosome-membrane junction during translocation requires the participation of a GTP-binding protein. *J. Cell Biol.* **103**, 2253–2261.
12. Meyer, H. A., Grau, H., Kraft, R., Kostka, S., Prehn, S., Kalies, K. U., and Hartmann, E. (2000). Mammalian Sec61 is associated with Sec62 and Sec63. *J. Biol. Chem.* **275**, 14550–14557.
13. Tyedmers, J., Lerner, M., Bies, C., Dudek, J., Skowronek, M. H., Haas, I. G., Heim, N., Nastainczyk, J., Volkmer, J., and Zimmermann, R. (2000). Homologs of the yeast Sec complex subunits Sec62p and Sec63p are abundant proteins in dog pancreas microsomes. *Proc. Natl. Acad. Sci. USA* **97**, 7214–7219.
14. Matlack, K. E., Misselwitz, B., Plath, K., and Rapoport, T. A. (1999). BiP acts as a molecular ratchet during posttranslational transport of prepro-alpha factor across the ER membrane. *Cell* **97**, 553–564.
15. Rapoport, T. A. (2007). Protein translocation across the eukaryotic endoplasmic reticulum and bacterial plasma membranes. *Nature* **450**, 663–669.

16. Whitley, P., Nilsson, I. M., and von Heijne, G. (1996). A nascent secretory protein may traverse the ribosome/endoplasmic reticulum translocase complex as an extended chain. *J. Biol. Chem.* **271,** 6241–6244.
17. Mingarro, I., Nilsson, I., Whitley, P., and von Heijne, G. (2000). Different conformations of nascent polypeptides during translocation across the ER membrane. *BMC Cell. Biol.* **1,** 3.
18. Kowarik, M., Kung, S., Martoglio, B., and Helenius, A. (2002). Protein folding during cotranslational translocation in the endoplasmic reticulum. *Mol. Cell* **10,** 769–778.
19. McCormick, P. J., Miao, Y., Shao, Y., Lin, J., and Johnson, A. E. (2003). Cotranslational protein integration into the ER membrane is mediated by the binding of nascent chains to translocon proteins. *Mol. Cell* **12,** 329–341.
20. Skach, W. R., and Lingappa, V. R. (1993). Amino-terminal assembly of human P-glycoprotein at the endoplasmic reticulum is directed by cooperative actions of two internal sequences. *J. Biol. Chem.* **268,** 23552–23561.
21. Meacock, S. L., Lecomte, F. J., Crawshaw, S. G., and High, S. (2002). Different transmembrane domains associate with distinct endoplasmic reticulum components during membrane integration of a polytopic protein. *Mol. Biol. Cell* **13,** 4114–4129.
22. Borel, A. C., and Simon, S. M. (1996). Biogenesis of polytopic membrane proteins: Membrane segments assemble within translocation channels prior to membrane integration. *Cell* **85,** 379–389.
23. Kida, Y., Morimoto, F., and Sakaguchi, M. (2007). Two translocating hydrophilic segments of a nascent chain span the ER membrane during multispanning protein topogenesis. *J. Cell Biol.* **179,** 1441–1452.
24. Skach, W. R. (2007). The expanding role of the ER translocon in membrane protein folding. *J. Cell Biol.* **179,** 1333–1335.
25. Morgan, D. G., Menetret, J. F., Radermacher, M., Neuhof, A., Akey, I. V., Rapoport, T. A., and Akey, C. W. (2000). A comparison of the yeast and rabbit 80 S ribosome reveals the topology of the nascent chain exit tunnel, inter-subunit bridges and mammalian rRNA expansion segments. *J. Mol. Biol.* **301,** 301–321.
26. Haas, I. G., and Wabl, M. (1983). Immunoglobulin heavy chain binding protein. *Nature* **306,** 387–389.
27. Blond-Elguindi, S., Cwirla, S. E., Dower, W. J., Lipshutz, R. J., Sprang, S. R., Sambrook, J. F., and Gething, M. J. (1993). Affinity panning of a library of peptides displayed on bacteriophages reveals the binding specificity of BiP. *Cell* **75,** 717–728.
28. Hendershot, L. M. (2004). The ER function BiP is a master regulator of ER function. *Mt Sinai J. Med.* **71,** 289–297.
29. Melnick, J., Dul, J. L., and Argon, Y. (1994). Sequential interaction of the chaperones BiP and GRP94 with immunoglobulin chains in the endoplasmic reticulum. *Nature* **370,** 373–375.
30. Nieland, T. J., Tan, M. C., Monne-van Muijen, M., Koning, F., Kruiseek, A. M., and van Bleek, G. M. (1996). Isolation of an immunodominant viral peptide that is endogenously bound to the stress protein GP96/GRP94. *Proc. Natl. Acad. Sci. USA* **93,** 6135–6139.
31. Argon, Y., and Simen, B. B. (1999). GRP94, an ER chaperone with protein and peptide binding properties. *Semin. Cell Dev. Biol.* **10,** 495–505.
32. Sitia, R., Neuberger, M., Alberini, C., Bet, P., Fra, A., Valetti, C., Williams, G., and Milstein, C. (1990). Developmental regulation of IgM secretion: The role of the carboxy-terminal cysteine. *Cell* **60,** 781–790.
33. Reddy, P. S., and Corley, R. B. (1998). Assembly, sorting, and exit of oligomeric proteins from the endoplasmic reticulum. *Bioessays* **20,** 546–554.
34. Molinari, M., and Helenius, A. (1999). Glycoproteins form mixed disulphides with oxidoreductases during folding in living cells. *Nature* **402,** 90–93.

35. Ellgaard, L., and Ruddock, L. W. (2005). The human protein disulphide isomerase family: substrate interactions and functional properties. *EMBO Rep.* **6**, 28–32.
36. Kiefhaber, T., Quaas, R., Hahn, U., and Schmid, F. X. (1990). Folding of ribonuclease T1. 2. Kinetic models for the folding and unfolding reactions. *Biochemistry* **29**, 3061–3070.
37. Ellgaard, L., Molinari, M., and Helenius, A. (1999). Setting the standards: Quality control in the secretory pathway. *Science* **286**, 1882–1888.
38. Meunier, L., Usherwood, Y. K., Chung, K. T., and Hendershot, L. M. (2002). A subset of chaperones and folding enzymes form multiprotein complexes in endoplasmic reticulum to bind nascent proteins. *Mol. Biol. Cell* **13**, 4456–4469.
39. Zhang, J., and Herscovitz, H. (2003). Nascent lipidated apolipoprotein B is transported to the Golgi as an incompletely folded intermediate as probed by its association with network of endoplasmic reticulum molecular chaperones, GRP94, ERp72, BiP, calreticulin, and cyclophilin B. *J. Biol. Chem.* **278**, 7459–7468.
40. Anfinsen, C. B. (1973). Principles that govern the folding of protein chains. *Science* **181**, 223–230.
41. Chen, W., Helenius, J., Braakman, I., and Helenius, A. (1995). Cotranslational folding and calnexin binding during glycoprotein synthesis. *Proc. Natl. Acad. Sci. USA* **92**, 6229–6233.
42. Molinari, M., Galli, C., Piccaluga, V., Pieren, M., and Paganetti, P. (2002). Sequential assistance of molecular chaperones and transient formation of covalent complexes during protein degradation from the ER. *J. Cell Biol.* **158**, 247–257.
43. Wahlman, J., DeMartino, G. N., Skach, W. R., Bulleid, N. J., Brodsky, J. L., and Johnson, A. E. (2007). Real-time fluorescence detection of ERAD substrate retrotranslocation in a mammalian *in vitro* system. *Cell* **129**, 943–955.
44. Tsai, B., Ye, Y., and Rapoport, T. A. (2002). Retro-translocation of proteins from the endoplasmic reticulum into the cytosol. *Nat. Rev. Mol. Cell Biol.* **3**, 246–255.
45. Bernardi, K. M., Forster, M. L., Lencer, W. I., and Tsai, B. (2008). Derlin-1 facilitates the retro-translocation of cholera toxin. *Mol. Biol. Cell* **19**, 877–884.
46. Helenius, A., and Aebi, M. (2004). Roles of N-Linked Glycans in the Endoplasmic Reticulum. *Annu. Rev. Biochem.* **73**, 1019–1049.
47. Chavan, M., Chen, Z., Li, G., Schindelin, H., Lennarz, W. J., and Li, H. (2006). Dimeric organization of the yeast oligosaccharyl transferase complex. *Proc. Natl. Acad. Sci. USA* **103**, 8947–8952.
48. Li, H., Chavan, M., Schindelin, H., Lennarz, W. J., and Li, H. (2008). Structure of the oligosaccharyl transferase complex at 12 a resolution. *Structure* **16**, 432–440.
49. Imperiali, B., and Hendrickson, T. L. (1995). Asparagine-linked glycosylation: specificity and function of oligosaccharyl transferase. *Bioorg. Med. Chem.* **3**, 1565–1578.
50. Parodi, A. J., and Leloir, L. F. (1979). The role of lipid intermediates in the glycosylation of proteins in the eucaryotic cell. *Biochim. Biophys. Acta* **559**, 1–37.
51. Parodi, A. J., Behrens, N. H., Leloir, L. F., and Carminatti, H. (1972). The role of polyprenol-bound saccharides as intermediates in glycoprotein synthesis in liver. *Proc. Natl. Acad. Sci. USA* **69**, 3268–3272.
52. Nilsson, I. M., and von Heijne, G. (1993). Determination of the distance between the oligosaccharyltransferase active site and the endoplasmic reticulum membrane. *J. Biol. Chem.* **268**, 5798–5801.
53. Molinari, M., and Helenius, A. (2000). Chaperone selection during glycoprotein translocation into the endoplasmic reticulum. *Science* **288**, 331–333.
54. Hammond, C., and Helenius, A. (1994). Folding of VSV G protein: Sequential interaction with BiP and calnexin. *Science* **266**, 456–458.

55. Pipe, S. W., Morris, J. A., Shah, J., and Kaufman, R. J. (1998). Differential interaction of coagulation factor VIII and factor V with protein chaperones calnexin and calreticulin. *J. Biol. Chem.* **273,** 8537–8544.
56. Tomita, Y., Yamashita, T., Sato, H., and Taira, H. (1999). Kinetics of interactions of sendai virus envelope glycoproteins, F and HN, with endoplasmic reticulum-resident molecular chaperones, BiP, calnexin, and calreticulin. *J. Biochem. (Tokyo)* **126,** 1090–1100.
57. Wang, N., Daniels, R., and Hebert, D. N. (2005). The cotranslational maturation of the type I membrane glycoprotein tyrosinase: The heat shock protein 70 system hands off to the lectin-based chaperone system. *Mol. Biol. Cell* **16,** 3740–3752.
58. Helenius, A. (1994). How N-linked oligosaccharides affect glycoprotein folding in the endoplasmic reticulum. *Mol. Biol. Cell* **5,** 253–265.
59. Baksh, S., and Michalak, M. (1991). Expression of calreticulin in *Escherichia coli* and identification of its Ca2 + binding domains. *J. Biol. Chem.* **266,** 21458–21465.
60. Schrag, J. D., Bergeron, J. J., Li, Y., Borisova, S., Hahn, M., Thomas, D. Y., and Cygler, M. (2001). The Structure of calnexin, an ER chaperone involved in quality control of protein folding. *Mol. Cell* **8,** 633–644.
61. Frickel, E. M., Riek, R., Jelesarov, I., Helenius, A., Wuthrich, K., and Ellgaard, L. (2002). TROSY-NMR reveals interaction between ERp57 and the tip of the calreticulin P-domain. *Proc. Natl. Acad. Sci. USA* **99,** 1954–1959.
62. Leach, M. R., Cohen-Doyle, M. F., Thomas, D. Y., and Williams, D. B. (2002). Localization of the lectin, ERp57 binding, and polypeptide binding sites of calnexin and calreticulin. *J. Biol. Chem.* **277,** 29686–29697.
63. Pollock, S., Kozlov, G., Pelletier, M. F., Trempe, J. F., Jansen, G., Sitnikov, D., Bergeron, J. J., Gehring, K., Ekiel, I., and Thomas, D. Y. (2004). Specific interaction of ERp57 and calnexin determined by NMR spectroscopy and an ER two-hybrid system. *EMBO J.* **23,** 1020–1029.
64. Russell, S. J., Ruddock, L. W., Salo, K. E., Oliver, J. D., Roebuck, Q. P., Llewellyn, D. H., Roderick, H. L., Koivunen, P., Myllyharju, J., and High, S. (2004). The primary substrate binding site in the b' domain of ERp57 is adapted for endoplasmic reticulum lectin association. *J. Biol. Chem.* **279,** 18861–18869.
65. Oliver, J. D., van der Wal, F. J., Bulleid, N. J., and High, S. (1997). Interaction of the thiol-dependent reductase ERp57 with nascent glycoproteins. *Science* **275,** 86–88.
66. Zapun, A., Darby, N. J., Tessier, D. C., Michalak, M., Bergeron, J. J., and Thomas, D. Y. (1998). Enhanced catalysis of ribonuclease B folding by the interaction of calnexin or calreticulin with ERp57. *J. Biol. Chem.* **273,** 6009–6012.
67. Deprez, P., Gautschi, M., and Helenius, A. (2005). More than one glycan is needed for ER glucosidase II to allow entry of glycoproteins into the calnexin/calreticulin cycle. *Mol. Cell* **19,** 183–195.
68. Trombetta, E. S., Simons, J. F., and Helenius, A. (1996). Endoplasmic reticulum glucosidase II is composed of a catalytic subunit, conserved from yeast to mammals, and a tightly bound noncatalytic HDEL-containing subunit. *J. Biol. Chem.* **271,** 27509–27516.
69. Petrescu, A. J., Butters, T. D., Reinkensmeier, G., Petrescu, S., Platt, F. M., Dwek, R. A., and Wormald, M. R. (1997). The solution NMR structure of glucosylated N-glycans involved in the early stages of glycoprotein biosynthesis and folding. *EMBO. J.* **16,** 4302–4310.
70. Hebert, D. N., Foellmer, B., and Helenius, A. (1996). Calnexin and calreticulin promote folding, delay oligomerization and suppress degradation of influenza hemagglutinin in microsomes. *EMBO J.* **15,** 2961–2968.
71. Wilson, I. A., Skehel, J. J., and Wiley, D. C. (1981). Structure of the haemagglutinin membrane glycoprotein of influenza virus at 3 A resolution. *Nature* **289,** 366–373.
72. Pieren, M., Galli, C., Denzel, A., and Molinari, M. (2005). The use of calnexin and calreticulin by cellular and viral glycoproteins. *J. Biol. Chem.* **280,** 28265–28271.

73. Degen, E., and Williams, D. B. (1991). Participation of a novel 88-kD protein in the biogenesis of murine class I histocompatibility molecules. *J. Cell. Biol.* **112,** 1099–1115.
74. Wada, I., Rindress, D., Cameron, P. H., Ou, W. J., Doherty, J. J., 2nd, Louvard, D., Bell, A. W., Dignard, D., Thomas, D. Y., and Bergeron, J. J. (1991). SSR alpha and associated calnexin are major calcium binding proteins of the endoplasmic reticulum membrane. *J. Biol. Chem.* **266,** 19599–19610.
75. Degen, E., Cohen-Doyle, M. F., and Williams, D. B. (1992). Efficient dissociation of the p88 chaperone from major histocompatibility complex class I molecules requires both beta 2-microglobulin and peptide. *J. Exp. Med.* **175,** 1653–1661.
76. Ou, W. J., Cameron, P. H., Thomas, D. Y., and Bergeron, J. J. (1993). Association of folding intermediates of glycoproteins with calnexin during protein maturation. *Nature* **364,** 771–776.
77. Galvin, K., Krishna, S., Ponchel, F., Frohlich, M., Cummings, D. E., Carlson, R., Wands, J. R., Isselbacher, K. J., Pillai, S., and Ozturk, M. (1992). The major histocompatibility complex class I antigen-binding protein p88 is the product of the calnexin gene. *Proc. Natl. Acad. Sci. USA* **89,** 8452–8456.
78. Hammond, C., Braakman, I., and Helenius, A. (1994). Role of N-linked oligosaccharide recognition, glucose trimming, and calnexin in glycoprotein folding and quality control. *Proc. Natl. Acad. Sci. USA* **91,** 913–917.
79. Suh, K., Bergmann, J. E., and Gabel, C. A. (1989). Selective retention of monoglucosylated high mannose oligosaccharides by a class of mutant vesicular stomatitis virus G proteins. *J. Cell. Biol.* **108,** 811–819.
80. Parodi, A. J., Mendelzon, D. H., Lederkremer, G. Z., and Martin-Barrientos, J. (1984). Evidence that transient glucosylation of protein-linked Man9GlcNAc2, Man8GlcNAc2, and Man7GlcNAc2 occurs in rat liver and Phaseolus vulgaris cells. *J. Biol. Chem.* **259,** 6351–6357.
81. Cabral, C. M., Liu, Y., Moremen, K. W., and Sifers, R. N. (2002). Organizational diversity among distinct glycoprotein endoplasmic reticulum-associated degradation programs. *Mol. Biol. Cell* **13,** 2639–2650.
82. Caramelo, J. J., Castro, O. A., de Prat-Gay, G., and Parodi, A. J. (2004). The endoplasmic reticulum glucosyltransferase recognizes nearly native glycoprotein folding intermediates. *J. Biol. Chem.* **279,** 46280–46285.
83. Kimura, T., Hosoda, Y., Sato, Y., Kitamura, Y., Ikeda, T., Horibe, T., and Kikuchi, M. (2005). Interactions among yeast protein-disulfide isomerase proteins and endoplasmic reticulum chaperone proteins influence their activities. *J. Biol. Chem.* **280,** 31438–31441.
84. Fanchiotti, S., Fernandez, F., D'Alessio, C., and Parodi, A. J. (1998). The UDP-Glc:Glycoprotein glucosyltransferase is essential for Schizosaccharomyces pombe viability under conditions of extreme endoplasmic reticulum stress. *J. Cell. Biol.* **143,** 625–635.
85. Gurkan, C., Stagg, S. M., Lapointe, P., and Balch, W. E. (2006). The COPII cage: Unifying principles of vesicle coat assembly. *Nat. Rev. Mol. Cell Biol.* **7,** 727–738.
86. Bonifacino, J. S., and Glick, B. S. (2004). The mechanisms of vesicle budding and fusion. *Cell* **116,** 153–166.
87. Pepperkok, R., Scheel, J., Horstmann, H., Hauri, H. P., Griffiths, G., and Kreis, T. E. (1993). Beta-COP is essential for biosynthetic membrane transport from the endoplasmic reticulum to the Golgi complex *in vivo*. *Cell* **74,** 71–82.
88. Aridor, M., Bannykh, S. I., Rowe, T., and Balch, W. E. (1995). Sequential coupling between COPII and COPI vesicle coats in endoplasmic reticulum to Golgi transport. *J. Cell Biol.* **131,** 875–893.
89. Appenzeller-Herzog, C., and Hauri, H. P. (2006). The ER-Golgi intermediate compartment (ERGIC): In search of its identity and function. *J. Cell Sci.* **119,** 2173–2183.
90. Barlowe, C. (2003). Signals for COPII-dependent export from the ER: What's the ticket out? *Trends Cell Biol.* **13,** 295–300.

91. Kamiya, Y., Kamiya, D., Yamamoto, K., Nyfeler, B., Hauri, H. P., and Kato, K. (2008). Molecular basis of sugar recognition by the human L-type lectins ERGIC-53, VIPL, and VIP36. *J. Biol. Chem.* **283,** 1857–1861.
92. Vollenweider, F., Kappeler, F., Itin, C., and Hauri, H. P. (1998). Mistargeting of the lectin ERGIC-53 to the endoplasmic reticulum of HeLa cells impairs the secretion of a lysosomal enzyme. *J. Cell Biol.* **142,** 377–389.
93. Appenzeller, C., Andersson, H., Kappeler, F., and Hauri, H. P. (1999). The lectin ERGIC-53 is a cargo transport receptor for glycoproteins. *Nat. Cell Biol.* **1,** 330–334.
94. Zhang, B., Cunningham, M. A., Nichols, W. C., Bernat, J. A., Seligsohn, U., Pipe, S. W., McVey, J. H., Schulte-Overberg, U., de Bosch, N. B., Ruiz-Saez, A., White, G. C., Tuddenham, E. G., Kaufman, R. J., and Ginsburg, D. (2003). Bleeding due to disruption of a cargo-specific ER-to-Golgi transport complex. *Nat. Genet.* **34,** 220–225.
95. Nyfeler, B., Reiterer, V., Wendeler, M. W., Stefan, E., Zhang, B., Michnick, S. W., and Hauri, H. P. (2008). Identification of ERGIC-53 as an intracellular transport receptor of alpha1-antitrypsin. *J. Cell. Biol.* **180,** 705–712.
96. Hebert, D. N., and Molinari, M. (2007). In and Out of the ER: Protein Folding, Quality Control, Degradation, and Related Human Diseases. *Physiol. Rev.* **87,** 1377–1408.
97. Ray, M. K., Yang, J., Sundaram, S., and Stanley, P. (1991). A novel glycosylation phenotype expressed by Lec23, a Chinese hamster ovary mutant deficient in alpha-glucosidase I. *J. Biol. Chem.* **266,** 22818–22825.
98. Reitman, M. L., Trowbridge, I. S., and Kornfeld, S. (1982). A lectin-resistant mouse lymphoma cell line is deficient in glucosidase II, a glycoprotein-processing enzyme. *J. Biol. Chem.* **257,** 10357–10363.
99. Denzel, A., Molinari, M., Trigueros, C., Martin, J. E., Velmurgan, S., Brown, S., Stamp, G., and Owen, M. J. (2002). Early postnatal death and motor disorders in mice congenitally deficient in calnexin expression. *Mol. Cell. Biol.* **22,** 7398–7404.
100. Mesaeli, N., Nakamura, K., Zvaritch, E., Dickie, P., Dziak, E., Krause, K. H., Opas, M., MacLennan, D. H., and Michalak, M. (1999). Calreticulin is essential for cardiac development. *J. Cell. Biol.* **144,** 857–868.
101. Garbi, N., Tanaka, S., Momburg, F., and Hammerling, G. J. (2006). Impaired assembly of the major histocompatibility complex class I peptide-loading complex in mice deficient in the oxidoreductase ERp57. *Nat. Immunol.* **7,** 93–102.
102. Molinari, M., Galli, C., Vanoni, O., Arnold, S. M., and Kaufman, R. J. (2005). Persistent Glycoprotein Misfolding Activates the Glucosidase II/UGT1-Driven Calnexin Cycle to Delay Aggregation and Loss of Folding Competence. *Mol. Cell* **20,** 503–512.
103. Hong, Y., Sundaram, S., Shin, D. J., and Stanley, P. (2004). The Lec23 Chinese hamster ovary mutant is a sensitive host for detecting mutations in alpha-glucosidase I that give rise to congenital disorder of glycosylation IIb (CDG IIb). *J. Biol. Chem.* **279,** 49894–49901.
104. Elbein, A. D. (1987). Inhibitors of the biosynthesis and processing of N-linked oligosaccharide chains. *Annu. Rev. Biochem.* **56,** 497–534.
105. Kearse, K. P., Williams, D. B., and Singer, A. (1994). Persistence of glucose residues on core oligosaccharides prevents association of TCR alpha and TCR beta proteins with calnexin and results specifically in accelerated degradation of nascent TCR alpha proteins within the endoplasmic reticulum. *EMBO. J.* **13,** 3678–3686.
106. Ora, A., and Helenius, A. (1995). Calnexin fails to associate with substrate proteins in glucosidase-deficient cell lines. *J. Biol. Chem.* **270,** 26060–26062.
107. Branza-Nichita, N., Petrescu, A. J., Dwek, R. A., Wormald, M. R., Platt, F. M., and Petrescu, S. M. (1999). Tyrosinase folding and copper loading *in vivo*: A crucial role for calnexin and alpha-glucosidase II. *Biochem. Biophys. Res. Commun.* **261,** 720–725.

108. Machold, R. P., and Ploegh, H. L. (1996). Intermediates in the assembly and degradation of class I major histocompatibility complex (MHC) molecules probed with free heavy chain-specific monoclonal antibodies. *J. Exp. Med.* **184**, 2251–2259.
109. Zhang, J. X., Braakman, I., Matlack, K. E., and Helenius, A. (1997). Quality control in the secretory pathway: The role of calreticulin, calnexin and BiP in the retention of glycoproteins with C-terminal truncations. *Mol. Biol. Cell* **8**, 1943–1954.
110. De Praeter, C. M., Gerwig, G. J., Bause, E., Nuytinck, L. K., Vliegenthart, J. F., Breuer, W., Kamerling, J. P., Espeel, J. P., Martin, J. J., De Paepe, A. M., Chan, N. W., Dacremont, G. A. et al. (2000). A novel disorder caused by defective biosynthesis of N-linked oligosaccharides due to glucosidase I deficiency. *Am. J. Hum. Genet.* **66**, 1744–1756.
111. Guo, L., Nakamura, K., Lynch, J., Opas, M., Olson, E. N., Agellon, L. B., and Michalak, M. (2002). Cardiac-specific expression of calcineurin reverses embryonic lethality in calreticulin-deficient mouse. *J. Biol. Chem.* **277**, 50776–50779.
112. Ikawa, M., Wada, I., Kominami, K., Watanabe, D., Toshimori, K., Nishimune, Y., and Okabe, M. (1997). The putative chaperone calmegin is required for sperm fertility. *Nature* **387**, 607–611.
113. Ikawa, M., Nakanishi, T., Yamada, S., Wada, I., Kominami, K., Tanaka, H., Nozaki, M., Nishimune, Y., and Okabe, M. (2001). Calmegin is required for fertilin alpha/beta heterodimerization and sperm fertility. *Dev. Biol.* **240**, 254–261.
114. Gao, B., Adhikari, R., Howarth, M., Nakamura, K., Gold, M. C., Hill, A. B., Knee, R., Michalak, M., and Elliott, T. (2002). Assembly and antigen-presenting function of MHC class I molecules in cells lacking the ER chaperone calreticulin. *Immunity* **16**, 99–109.
115. Molinari, M., Eriksson, K. K., Calanca, V., Galli, C., Cresswell, P., Michalak, M., and Helenius, A. (2004). Contrasting Functions of Calreticulin and Calnexin in Glycoprotein Folding and ER Quality Control. *Mol. Cell* **13**, 125–135.
116. Danilczyk, U. G., Cohen-Doyle, M. F., and Williams, D. B. (2000). Functional relationship between calreticulin, calnexin, and the endoplasmic reticulum luminal domain of calnexin. *J. Biol. Chem.* **275**, 13089–13097.
117. Peterson, J. R., Ora, A., Van, P. N., and Helenius, A. (1995). Transient, lectin-like association of calreticulin with folding intermediates of cellular and viral glycoproteins. *Mol. Biol. Cell* **6**, 1173–1184.
118. Halaban, R., Cheng, E., Zhang, Y., Moellmann, G., Hanlon, D., Michalak, M., Setaluri, V., and Hebert, D. N. (1997). Aberrant retention of tyrosinase in the endoplasmic reticulum mediates accelerated degradation of the enzyme and contributes to the dedifferentiated phenotype of amelanotic melanoma cells. *Proc. Natl. Acad. Sci. USA* **94**, 6210–6215.
119. Keller, S. H., Lindstrom, J., and Taylor, P. (1998). Inhibition of glucose trimming with castanospermine reduces calnexin association and promotes proteasome degradation of the alpha-subunit of the nicotinic acetylcholine receptor. *J. Biol. Chem.* **273**, 17064–17072.
120. Van Leeuwen, J. E. M., and Kearse, K. P. (1996). The related molecular chaperones calnexin and calreticulin differentially associate with nascent T cell antigen receptor proteins within the endoplasmic reticulum. *J. Biol. Chem.* **271**, 25345–25349.
121. Otteken, A., and Moss, B. (1996). Calreticulin interacts with newly synthesized human immunodeficiency virus type 1 envelope glycoprotein, suggesting a chaperone function similar to that of calnexin. *J. Biol. Chem.* **271**, 97–103.
122. Gruters, R. A., Neefjes, J. J., Tersmette, M., de Goede, R. E., Tulp, A., Huisman, H. G., Miedema, F., and Ploegh, H. L. (1987). Interference with HIV-induced syncytium formation and viral infectivity by inhibitors of trimming glucosidase. *Nature* **330**, 74–77.
123. Fischer, P. B., Karlsson, G. B., Butters, T. D., Dwek, R. A., and Platt, F. M. (1996). N-butyldeoxynojirimycin-mediated inhibition of human immunodeficiency virus entry correlates with changes in antibody recognition of the V1/V2 region of gp 120. *J. Virol.* **70**, 7143–7152.

124. Mehta, A., Lu, X., Block, T. M., Blumberg, B. S., and Dwek, R. A. (1997). Hepatitis B virus (HBV) envelope glycoproteins vary drastically in their sensitivity to glycan processing: Evidence that alteration of a single N-linked glycosylation site can regulate HBV secretion. *Proc. Natl. Acad. Sci. USA* **94,** 1822–1827.
125. Mehta, A., Zitzmann, N., Rudd, P. M., Block, T. M., and Dwek, R. A. (1998). Alpha-glucosidase inhibitors as potential broad based anti-viral agents. *FEBS Lett.* **430,** 17–22.
126. Ouzounov, S., Mehta, A., Dwek, R. A., Block, T. M., and Jordan, R. (2002). The combination of interferon alpha-2b and n-butyl deoxynojirimycin has a greater than additive antiviral effect upon production of infectious bovine viral diarrhea virus (BVDV) *in vitro*: Implications for hepatitis C virus (HCV) therapy. *Antiviral Res.* **55,** 425–435.
127. Wu, S. F., Lee, C. J., Liao, C. L., Dwek, R. A., Zitzmann, N., and Lin, Y. L. (2002). Antiviral effects of an iminosugar derivative on flavivirus infections. *J. Virol.* **76,** 3596–3604.
128. Soldà, T., Garbi, N., Hammerling, G. J., and Molinari, M. (2006). Consequences of ERp57 Deletion on Oxidative Folding of Obligate and Facultative Clients of the Calnexin Cycle. *J. Biol. Chem.* **281,** 6219–6226.
129. Hughes, E. A., and Cresswell, P. (1998). The thiol oxidoreductase ERp57 is a component of the MHC class I peptide-loading complex. *Curr. Biol.* **8,** 709–712.
130. Garbi, N., Hammerling, G., and Tanaka, S. (2007). Interaction of ERp57 and tapasin in the generation of MHC class I-peptide complexes. *Curr. Opin. Immunol.* **19,** 99–105.
131. Peaper, D. R., Wearsch, P. A., and Cresswell, P. (2005). Tapasin and ERp57 form a stable disulfide-linked dimer within the MHC class I peptide-loading complex. *EMBO J.* **24,** 3613–3623.
132. Kienast, A., Preuss, M., Winkler, M., and Dick, T. P. (2007). Redox regulation of peptide receptivity of major histocompatibility complex class I molecules by ERp57 and tapasin. *Nat. Immunol.* **8,** 864–872.
133. Wearsch, P. A., and Cresswell, P. (2007). Selective loading of high-affinity peptides onto major histocompatibility complex class I molecules by the tapasin-ERp57 heterodimer. *Nat. Immunol.* **8,** 873–881.
134. Jessop, C. E., Chakravarthi, S., Garbi, N., Hammerling, G. J., Lovell, S., and Bulleid, N. J. (2007). ERp57 is essential for efficient folding of glycoproteins sharing common structural domains. *EMBO J.* **26,** 28–40.
135. Mezghrani, A., Fassio, A., Benham, A., Simmen, T., Braakman, I., and Sitia, R. (2001). Manipulation of oxidative protein folding and PDI redox state in mammalian cells. *EMBO J.* **20,** 6288–6296.
136. Antoniou, A. N., and Powis, S. J. (2003). Characterization of the ERp57-Tapasin complex by rapid cellular acidification and thiol modification. *Antioxid Redox Signal* **5,** 375–379.
137. Jessop, C. E., and Bulleid, N. J. (2004). Glutathione directly reduces an oxidoreductase in the endoplasmic reticulum of mammalian cells. *J. Biol. Chem.* **279,** 55341–55347.
138. Labunskyy, V. M., Hatfield, D. L., and Gladyshev, V. N. (2007). The Sep 15 protein family: Roles in disulfide bond formation and quality control in the endoplasmic reticulum. *IUBMB Life* **59,** 1–5.
139. Soldà, T., Galli, C., Kaufman, R. J., and Molinari, M. (2007). Substrate-specific requirements for UGT1-dependent release from calnexin. *Mol. Cell* **27,** 238–249.
140. Williams, D. B. (2006). Beyond lectins: the calnexin/calreticulin chaperone system of the endoplasmic reticulum. *J. Cell Sci.* **119,** 615–623.
141. Gallione, C. J., and Rose, J. K. (1985). A single amino acid substitution in a hydrophobic domain causes temperature-sensitive cell-surface transport of a mutant viral glycoprotein. *J. Virol.* **54,** 374–382.
142. Wada, I., Kai, M., Imai, S., Sakane, F., and Kanoh, H. (1997). Promotion of transferrin folding by cyclic interactions with calnexin and calreticulin. *EMBO J.* **16,** 5420–5432.

143. Garbi, N., Tan, P., Diehl, A. D., Chambers, B. J., Ljunggren, H. G., Momburg, F., and Hammerling, G. J. (2000). Impaired immune responses and altered peptide repertoire in tapasin-deficient mice. *Nat. Immunol.* **1,** 234–238.
144. Nagai, N., Hosokawa, M., Itohara, S., Adachi, E., Matsushita, T., Hosokawa, N., and Nagata, K. (2000). Embryonic lethality of molecular chaperone hsp47 knockout mice is associated with defects in collagen biosynthesis. *J. Cell. Biol.* **150,** 1499–1506.
145. Nakanishi, T., Isotani, A., Yamaguchi, R., Ikawa, M., Baba, T., Suarez, S. S., and Okabe, M. (2004). Selective passage through the uterotubal junction of sperm from a mixed population produced by chimeras of calmegin-knockout and wild-type male mice. *Biol. Reprod.* **71,** 959–965.
146. Wanderling, S., Simen, B. B., Ostrovsky, O., Ahmed, N. T., Vogen, S. M., Gidalevitz, T., and Argon, Y. (2007). GRP94 is essential for mesoderm induction and muscle development because it regulates insulin-like growth factor secretion. *Mol. Biol. Cell* **18,** 3764–3775.
147. Luo, S., Mao, C., Lee, B., and Lee, A. S. (2006). GRP78/BiP Is Required for Cell Proliferation and Protecting the Inner Cell Mass from Apoptosis during Early Mouse Embryonic Development. *Mol. Cell Biol.* **26,** 5688–5697.
148. Montecucco, C., and Molinari, M. (2006). Microbiology: Death of a chaperone. *Nature* **443,** 511–512.
149. Anelli, T., and Sitia, R. (2008). Protein quality control in the early secretory pathway. *EMBO J.* **27,** 315–327.
150. Sekijima, Y., Wiseman, R. L., Matteson, J., Hammarstrom, P., Miller, S. R., Sawkar, A. R., Balch, W. E., and Kelly, J. W. (2005). The biological and chemical basis for tissue-selective amyloid disease. *Cell* **121,** 73–85.
151. Johnson, S. M., Wiseman, R. L., Sekijima, Y., Green, N. S., Adamski-Werner, S. L., and Kelly, J. W. (2005). Native state kinetic stabilization as a strategy to ameliorate protein misfolding diseases: A focus on the transthyretin amyloidoses. *Acc. Chem. Res.* **38,** 911–921.
152. Jiricny, J. (2006). The multifaceted mismatch-repair system. *Nat. Rev. Mol. Cell Biol.* **7,** 335–346.
153. Branzei, D., and Foiani, M. (2008). Regulation of DNA repair throughout the cell cycle. *Nat. Rev. Mol. Cell Biol.*
154. Kopito, R. R. (1999). Biosynthesis and degradation of CFTR. *Physiol. Rev.* **79,** S167–173.
155. Schubert, U., Anton, L. C., Gibbs, J., Norbury, C. C., Yewdell, J. W., and Bennink, J. R. (2000). Rapid degradation of a large fraction of newly synthesized proteins by proteasomes. *Nature* **404,** 770–774.
156. Vabulas, R. M., and Hartl, F. U. (2005). Protein synthesis upon acute nutrient restriction relies on proteasome function. *Science* **310,** 1960–1963.
157. Molinari, M., and Sitia, R. (2005). The secretory capacity of a cell depends on the efficiency of endoplasmic reticulum-associated degradation. *Curr. Top Microbiol. Immunol.* **300,** 1–15.
158. Moremen, K. W., and Molinari, M. (2006). N-linked glycan recognition and processing: the molecular basis of endoplasmic reticulum quality control. *Curr. Opin. Struct. Biol.*592–599.
159. Molinari, M. (2007). N-glycan structure dictates extension of protein folding or onset of disposal. *Nat. Chem. Biol.* **3,** 313–320.
160. Wiertz, E., Hill, A., Tortorella, D., and Ploegh, H. (1997). Cytomegaloviruses use multiple mechanisms to elude the host immune response. *Immunol. Lett.* **57,** 213–216.
161. Willey, R. L., Maldarelli, F., Martin, M. A., and Strebel, K. (1992). Human immunodeficiency virus type 1 Vpu protein induces rapid degradation of CD4. *J. Virol.* **66,** 7193–7200.
162. Gil, G., Faust, J. R., Chin, D. J., Goldstein, J. L., and Brown, M. S. (1985). Membrane-bound domain of HMG CoA reductase is required for sterol-enhanced degradation of the enzyme. *Cell* **41,** 249–258.

163. Chin, D. J., Gil, G., Faust, J. R., Goldstein, J. L., Brown, M. S., and Luskey, K. L. (1985). Sterols accelerate degradation of hamster 3-hydroxy-3-methylglutaryl coenzyme A reductase encoded by a constitutively expressed cDNA. *Mol. Cell Biol.* **5**, 634–641.
164. Hampton, R. Y. (2002). Proteolysis and sterol regulation. *Annu. Rev. Cell Dev. Biol.* **18**, 345–378.
165. Conn, P. M., Ulloa-Aguirre, A., Ito, J., and Janovick, J. A. (2007). G protein-coupled receptor trafficking in health and disease: Lessons learned to prepare for therapeutic mutant rescue in vivo. *Pharmacol. Rev.* **59**, 225–250.
166. Aridor, M. (2007). Visiting the ER: The endoplasmic reticulum as a target for therapeutics in traffic related diseases. *Adv. Drug Deliv. Rev.* **59**, 759–781.
167. Yoshida, H. (2007). ER stress and diseases. *FEBS J.* **274**, 630–658.
168. Wu, Y., Whitman, I., Molmenti, E., Moore, K., Hippenmeyer, P., and Perlmutter, D. H. (1994). A lag in intracellular degradation of mutant alpha 1-antitrypsin correlates with the liver disease phenotype in homozygous PiZZ alpha 1-antitrypsin deficiency. *Proc. Natl. Acad. Sci. USA* **91**, 9014–9018.
169. Ward, C. L., and Kopito, R. R. (1994). Intracellular turnover of cystic fibrosis transmembrane conductance regulator. Inefficient processing and rapid degradation of wild-type and mutant proteins. *J. Biol. Chem.* **269**, 25710–25718.
170. Liu, Y., Choudhury, P., Cabral, C. M., and Sifers, R. N. (1997). Intracellular disposal of incompletely folded human alpha 1-antitryspin involves release from calnexin and post-translational trimming of asparagine-linked oligosaccharides. *J. Biol. Chem.* **272**, 7946–7951.
171. Wikstrom, L., and Lodish, H. F. (1991). Nonlysosomal, pre-Golgi degradation of unassembled asialoglycoprotein receptor subunits: A TLCK- and TPCK-sensitive cleavage within the ER. *J. Cell. Biol.* **113**, 997–1007.
172. Lippincott-Schwartz, J., Bonifacino, J. S., Yuan, L. C., and Klausner, R. D. (1988). Degradation from the endoplasmic reticulum: Disposing of newly synthesized proteins. *Cell* **54**, 209–220.
173. Knittler, M. R., Dirks, S., and Haas, I. G. (1995). Molecular chaperones involved in protein degradation in the endoplasmic reticulum: Quantitative interaction of the heat shock cognate protein BiP with partially folded immunoglobulin light chains that are degraded in the endoplasmic reticulum. *Proc. Natl. Acad. Sci. USA* **92**, 1764–1768.
174. Okuda-Shimizu, Y., and Hendershot, L. M. (2007). Characterization of an ERAD pathway for nonglycosylated BiP substrates, which require Herp. *Mol. Cell* **28**, 544–554.
175. Wolf, D. H., and Schafer, A. (2005). CPY° and the power of yeast genetics in the elucidation of quality control and associated protein degradation of the endoplasmic reticulum. *Curr. Top. Microbiol. Immunol.* **300**, 41–56.
176. Le, A., Graham, K. S., and Sifers, R. N. (1990). Intracellular degradation of the transport-impaired human PiZ alpha 1-antitrypsin variant. Biochemical mapping of the degradative event among compartments of the secretory pathway. *J. Biol. Chem.* **265**, 14001–14007.
177. Amara, J. F., Lederkremer, G., and Lodish, H. F. (1989). Intracellular degradation of unassembled asialoglycoprotein receptor subunits: A pre-Golgi, nonlysosomal endoproteolytic cleavage. *J. Cell. Biol.* **109**, 3315–3324.
178. Fagioli, C., and Sitia, R. (2001). Glycoprotein quality control in the endoplasmic reticulum. Mannose trimming by endoplasmic reticulum mannosidase I times the proteasomal degradation of unassembled immunoglobulin subunits. *J. Biol. Chem.* **276**, 12885–12892.
179. de Virgilio, M., Kitzmuller, C., Schwaiger, E., Klein, M., Kreibich, G., and Ivessa, N. E. (1999). Degradation of a short-lived glycoprotein from the lumen of the endoplasmic reticulum: The role of N-linked glycans and the unfolded protein response. *Mol. Biol. Cell* **10**, 4059–4073.

180. Mancini, R., Aebi, M., and Helenius, A. (2003). Multiple endoplasmic reticulum-associated pathways degrade mutant yeast carboxypeptidase Y in mammalian cells. *J. Biol. Chem.* **278**, 46895–46905.
181. Cabral, C. M., Liu, Y., and Sifers, R. N. (2001). Dissecting glycoprotein quality control in the secretory pathway. *Trends Biochem. Sci.* **26**, 619–624.
182. Lederkremer, G. Z., and Glickman, M. H. (2005). A window of opportunity: Timing protein degradation by trimming of sugars and ubiquitins. *Trends Biochem. Sci.* **30**, 297–303.
183. Olivari, S., and Molinari, M. (2007). Glycoprotein folding and the role of EDEM1, EDEM2 and EDEM3 in degradation of folding-defective glycoproteins. *FEBS Lett.* **581**, 3658–3664.
184. Spiro, R. G., Zhu, Q., Bhoyroo, V., and Soling, H. D. (1996). Definition of the lectin-like properties of the molecular chaperone, calreticulin, and demonstration of its copurification with endomannosidase from rat liver Golgi. *J. Biol. Chem.* **271**, 11588–11594.
185. Totani, K., Ihara, Y., Matsuo, I., and Ito, Y. (2006). Substrate specificity analysis of endoplasmic reticulum glucosidase II using synthetic high-mannose-type glycans. *J. Biol. Chem.* **281**, 31502–31508.
186. Grinna, L. S., and Robbins, P. W. (1980). Substrate specificities of rat liver microsomal glucosidases which process glycoproteins. *J. Biol. Chem.* **255**, 2255–2258.
187. Sousa, M. C., Ferrero-Garcia, M. A., and Parodi, A. J. (1992). Recognition of the oligosaccharide and protein moieties of glycoproteins by the UDP-Glc:glycoprotein glucosyltransferase. *Biochemistry* **31**, 97–105.
188. Liu, Y., Choudhury, P., Cabral, C. M., and Sifers, R. N. (1999). Oligosaccharide modification in the early secretory pathway directs the selection of a misfolded glycoprotein for degradation by the proteasome. *J. Biol. Chem.* **274**, 5861–5867.
189. Tokunaga, F., Brostrom, C., Koide, T., and Arvan, P. (2000). Endoplasmic reticulum (ER)-associated degradation of misfolded N-linked glycoproteins is suppressed upon inhibition of ER mannosidase I. *J. Biol. Chem.* **275**, 40757–40764.
190. Su, K., Stoller, T., Rocco, J., Zemsky, J., and Green, R. (1993). Pre-Golgi degradation of yeast prepro-alpha-factor expressed in a mammalian cell. Influence of cell type-specific oligosaccharide processing on intracellular fate. *J. Biol. Chem.* **268**, 14301–14309.
191. Fernandez, F. S., Trombetta, S. E., Hellman, U., and Parodi, A. J. (1994). Purification to homogeneity of UDP-glucose:glycoprotein glucosyltransferase from Schizosaccharomyces pombe and apparent absence of the enzyme fro Saccharomyces cerevisiae. *J. Biol. Chem.* **269**, 30701–30706.
192. Jakob, C. A., Burda, P., Roth, J., and Aebi, M. (1998). Degradation of misfolded endoplasmic reticulum glycoproteins in Saccharomyces cerevisiae is determined by a specific oligosaccharide structure. *J. Cell Biol.* **142**, 1223–1233.
193. Foulquier, F., Harduin-Lepers, A., Duvet, S., Marchal, I., Mir, A. M., Delannoy, P., Chirat, F., and Cacan, R. (2002). The unfolded protein response in a dolichyl phosphate mannose-deficient Chinese hamster ovary cell line points out the key role of a demannosylation step in the quality-control mechanism of N-glycoproteins. *Biochem. J.* **362**, 491–498.
194. Kitzmuller, C., Caprini, A., Moore, S. E., Frenoy, J. P., Schwaiger, E., Kellermann, O., Ivessa, N. E., and Ermonval, M. (2003). Processing of N-linked glycans during endoplasmic-reticulum-associated degradation of a short-lived variant of ribophorin I. *Biochem. J.* **376**, 687–696.
195. Foulquier, F., Duvet, S., Klein, A., Mir, A. M., Chirat, F., and Cacan, R. (2004). Endoplasmic reticulum-associated degradation of glycoproteins bearing Man5GlcNAc2 and Man9-GlcNAc2 species in the MI8-5 CHO cell line. *Eur. J. Biochem.* **271**, 398–404.
196. Frenkel, Z., Gregory, W., Kornfeld, S., and Lederkremer, G. Z. (2003). Endoplasmic reticulum-associated degradation of mammalian glycoproteins involves sugar chain trimming to Man6–5GlcNAc2. *J. Biol. Chem.* **278**, 34119–34124.

197. Hosokawa, N., Tremblay, L. O., You, Z., Herscovics, A., Wada, I., and Nagata, K. (2003). Enhancement of endoplasmic reticulum (ER) degradation of misfolded Null Hong Kong alpha1-antitrypsin by human ER mannosidase I. *J. Biol. Chem.* **278**, 26287–26294.
198. Helenius, A., and Aebi, M. (2001). Intracellular functions of N-linked glycans. *Science* **291**, 2364–2369.
199. Cacan, R., Villers, C., Belard, M., Kaiden, A., Krag, S. S., and Verbert, A. (1992). Different fates of the oligosaccharide moieties of lipid intermediates. *Glycobiology* **2**, 127–136.
200. Ermonval, M., Kitzmuller, C., Mir, A. M., Cacan, R., and Ivessa, N. E. (2001). N-glycan structure of a short-lived variant of ribophorin I expressed in the MadIA214 glycosylation-defective cell line reveals the role of a mannosidase that is not ER mannosidase I in the process of glycoprotein degradation. *Glycobiology* **11**, 565–576.
201. Olivari, S., Cali, T., Salo, K. E., Paganetti, P., Ruddock, L. W., and Molinari, M. (2006). EDEM1 regulates ER-associated degradation by accelerating de-mannosylation of folding-defective polypeptides and by inhibiting their covalent aggregation. *Biochem. Biophys. Res. Commun.* **349**, 1278–1284.
202. Herscovics, A., Romero, P. A., and Tremblay, L. O. (2002). The specificity of the yeast and human class I ER alpha 1,2-mannosidases involved in ER quality control is not as strict previously reported. *Glycobiology* **12**, 14G–15G.
203. Avezov, E., Frenkel, Z., Ehrlich, M., Herscovics, A., and Lederkremer, G. Z. (2008). Endoplasmic Reticulum (ER) Mannosidase I Is Compartmentalized and Required for N-Glycan Trimming to Man5 6G1cNAc2 in Glycoprotein ER-associated Degradation. *Mol. Biol. Cell* **19**, 216–225.
204. Romero, P. A., Vallee, F., Howell, P. L., and Herscovics, A. (2000). Mutation of Arg(273) to Leu alters the specificity of the yeast N-glycan processing class I alpha 1,2-mannosidase. *J. Biol. Chem.* **275**, 11071–11074.
205. Lobsanov, Y. D., Vallee, F., Imberty, A., Yoshida, T., Yip, P., Herscovics, A., and Howell, P. L. (2002). Structure of Penicillium citrinum alpha 1,2-mannosidase reveals the basis for differences in specificity of the endoplasmic reticulum and Golgi class I enzymes. *J. Biol. Chem.* **277**, 5620–5630.
206. Karaveg, K., Siriwardena, A., Tempel, W., Liu, Z. J., Glushka, J., Wang, B. C., and Moremen, K. W. (2005). Mechanism of class 1 (glycosylhydrolase family 47) {alpha}-mannosidases involved in N-glycan processing and endoplasmic reticulum quality control. *J. Biol. Chem.* **280**, 16197–16207.
207. Hirao, K., Natsuka, Y., Tamura, T., Wada, I., Morito, D., Natsuka, S., Romero, P., Sleno, B., Tremblay, L. O., Herscovics, A., Nagata, K., and Hosokawa, N. (2006). EDEM3, a soluble EDEM homolog, enhances glycoprotein ERAD and mannose trimming. *J. Biol. Chem.* **281**, 9650–9658.
208. Banerjee, S., Vishwanath, P., Cui, J., Kelleher, D. J., Gilmore, R., Robbins, P. W., and Samuelson, J. (2007). The evolution of N-glycan-dependent endoplasmic reticulum quality control factors for glycoprotein folding and degradation. *Proc. Natl. Acad. Sci. USA* **104**, 11676–11681.
209. Kabani, M., Kelley, S. S., Morrow, M. W., Montgomery, D. L., Sivendran, R., Rose, M. D., Gierasch, L. M., and Brodsky, J. L. (2003). Dependence of endoplasmic reticulum-associated degradation on the peptide binding domain and concentration of BiP. *Mol. Biol. Cell* **14**, 3437–3448.
210. Nishikawa, S., Fewell, S. W., Kato, Y., Brodsky, J. L., and Endo, T. (2001). Molecular chaperones in the yeast endoplasmic reticulum maintain the solubility of proteins for retrotranslocation and degradation. *J. Cell Biol.* **153**, 1061–1070.
211. Tsai, B., Rodighiero, C., Lencer, W. I., and Rapoport, T. A. (2001). Protein disulfide isomerase acts as a redox-dependent chaperone to unfold cholera toxin. *Cell* **104**, 937–948.

212. Gillece, P., Luz, J. M., Lennarz, W. J., de La Cruz, F. J., and Romisch, K. (1999). Export of a cysteine-free misfolded secretory protein from the endoplasmic reticulum for degradation requires interaction with protein disulfide isomerase. *J. Cell Biol.* **147,** 1443–1456.
213. Oda, Y., Okada, T., Yoshida, H., Kaufman, R. J., Nagata, K., and Mori, K. (2006). Derlin-2 and Derlin-3 are regulated by the mammalian unfolded protein response and are required for ER-associated degradation. *J. Cell Biol.* **172,** 383–393.
214. Kokame, K., Agarwala, K. L., Kato, H., and Miyata, T. (2000). Herp, a new ubiquitin-like membrane protein induced by endoplasmic reticulum stress. *J. Biol. Chem.* **275,** 32846–32853.
215. Vashist, S., and Ng, D. T. (2004). Misfolded proteins are sorted by a sequential checkpoint mechanism of ER quality control. *J. Cell Biol.* **165,** 41–52.
216. Ismail, N., and Ng, D. T. (2006). Have you HRD? Understanding ERAD Is DOAble! *Cell* **126,** 237–239.
217. Kostova, Z., Tsai, Y. C., and Weissman, A. M. (2007). Ubiquitin ligases, critical mediators of endoplasmic reticulum-associated degradation. *Semin. Cell Dev. Biol.* **18,** 770–779.
218. Nakatsukasa, K., Huyer, G., Michaelis, S., and Brodsky, J. L. (2008). Dissecting the ER-associated degradation of a misfolded polytopic membrane protein. *Cell* **132,** 101–112.
219. Carvalho, P., Goder, V., and Rapoport, T. A. (2006). Distinct ubiquitin-ligase complexes define convergent pathways for the degradation of ER proteins. *Cell* **126,** 361–373.
220. Li, W., Tu, D., Brunger, A. T., and Ye, Y. (2007). A ubiquitin ligase transfers preformed polyubiquitin chains from a conjugating enzyme to a substrate. *Nature* **446,** 333–337.
221. Denic, V., Quan, E. M., and Weissman, J. S. (2006). A luminal surveillance complex that selects misfolded glycoproteins for ER-associated degradation. *Cell* **126,** 349–359.
222. Gauss, R., Jarosch, E., Sommer, T., and Hirsch, C. (2006). A complex of Yos9p and the HRD ligase integrates endoplasmic reticulum quality control into the degradation machinery. *Nat. Cell Biol.* **8,** 849–854.
223. Yagishita, N., Yamasaki, S., Nishioka, K., and Nakajima, T. (2008). Synoviolin, protein folding and the maintenance of joint homeostasis. *Nat. Clin. Pract. Rheumatol* **4,** 91–97.
224. Fang, S., Ferrone, M., Yang, C., Jensen, J. P., Tiwari, S., and Weissman, A. M. (2001). The tumor autocrine motility factor receptor, gp78, is a ubiquitin protein ligase implicated in degradation from the endoplasmic reticulum. *Proc. Natl. Acad. Sci. USA* **98,** 14422–14427.
225. Hassink, G., Kikkert, M., van Voorden, S., Lee, S. J., Spaapen, R., van Laar, T., Coleman, C. S., Bartee, E., Fruh, K., Chau, V., and Wiertz, E. (2005). TEB4 is a C4HC3 RING finger-containing ubiquitin ligase of the endoplasmic reticulum. *Biochem. J.* **388,** 647–655.
226. Morito, D., Hirao, K., Oda, Y., Hosokawa, N., Tokunaga, F., Cyr, D. M., Tanaka, K., Iwai, K., and Nagata, K. (2008). Gp78 Cooperates with RMA1 in ER-associated Degradation of CFTR {Delta} F508. *Mol. Biol. Cell* **19,** 1328–1336.
227. Imai, Y., Soda, M., Inoue, H., Hattori, N., Mizuno, Y., and Takahashi, R. (2001). An unfolded putative transmembrane polypeptide, which can lead to endoplasmic reticulum stress, is a substrate of Parkin. *Cell* **105,** 891–902.
228. Younger, J. M., Chen, L., Ren, H. Y., Rosser, M. F., Turnbull, E. L., Fan, C. Y., Patterson, C., and Cyr, D. M. (2006). Sequential quality-control checkpoints triage misfolded cystic fibrosis transmembrane conductance regulator. *Cell* **126,** 571–582.
229. Yoshida, Y., Chiba, T., Tokunaga, F., Kawasaki, H., Iwai, K., Suzuki, T., Ito, Y., Matsuoka, K., Yoshida, M., Tanaka, K., and Tai, T. (2002). E3 ubiquitin ligase that recognizes sugar chains. *Nature* **418,** 438–442.
230. Ploegh, H. L. (2007). A lipid-based model for the creation of an escape hatch from the endoplasmic reticulum. *Nature* **448,** 435–438.

231. Wiertz, E. J., Tortorella, D., Bogyo, M., Yu, J., Mothes, W., Jones, T. R., Rapoport, T. A., and Ploegh, H. L. (1996). Sec61-mediated transfer of a membrane protein from the endoplasmic reticulum to the proteasome for destruction. *Nature* **384**, 432–438.
232. Huyer, G., Piluek, W. F., Fansler, Z., Kreft, S. G., Hochstrasser, M., Brodsky, J. L., and Michaelis, S. (2004). Distinct machinery is required in Saccharomyces cerevisiae for the endoplasmic reticulum-associated degradation of a multispanning membrane protein and a soluble luminal protein. *J. Biol. Chem.* **279**, 38369–38378.
233. Pilon, M., Schekman, R., and Romisch, K. (1997). Sec61p mediates export of a misfolded secretory protein from the endoplasmic reticulum to the cytosol for degradation. *EMBO J.* **16**, 4540–4548.
234. Plemper, R. K., Bohmler, S., Bordallo, J., Sommer, T., and Wolf, D. H. (1997). Mutant analysis links the translocon and BiP to retrograde protein transport for ER degradation. *Nature* **388**, 891–895.
235. Walter, J., Urban, J., Volkwein, C., and Sommer, T. (2001). Sec61p-independent degradation of the tail-anchored ER membrane protein Ubc6p. *EMBO J.* **20**, 3124–3131.
236. Simpson, J. C., Roberts, L. M., Romisch, K., Davey, J., Wolf, D. H., and Lord, J. M. (1999). Ricin A chain utilises the endoplasmic reticulum-associated protein degradation pathway to enter the cytosol of yeast. *FEBS Lett.* **459**, 80–84.
237. Kalies, K. U., Allan, S., Sergeyenko, T., Kroger, H., and Romisch, K. (2005). The protein translocation channel binds proteasomes to the endoplasmic reticulum membrane. *EMBO J.* **24**, 2284–2293.
238. Lilley, B. N., and Ploegh, H. L. (2004). A membrane protein required for dislocation of misfolded proteins from the ER. *Nature* **429**, 834–840.
239. Lilley, B. N., and Ploegh, H. L. (2005). Multiprotein complexes that link dislocation, ubiquitination, and extraction of misfolded proteins from the endoplasmic reticulum membrane. *Proc. Natl. Acad. Sci. USA* **102**, 14296–14301.
240. Ye, Y., Shibata, Y., Kikkert, M., van Voorden, S., Wiertz, E., and Rapoport, T. A. (2005). Inaugural Article: Recruitment of the p97 ATPase and ubiquitin ligases to the site of retrotranslocation at the endoplasmic reticulum membrane. *Proc. Natl. Acad. Sci. USA* **102**, 14132–14138.
241. Ye, Y., Shibata, Y., Yun, C., Ron, D., and Rapoport, T. A. (2004). A membrane protein complex mediates retro-translocation from the ER lumen into the cytosol. *Nature* **429**, 841–847.
242. Buschhorn, B. A., Kostova, Z., Medicherla, B., and Wolf, D. H. (2004). A genome-wide screen identifies Yos9p as essential for ER-associated degradation of glycoproteins. *FEBS Lett.* **577**, 422–426.
243. Bhamidipati, A., Denic, V., Quan, E. M., and Weissman, J. S. (2005). Exploration of the topological requirements of ERAD identifies Yos9p as a lectin sensor of misfolded glycoproteins in the ER lumen. *Mol. Cell* **19**, 741–751.
244. Kim, W., Spear, E. D., and Ng, D. T. (2005). Yos9p detects and targets misfolded glycoproteins for ER-associated degradation. *Mol. Cell* **19**, 753–764.
245. Szathmary, R., Bielmann, R., Nita-Lazar, M., Burda, P., and Jakob, C. A. (2005). Yos9 protein is essential for degradation of misfolded glycoproteins and may function as lectin in ERAD. *Mol. Cell* **19**, 765–775.
246. Litovchick, L., Friedmann, E., and Shaltiel, S. (2002). A selective interaction between OS-9 and the carboxyl-terminal tail of meprin beta. *J. Biol. Chem.* **277**, 34413–34423.
247. Baek, J. H., Mahon, P. C., Oh, J., Kelly, B., Krishnamachary, B., Pearson, M., Chan, D. A., Giaccia, A. J., and Semenza, G. L. (2005). OS-9 interacts with hypoxia-inducible factor 1alpha and prolyl hydroxylases to promote oxygen-dependent degradation of HIF-1 alpha. *Mol. Cell* **17**, 503–512.

248. Wang, Y., Fu, X., Gaiser, S., Kottgen, M., Kramer-Zucker, A., Walz, G., and Wegierski, T. (2007). OS-9 Regulates the Transit and Polyubiquitination of TRPV4 in the Endoplasmic Reticulum. *J. Biol. Chem.* **282**, 36561–36570.
249. Christianson, J. C., Shaler, T. A., Tyler, R. E., and Kopito, R. R. (2008). OS-9 and GRP94 deliver mutant alpha 1-antitrypsin to the Hrd1?SEL1L ubiquitin ligase complex for ERAD *Nat. Cell Biol.* **10**, 272–282.
250. Kaneko, M., Yasui, S., Niinuma, Y., Arai, K., Omura, T., Okuma, Y., and Nomura, Y. (2007). A different pathway in the endoplasmic reticulum stress-induced expression of human HRD1 and SEL1 genes. *FEBS Lett.* **581**, 5355–5360.
251. Shintani, T., and Klionsky, D. J. (2004). Autophagy in health and disease: a double-edged sword. *Science* **306**, 990–995.
252. Cuervo, A. M. (2004). Autophagy: In sickness and in health. *Trends Cell Biol.* **14**, 70–77.
253. Mizushima, N. (2005). The pleiotropic role of autophagy: From protein metabolism to bactericide. *Cell Death Differ* **12**(Suppl 2), 1535–1541.
254. Yoshimori, T. (2007). Autophagy: Paying Charon's Toll. *Cell* **128**, 833–836.
255. Yorimitsu, T., Nair, U., Yang, Z., and Klionsky, D. J. (2006). Endoplasmic reticulum stress triggers autophagy. *J. Biol. Chem.* **281**, 30299–30304.
256. Ogata, M., Hino, S., Saito, A., Morikawa, K., Kondo, S., Kanemoto, S., Murakami, T., Taniguchi, M., Tanii, I., Yoshinaga, K., Shiosaka, S., Hammarback, J. A. *et al.* (2006). Autophagy is activated for cell survival after endoplasmic reticulum stress. *Mol. Cell Biol.* **26**, 9220–9231.
257. Bernales, S., McDonald, K. L., and Walter, P. (2006). Autophagy counterbalances endoplasmic reticulum expansion during the unfolded protein response. *PLoS Biol.* **4**, e423.
258. Mizushima, N., Yamamoto, A., Hatano, M., Kobayashi, Y., Kabeya, Y., Suzuki, K., Tokuhisa, T., Ohsumi, Y., and Yoshimori, T. (2001). Dissection of autophagosome formation using Apg5-deficient mouse embryonic stem cells. *J. Cell Biol.* **152**, 657–668.
259. Hara, T., Nakamura, K., Matsui, M., Yamamoto, A., Nakahara, Y., Suzuki-Migishima, R., Yokoyama, M., Mishima, K., Saito, I., Okano, H., and Mizushima, N. (2006). Suppression of basal autophagy in neural cells causes neurodegenerative disease in mice. *Nature* **441**, 885–889.
260. Komatsu, M., Waguri, S., Chiba, T., Murata, S., Iwata, J., Tanida, I., Ueno, T., Koike, M., Uchiyama, Y., Kominami, E., and Tanaka, K. (2006). Loss of autophagy in the central nervous system causes neurodegeneration in mice. *Nature* **441**, 880–884.
261. Kamimoto, T., Shoji, S., Hidvegi, T., Mizushima, N., Umebayashi, K., Perlmutter, D. H., and Yoshimori, T. (2006). Intracellular inclusions containing mutant alpha1-antitrypsin Z are propagated in the absence of autophagic activity. *J. Biol. Chem.* **281**, 4467–4476.
262. Kouroku, Y., Fujita, E., Tanida, I., Ueno, T., Isoai, A., Kumagai, H., Ogawa, S., Kaufman, R. J., Kominami, E., and Momoi, T. (2007). ER stress (PERK/eIF2alpha phosphorylation) mediates the polyglutamine-induced LC3 conversion, an essential step for autophagy formation. *Cell Death Differ* **14**, 230–239.
263. Fujita, E., Kouroku, Y., Isoai, A., Kumagai, H., Misutani, A., Matsuda, C., Hayashi, Y. K., and Momoi, T. (2007). Two endoplasmic reticulum-associated degradation (ERAD) systems for the novel variant of the mutant dysferlin: Ubiquitin/proteasome ERAD(I) and autophagy/lysosome ERAD(II). *Hum. Mol. Genet.* **16**, 618–629.
264. Yorimitsu, T., and Klionsky, D. J. (2007). Eating the endoplasmic reticulum: Quality control by autophagy. *Trends Cell Biol.* **17**, 279–285.
265. Klionsky, D. J. (2007). Autophagy: From phenomenology to molecular understanding in less than a decade. *Nat. Rev. Mol. Cell Biol.* **8**, 931–937.
266. Ruddock, L. W., and Molinari, M. (2006). N-glycan processing in ER quality control. *J. Cell Sci.* **119**, 4373–4380.

267. Meusser, B., Hirsch, C., Jarosch, E., and Sommer, T. (2005). ERAD: The long road to destruction. *Nat. Cell Biol.* **7**, 766–772.
268. Romisch, K. (2005). Endoplasmic reticulum-associated degradation. *Annu. Rev. Cell Dev. Biol.* **21**, 435–456.
269. Klausner, R. D., and Sitia, R. (1990). Protein degradation in the endoplasmic reticulum. *Cell* **62**, 611–614.
270. Wu, Y., Termine, D. J., Swulius, M. T., Moremen, K. W., and Sifers, R. N. (2007). Human endoplasmic reticulum mannosidase I is subject to regulated proteolysis. *J. Biol. Chem.* **282**, 4841–4849.
271. Zuber, C., Cormier, J. H., Guhl, B., Santimaria, R., Hebert, D. N., and Roth, J. (2007). EDEMI reveals a quality control vesicular transport pathway out of the endoplasmic reticulum not involving the COPII exit sites. *Proc. Natl. Acad. Sci. USA* **104**, 4407–4412.
272. Brodsky, J. L. (2007). The protective and destructive roles played by molecular chaperones during ERAD (endoplasmic-reticulum-associated degradation). *Biochem. J.* **404**, 353–363.
273. Wiseman, R. L., Powers, E. T., Buxbaum, J. N., Kelly, J. W., and Balch, W. E. (2007). An adaptable standard for protein export from the endoplasmic reticulum. *Cell* **131**, 809–821.
274. Brightman, S. E., Blatch, G. L., and Zetter, B. R. (1995). Isolation of a mouse cDNA encoding MTJ1, a new murine member of the DnaJ family of proteins. *Gene* **153**, 249–254.
275. Chevalier, M., Rhee, H., Elguindi, E. C., and Blond, S. Y. (2000). Interaction of murine BiP/GRP78 with the DnaJ homologue MTJ1. *J. Biol. Chem.* **275**, 19620–19627.
276. Skowronek, M. H., Rotter, M., and Haas, I. G. (1999). Molecular characterization of a novel mammalian DnaJ-like Sec63p homolog. *Biol. Chem.* **380**, 1133–1138.
277. Bies, C., Guth, S., Janoschek, K., Nastainczyk, W., Volkmer, J., and Zimmermann, R. (1999). A Scj 1p homolog and folding catalysts present in dog pancreas microsomes. *Biol. Chem.* **380**, 1175–1182.
278. Yu, M., Haslam, R. H., and Haslam, D. B. (2000). HEDJ, an Hsp40 co-chaperone localized to the endoplasmic reticulum of human cells. *J. Biol. Chem.* **275**, 24984–24992.
279. Lau, P. P., Villanueva, H., Kobayashi, K., Nakamuta, M., Chang, B. H., and Chan, L. (2001). A DnaJ protein, apobec-1-binding protein-2, modulates apolipoprotein B mRNA editing. *J. Biol. Chem.* **276**, 46445–46452.
280. Prols, F., Mayer, M. P., Renner, O., Czarnecki, P. G., Ast, M., Gassler, C., Wilting, J., Kurz, H., and Christ, B. (2001). Upregulation of the cochaperone Mdg 1 in endothelial cells is induced by stress and during *in vitro* angiogenesis. *Exp. Cell Res.* **269**, 42–53.
281. Shen, Y., and Hendershot, L. M. (2005). ERdj3, a stress-inducible endoplasmic reticulum DnaJ homologue, serves as a cofactor for BiP's interactions with unfolded substrates. *Mol. Biol. Cell* **16**, 40–50.
282. Shen, Y., Meunier, L., and Hendershot, L. M. (2002). Identification and characterization of a novel endoplasmic reticulum (ER) DnaJ homologue, which stimulates ATPase activity of BiP *in vitro* and is induced by ER stress. *J. Biol. Chem.* **277**, 15947–15956.
283. Cunnea, P. M., Miranda-Vizuete, A., Bertoli, G., Simmen, T., Damdimopoulos, A. E., Hermann, S., Leinonen, S., Huikko, M. P., Gustafsson, J. A., Sitia, R., and Spyrou, G. (2003). ERdj5, an endoplasmic reticulum (ER)-resident protein containing DnaJ and thioredoxin domains, is expressed in secretory cells or following ER stress. *J. Biol. Chem.* **278**, 1059–1066.
284. Hosoda, A., Kimata, Y., Tsuru, A., and Kohno, K. (2003). JPDI, a novel endoplasmic reticulum-resident protein containing both a BiP-interacting J-domain and thioredoxin-like motifs. *J. Biol. Chem.* **278**, 2669–2676.
285. Chung, K. T., Shen, Y., and Hendershot, L. M. (2002). BAP, a mammalian BiP-associated protein, is a nucleotide exchange factor that regulates the ATPase activity of BiP. *J. Biol. Chem.* **277**, 47557–47563.

286. Weitzmann, A., Volkmer, J., and Zimmermann, R. (2006). The nucleotide exchange factor activity of Grp 170 may explain the non-lethal phenotype of loss of Sill function in man and mouse. *FEBS Lett.* **580**, 5237–5240.
287. Steel, G. J., Fullerton, D. M., Tyson, J. R., and Stirling, C. J. (2004). Coordinated activation of Hsp70 chaperones. *Science* **303**, 98–101.
288. Smith, T., Ferreira, L. R., Hebert, C., Norris, K., and Sauk, J. J. (1995). Hsp47 and cyclophilin B traverse the endoplasmic reticulum with procollagen into pre-Golgi intermediate vesicles. A role for Hsp47 and cyclophilin B in the export of procollagen from the endoplasmic reticulum. *J. Biol. Chem.* **270**, 18323–18328.
289. Bush, K. T., Hendrickson, B. A., and Nigam, S. K. (1994). Induction of the FK506-binding protein, FKBP13, under conditions which misfold proteins in the endoplasmic reticulum. *Biochem. J.* **303**(Pt 3), 705–708.
290. Zhang, X., Wang, Y., Li, H., Zhang, W., Wu, D., and Mi, H. (2004). The mouse FKBP23 binds to BiP in ER and the binding of C-terminal domain is interrelated with Ca2 + concentration. *FEBS Lett.* **559**, 57–60.
291. Wang, Y., Han, R., Wu, D., Li, J., Chen, C., Ma, H., and Mi, H. (2007). The binding of FKBP23 to BiP modulates BiP's ATPase activity with its PPIase activity. *Biochem. Biophys. Res. Commun.* **354**, 315–320.
292. Davis, E. C., Broekelmann, T. J., Ozawa, Y., and Mecham, R. P. (1998). Identification of tropoelastin as a ligand for the 65-kD FK506-binding protein, FKBP65, in the secretory pathway. *J. Cell Biol.* **140**, 295–303.
293. Calì, T., Galli, C., Olivari, S., and Molinari, M. (2008). Segregation and Rapid Turnover of EDEM1 Modulates Standard ERAD and Folding Activities. *Biochem. Biophys. Res. Commun.* **371**, 405–410.
294. Bernasconi, R., Pertel, T., Luban, J., and Molinari, M. (2008). A Dual Task for the Xbp1-Responsive OS-9 Variants in the Mammalian ER: Inhibiting Secretion of Misfolded Protein Conformers and Enhancing Their Disposal. *J. Biol. Chem.* **283**, 16446–16454..

All-Atom Protein Folding with Free-Energy Forcefields

A. Verma[*], S. M. Gopal[†],
A. Schug[¶], T. Herges[‡],
K. Klenin[‡], and W. Wenzel[‡]

[*]*Forschungszentrum Karlsruhe, Institute for Scientific Computing, Karlsruhe, Germany*

[†]*Desp. of Biochemistry, Michigan State University, East Lansing, MI, USA*

[‡]*Forschungszentrum Karlsruhe, Institute for Nanotechnology, Karlsruhe, Germany*

[¶]Center for Theoretical Biological Physics, UC San Diego, CA, USA

I. Introduction	182
II. Protein Structure and Folding	186
A. Protein Composition	186
B. Protein Structure	190
C. Dominant Forces in Protein Folding	193
D. The Protein-Folding Problem	195
E. Early Models for Protein Folding	197
III. Kinetic Folding Methods	199
A. Molecular Dynamics	199
B. Forcefields	203
C. Protein-Folding Simulations	205
IV. Free-Energy Forcefields and Simulation Methods	205
A. Anfinsen's Hypothesis	205
B. A Biophysical All-Atom Free-Energy Forcefield	206
C. Stochastic Simulation Methods	209
V. Free-Energy Protein Folding	216
A. Helical Proteins	216
B. Hairpins	224
C. Three-Stranded Sheet (GSGS Peptide)	236
D. Folding of a DNA-Binding Zinc Finger Motif	237
VI. Summary	242
VII. Outlook	246
References	247

"If you want to understand function, study structure."
(Francis Crick, Nobel Prize in Medicine, 1962)

I. Introduction

Proteins are the workhorses of all cellular life. They constitute the building blocks and the machinery of all cells. DNA carries the genetic information which encodes the production of protein molecules. To produce a protein, the corresponding gene is first transcribed into mRNA and then translated into a polypeptide chain of amino acids in the ribosome.

Proteins perform a variety of roles in the cell: structural proteins constitute the building blocks for cells and tissues, enzymes, like pepsin, catalyze complex reactions, signaling proteins, like insulin, transfer signals between or within the cells. Transport proteins, like hemoglobin, carry small molecules or ions, while receptor proteins like rhodopsin generate response to stimuli. The mechanisms of all these biophysical processes depend on the precise folding of their respective polypeptide chains.

From the work of C. B. Anfinsen and coworkers in the 1960s, we now know that the amino acid sequence of a polypeptide chain in the appropriate physiological environment can fully determine its folding into a so-called native conformation. Unlike man-made polymers of similar length, functional proteins assume unique three-dimensional structures under physiological conditions and there must be rules governing this sequence-to-structure transition. Protein structures can be determined experimentally, by X-ray crystallography or NMR methods, but these experiments are still challenging and do not work for all proteins. From the theoretical standpoint, it is still not possible to reliably predict the native three-dimensional conformation of most proteins given their amino acid sequence alone.

The triplet genetic code by which the DNA sequence determines the amino acid sequence of polypeptide chains is well understood. However, unfolded polypeptide chains lack most of the properties needed for their biological function. The chain must fold into its native three-dimensional conformation in order to perform its function. Despite much research in this direction and the emergence of novel folding paradigms during the last decade, much of the mechanism by which the protein performs this auto-induced folding reaction is still unclear.

To perform their biological function polypeptide chains interact with their aqueous or lipid environment to fold into discrete, highly organized three-dimensional structures. Because of great advancement in sequencing techniques for proteins and nucleotides compared to structure determination methods, the number of known protein structures lags far behind the number of known sequences. Various genome projects have rapidly increased the number of known sequences. Entire genomes are reported for the human, the mouse, the chicken, the fruit fly, and many fungi. Currently, over one million protein sequences are known, compared to about 40,000 structures deposited in the Protein Data Bank (the worldwide database of protein structures). Reliable theoretical methods for protein structure prediction could help to reduce this gap between sequence and structural databases and elucidate the biological information in structurally unresolved sequences.

Therefore it would be very helpful to develop methods for protein structure prediction on the basis of the amino acid sequence alone. Even if this goal it is not fully realized, methods that can complete partially resolved experimental protein structures would be very helpful to determine the structure of proteins where neither theoretical methods nor experimental techniques alone can succeed. For the transmembrane family of proteins, present day experimental methods fail, which are responsible for the entire communication of the cell with its environment. Theoretical methods would be very helpful to investigate these proteins.

There are large number of related questions, for instance, regarding the interactions of a given protein with a large variety of other proteins, where theoretical methods could also contribute to our understanding of biological function. Protein–protein interactions govern the cell signaling processes and are very important for the assembly of large protein structures in the cell. Because it is known that proteins change their shape upon binding to other proteins, the structure of the isolated constituents is only an approximation to the structure found in the complex in which the proteins ultimately function. In order to address these questions, it is important to develop accurate atomistic models for protein structure prediction. To use a protein structure for emerging methods of computer aided drug design, the resolution of the protein structure must be below 1 Å. In order to predict the binding sites or interacting complex of two proteins, a resolution between 3 and 5 Å is desirable.

Related to the question of protein structure prediction is the question of how the proteins attain their final conformation—the so-called protein-folding problem. It remains one of the astonishing mysteries responsible for the evolution of life how these complex molecules can attain a unique native conformation with such precision. No man-made polymer of similar size is able to assemble into a predetermined structure with the precision encountered in the proteins that have evolved in nature.

Given its complexity, it is not surprising that the protein-folding process occasionally fails, and many of such failures are related to cellular dysfunction or disease. Therefore it is important not only to be able to predict the final structure of proteins but also very desirable to understand the mechanisms by which proteins fold.

Experiments of C. B. Anfinsen and coworkers showed convincingly that many proteins can indeed adopt their native conformation spontaneously, that is, sequence determines structure. This led to the "thermodynamic hypothesis" which states that the native three-dimensional structure of a native protein in its normal physiological milieu (solvent, pH, ionic strength, presence of other components such as metal ions or prosthetic groups, temperature, etc.) minimizes the Gibbs free energy of the whole system. The native conformation is determined by the totality of interatomic interactions and hence by the amino acid sequence in a given environment. This led to the "Levinthal paradox" which suggested that there must be pathways for protein folding, as a simple protein with a 100 amino acids is estimated to have a vast configurational space of the order of 2^{100} ($\sim 10^{30}$) possible conformations. Unless there is a specific mechanism, such a protein will need more than the age of the universe to locate its global free-energy minimum in an exhaustive search of this configurational space. For this reason, Levinthal stipulated that there must be a specific, multistep-folding reaction that leads to the native conformation. Unfortunately, only very few of the proposed multitude of intermediates in this folding path were ever found. Instead, a family of small proteins was detected that folds in a two-state fashion, in which no discernable folding intermediates exist at all. Protein folding in this scenario appears as a single reaction between folded and unfolded state of the protein with only one intervening energy barrier. For some proteins even barrierless folding was observed.

Levinthal's paradox can be resolved by the funnel paradigm of protein folding. In this paradigm, proteins are generally thought to have globally funneled energy landscapes with a small gradient directed toward the native state. This "folding funnel" landscape allows the protein to fold to the native state through any of a large number of pathways and intermediates, rather than being restricted to a single mechanism. Single molecule experiments, such as atomic force microscopy and optical tweezers have confirmed the existence of such funnels in protein folding.

The ultimate, very long-range goal of protein structure theory would be the development of methods to design proteins for a specific function. This would be very helpful for medical purposes and technological applications in nanobiology, but will require an understanding of various factors that influence the folding of the polypeptide and their sequence determinants. It is currently possible to modify existing proteins and also to generate a variety of hybrids. But the ability to design completely new proteins to carry out novel functions requires a much more profound understanding of how sequences determine folding.

Many theories and increasingly computational methods have been developed to understand the folding process. Simplified models have been applied to understand its physical principles. Lattice-based methods were among the first models that allowed efficient sampling of conformational space. The lattice models, either 2D square or 3D cubic, were used to study protein folding and unfolding, but they were too simplified for protein structure prediction. Subsequently "Gō models" were developed, where only native contacts interact favorably, and were useful to characterize some aspects of the folding of small proteins. The success of these models is limited by the fact that all residues interact in the same way. Further development led to statistically obtained knowledge-based potentials. These potentials were obtained and parameterized on the structures available from the Protein Data Bank. The knowledge-based potentials are mostly used for fold recognition or protein structure prediction.

With the increase in computational resources and speed, all-atom molecular dynamics (MD) simulations of protein folding have been undertaken. For most proteins, it is still not feasible to determine the protein structure from extended conformations using a single MD simulation. This is due to the fact that at the all-atom level, the typical time step in a MD simulation is about 1 fs while the protein folding occurs at a millisecond timescale. A single simulation would need years to complete. Replica exchange molecular dynamics (REMD) simulations have been successful in folding proteins from extended conformations, but are still limited to the size of 20–30 amino acids.

In this review, we explore an alternate approach for protein structure predication and folding that is based on the Anfinsen's hypothesis that most proteins are in thermodynamic equilibrium with their environment in their native state. For proteins of this class, the native conformation corresponds to the global optimum of the free energy of the protein. We know from many problems in physics and chemistry that the global optimum of a complex energy landscape can be obtained with high efficiency using stochastic optimization methods. These methods map the folding process found in nature onto a fictitious dynamical process that explores the free-energy surface of the protein. By construction, these fictitious dynamical processes not only find the conformation of lowest energy, but typically characterize the entire low-energy ensemble of competing metastable states. Since the total free-energy change for protein folding under physiological conditions is small, often only a few kcal mol^{-1}, a characterization of the low-energy ensemble of thermodynamically accessible protein conformations may be sufficient not only to predict the structure of the protein, but also to characterize the folding process. There are two important ingredients to this approach: first we need an accurate atomically resolved free-energy forcefield for proteins. Second we need a set of simulation methods that can reliably explore and characterize the low-energy ensemble of protein conformations.

This review is structured as follows: The next section introduces protein composition and structure and the problem of protein folding. Section II deals with kinetic biomolecular simulations that have attempted to elucidate protein folding for more than three decades. In Section III, we describe the fundamentals of the free-energy approach, the protein free-energy forcefield PFF02, and methods to efficiently explore the protein free-energy surface with stochastic simulation methods. In Section IV, we review all-atom folding simulations for various proteins with the free-energy approach. The key results of these investigations and opportunities for further work are outlined in the last section.

II. Protein Structure and Folding

The proteins we observe in nature have evolved to perform specific functions, such as catalyzing various reactions and carrying ions or other small molecules to various parts of the body. The functional property of a protein depends upon its three-dimensional structure. Under physiological conditions, a particular sequence of amino acids in a polypeptide chain folds into a compact three-dimensional structure. This three-dimensional structure, due to the specific properties, makes a protein perform a specific biological function. These single chains, which are folded into a respective three-dimensional structure, can still assemble together to form more complex functional units.

To understand the biological function of a protein, one needs to measure or predict its three-dimensional structure from its amino acid sequence. This prediction problem is still unsolved and remains one of the most basic challenges in biophysical chemistry (1, 2).

A. Protein Composition

1. AMINO ACIDS

The basic monomeric unit of a protein is an amino acid. There are 20 naturally occurring amino acids. All of the 20 amino acids have a central carbon atom (C_α), to which are attached a hydrogen atom, an amino group (NH_2), and a carboxyl group (COOH). The side chain which is attached as the fourth valency to the C_α differentiates the various amino acids. There are 20 different naturally occurring amino acids specified by the genetic code (there are very rare occurrences of some other amino acids in proteins). The 20 different side chains that occur in natural proteins are shown in Fig. 1. Their names are commonly abbreviated with either a three-letter code or a one-letter code.

Amino acids are linked together by the formation of peptide bonds to form a chain. A peptide bond is formed when the carboxyl group of the first amino acid reacts with the amino group of the next to eliminate water, as shown in Fig. 2. This process is repeated until the whole protein chain is synthesized. At the ends of the polypeptide chain, the amino group of the first amino acid and the carboxyl group of the last amino acid still remain intact. Thus the chain is generally referred as to run from amino (N)-terminus to carboxy (C)-terminus.

The formation of a succession of peptide bonds generates a "main chain" or "backbone" from which various "side chains" project outward.

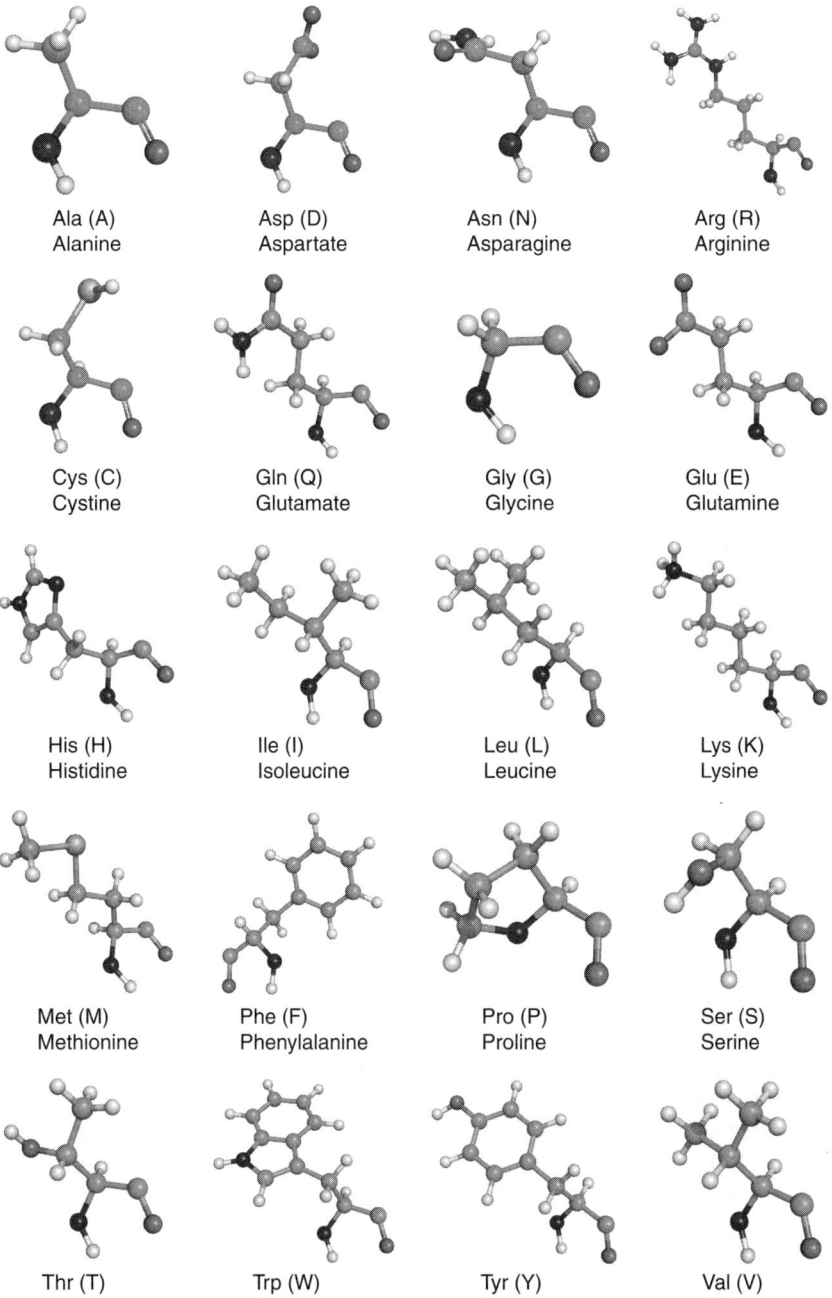

Fig. 1. Twenty naturally occurring amino acids: their structure, name, three-letter code and one-letter code. The structures are color coded with carbon (green), nitrogen (blue), oxygen (red), hydrogen (white), and sulfur (orange). (See Color Insert.)

$$H_3{}^+N-\underset{\underset{H}{|}}{\overset{\overset{R_1}{|}}{C}}-\overset{O}{\overset{\|}{C}}-OH + H-\underset{H}{\overset{H}{N}}-\underset{\underset{H}{|}}{\overset{\overset{R_2}{|}}{C}}-\overset{O}{\overset{\|}{C}}-O^- \xrightarrow{H_2O} H_3{}^+N-\underset{\underset{H}{|}}{\overset{\overset{R_1}{|}}{C}}-\overset{O}{\overset{\|}{C}}-\underset{H}{\overset{H}{N}}-\underset{\underset{H}{|}}{\overset{\overset{R_2}{|}}{C}}-\overset{O}{\overset{\|}{C}}-O^-$$

FIG. 2. Formation of a peptide bond.

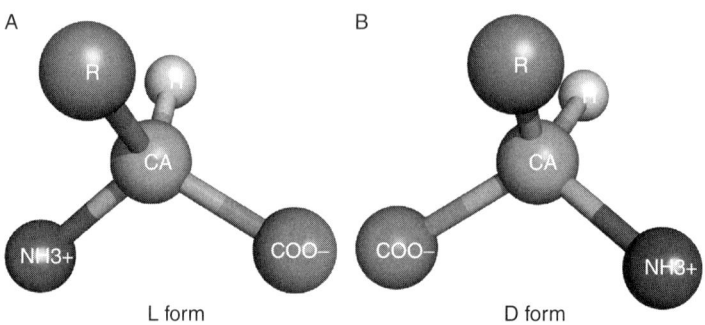

FIG. 3. The L- and D-chiral forms of amino acids. R (magenta) represents the side chain. (See Color Insert.)

The main chain atoms of a polypeptide chain are a carbon atom C_α to which the side chains is attached, a NH group bound to C_α, and a carbonyl group C=O, where the carbon atom C is attached to C_α. These units are called residues and are linked into a polypeptide chain by peptide bonds between the C atom of one residue and the nitrogen atom of the next. The basic repeating unit along the main chain is thus (NH–C_αH–CO), which is the residue of the common parts of amino acids after peptide bonds have been formed.

At the fourth valency of C_α is the side chain and depending upon the chemical structure of the side chain, the amino acids are divided into three different classes (1). The first class comprises those with strictly hydrophobic side chains Ala(A), Val(V), Leu(L), Ile(I), Phe(F), Pro(P), and Met(M). The second class includes four charged residues Asp(D), Glu(E), Lys(K), and Arg(R) and the third class comprises those with polar side chains Ser(S), Thr(T), Cys(C), Asn (N), Gln(Q), His(H), Tyr(Y), and Trp(W). The amino acid glycine (G) has only a hydrogen atom as the side chain and thus is the simplest of all the 20 amino acids. The amino acid proline (P) is also different from the rest as it is the only amino acid where both ends of the side chain are covalently bound to the main chain.

All amino acids (except glycine) are chiral molecules which can exist in two different forms with different hands, L- or D-form (see Fig. 3). Biological systems depend on specific detailed recognition of molecules involving differentiation between chiral forms. Amino acids are found in only one of the chiral forms, the L-form, during protein synthesis. There is, however, no obvious reason why the L-form was chosen during the evolution and not the D-form (3, 4).

2. THE POLYPEPTIDE CHAIN

The linkage of amino acids produces a polypeptide chain, with the backbone atoms linked through the peptide bond which does not change in its chemical structure during folding. The folding pattern of the polypeptide chain can be described in terms of angles of internal rotation around the bonds in the main chain. The bonds in the polypeptide backbone between N and C_α and between the C_α and C, are single bonds. Internal rotations around these bonds are not restricted by the electronic structure of the bond, but only by possible steric collisions in the conformations produced. In contrast, the peptide bond itself has a partial double bond character, with restricted internal rotation (5). This means that the NH and CO along with the two C_αs always remain in a peptide plane (see yellow regions in Fig. 4).

The peptide group can occur in both *cis* and *trans* forms, with the *trans* isomer being the more stable. For all the amino acids except proline, the energy difference between *cis* and *trans* states is very large (6). For proline, the energy difference is only about 1.2 kcal mol^{-1} (5). Taking a section of three peptide units having the sequences *trans–trans–trans* and *trans–cis–trans*, conformational energy calculations indicate that the latter can occur only to an extent of 0.1%, unless there occurs the sequence X-Pro, in which case it is of the order of 30%. This explains the extreme rarity of *cis* peptide units in proteins. As a result, virtually all the *cis* peptides in proteins appear between a proline and the residue preceding it in the chain, however, it follows that even with nonprolyl residues, *cis* peptide units are not forbidden, but can occur in some rare examples (6).

As most residues in proteins have *trans* peptide bonds, the main chain conformation of each residue is determined by two angles, commonly named as ϕ and ψ. The dihedral angle around the bond N–C_α is known as ϕ and the dihedral angle around the bond C_α–C is known as ψ. As ϕ involves a previous amino acid and ψ involves the next, the first amino acid and the last amino acid in the polypeptide chain have only one angle of rotation (ψ and ϕ, respectively). The angles of rotation are shown in Fig. 5B. Many combinations of ϕ and ψ produce sterically disallowed conformations. V. Sasisekharan, C. Ramakrishnan,

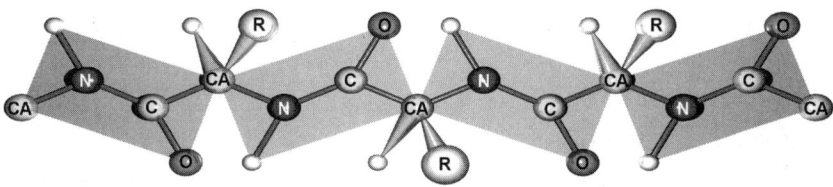

FIG. 4. Peptide planes of a polypeptide chain. R represents the side chain and white spheres are hydrogens.

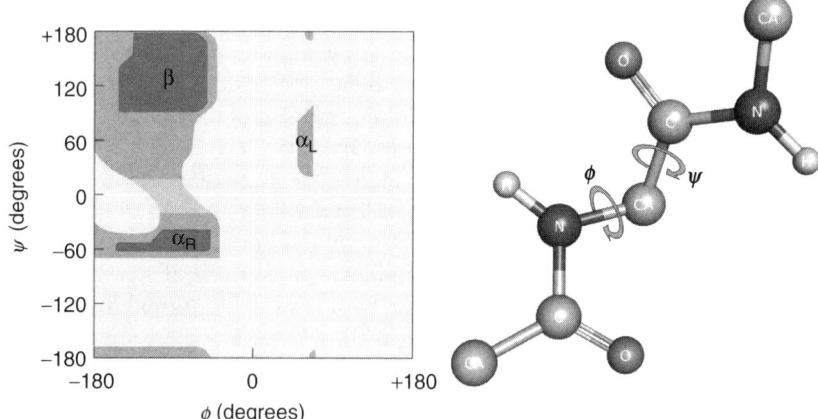

FIG. 5. Sasisekharan–Ramakrishnan–Ramachandran plot (the figure for Ramachandran plot was taken from the internet from http://www.bifi.unizar.es/). (See Color Insert.)

and G. N. Ramachandran first plotted the "allowed" regions in a graph of ϕ and ψ (7). The plot is generally known as the Ramachandran plot, shown in Fig. 5A. There are two main allowed regions, one around $\phi = -57°$, $\psi = -47°$ (denoted α_R) and the around $\phi = -125°$, $\psi = +125°$ (denoted β) with a neck like region between them. The mirror image of α_R, denoted α_L, is allowed equally for glycine residues only because glycine is achiral.

The two major allowed regions correspond to the two major types of secondary structures found in proteins, helix and sheet. A continuous stretch of residues, with all conformations in the α_R region, would form a right-handed helix (a helix formed by a stretch of residues in the α_L region would form the corresponding left-handed helix). In the β region, the chain is nearly fully extended. A continuous stretch of residues, with all the conformations in the β region, would form a single strand of a sheet (both helix and sheet are discussed in detail in the next section). The conformations that correspond to low-energy states of individual residues also permit the formation of structures with extensive main chain hydrogen bonding. The two effects thereby cooperate to lower the energy of the native state.

B. Protein Structure

Proteins are made up of unique sequences of the 20 naturally occurring amino acids. The protein structures are classified into four categories depending upon the amount of information known.

1. Primary Structure

Primary structure describes the sequence of amino acids starting from amino (N)-terminus to carbonyl (C)-terminus. The primary sequence is written in either the one-letter code or three-letter code, for example, SWTWEGNKWTWK or SER–TRP–THR–TRP–GLU–GLY–ASN–LYS–TRP–THR–TRP–LYS (three-letter code and one-letter code as described in Fig. 1) for the tryptophan zipper protein (PDB code: 1LE0).

2. Secondary Structure

Due to the "allowed" regions of the Ramachandran plot, polypeptide chains fold themselves into regularly repeating structures. In 1951, L. Pauling and R. Corey proposed two periodic structures called the α-helix and the β-pleated sheet. Later, other structures such as the β-turn and Ω-loop were also identified. Although not periodic, these common turn or loop structures were well defined and contribute along with α-helices and β-sheets to form the final protein structure:

- *α-Helix*. The α-helix is a spring like structure where tightly coiled backbone forms the inner part of the helix and the side chains project outward in a helical array (see Fig. 6A). The α-helix is stabilized by hydrogen bonds between the NH and CO groups of the main chain. In particular, the CO group of each amino acid forms a hydrogen bond with the NH group of the amino acid which is situated four residues ahead in sequence. Thus, except the amino acids near the ends of an α-helix, all the

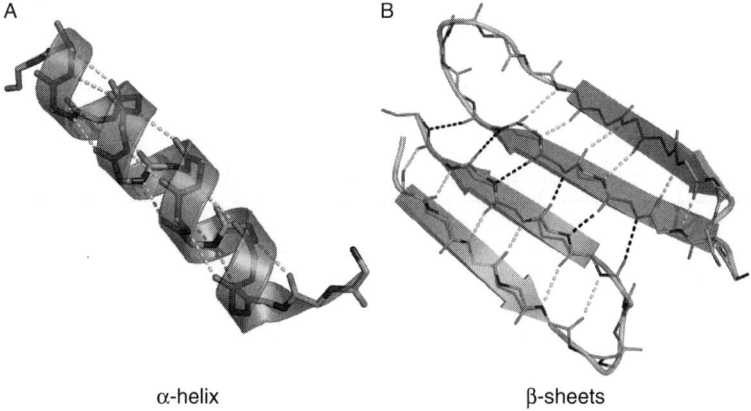

FIG. 6. Secondary structural elements. Dashed lines indicate the presence of hydrogen bonds. In (b), yellow/black bonds are shown for antiparallel/parallel β-sheets, respectively. (See Color Insert.)

main chain CO and NH groups are hydrogen bonded. Each residue is related to the next one by a rise of 1.5 Å along the helix axis and a rotation of 100°, which gives 3.6 amino acid residues per turn of helix. The screw sense of a helix can be right handed or left handed. The Ramachandran plot reveals that both the right-handed and left-handed helices are among the allowed conformations. However, right-handed helices are energetically more favorable because there is less steric clash between the side chains and the backbone. Essentially all α-helices found in proteins are right handed. There are also other types of helices, such as a 3_{10}-helix, a π-helix and polyproline II helix. The ideal parameters of these are given in Table I.

- *β-Sheet*. The β-sheet differs remarkably from the spring like α-helix. A polypeptide chain, called a β-strand, in a β-sheet is almost fully extended rather than being tightly coiled as in a helix. The distance between adjacent amino acids along a β-strand is approximately 3.5 Å in contrast with a distance of 1.5 Å along an α-helix (5). The side chains of adjacent amino acids point in opposite directions. A β-sheet is formed by linking two or more β-strands by hydrogen bonds. Adjacent chains in a β-sheet can run in opposite directions (antiparallel β-sheet) or in the same direction (parallel β-sheet), shown in Fig. 6b. In the antiparallel arrangement, the NH group and the CO group of each amino acid are, respectively, bonded to the CO and NH group of a partner on the adjacent chain. In the parallel arrangement, for each amino acid, the NH group is hydrogen bonded to the CO group of one amino acid on the adjacent strand, whereas the CO group is hydrogen bonded to the NH group on the amino acid two residues further along the chain. Hydrogen bonding of parallel and antiparallel β-strands are shown in Fig. 6b, in black and yellow, respectively. The ideal parameters are given in Table I. Many strands come together to form β-sheets with minimum being two for a β-hairpin and as many as 10 in β-barrel proteins. Such β-sheets can be purely antiparallel, purely parallel, or mixed.

TABLE I
STRUCTURAL PARAMETERS FOR PROTEIN SECONDARY STRUCTURES

Structure	ϕ	ψ	n	d (Å)
α-Helix	−57	−47	3.6	1.5
3_{10}-Helix	−49	−26	3.0	2.0
π-Helix	−57	−70	4.4	1.1
Polyproline II helix	−79	149	3.0	3.1
Parallel β-strand	−119	113	2.0	3.2
Antiparallel β-strand	−139	135	2.0	3.4

ϕ and ψ are the conformational angles of the main chain, n is the number of residues per turn, and d is the displacement between successive residues along the axis.

3. Tertiary Structure

The tertiary structure is formed by the assembly of secondary structural elements along with turns and loops into a three-dimensional arrangement. The tertiary structure mainly has a hydrophobic core with charged residues on the surface of protein. The charged residues on the surface, give the protein its biological activity and are thus responsible for its biological function.

4. Quaternary Structure

Tertiary structures of proteins (independent folding chains) can still assemble themselves under physiological conditions in order to perform specific functions. These are termed as quaternary structure. For example, four identical chains come together to form the hemoglobin complex. Figure 7 shows all four kinds of structure for a gene regulating protein (PDB code: 5CRO).

C. Dominant Forces in Protein Folding

During folding, different sets of residues come in proximity of each other in different possible conformations of the same polypeptide chain. The interactions of side chains and main chain, with one another and with the solvent and with

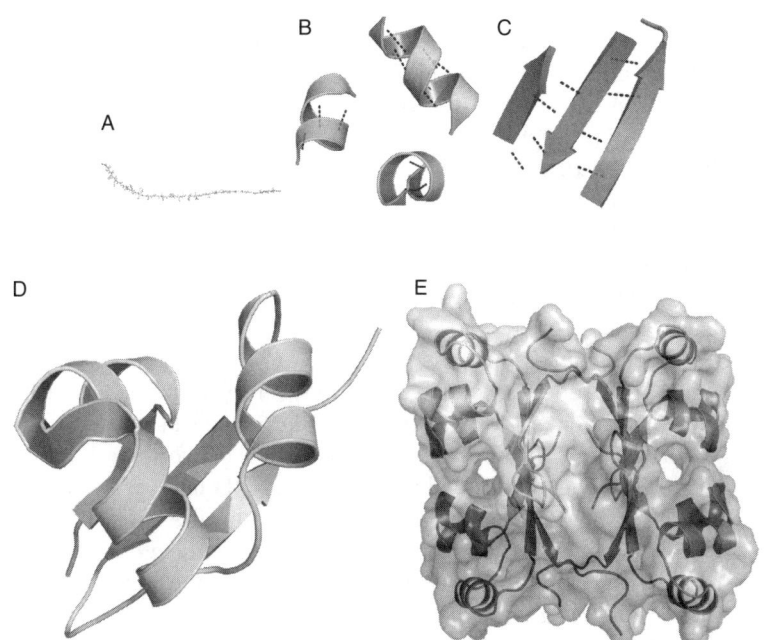

FIG. 7. Protein structure classifications: (A) the primary structure, (B) and (C) the secondary structural elements, helices and sheets, respectively, (D) the tertiary structure, and (E) the quaternary structure of a protein. (See Color Insert.)

other surrounding proteins or ligands, determine the energy of the conformation. Proteins have evolved so that one folding arrangement of the backbone and its side chain produces a set of interactions that is significantly more favorable than all other possible conformations. This conformation is called the native state of a protein. The experiments of C. B. Anfinsen and coworkers showed that for many proteins, the protein structure is determined by the amino acid sequence alone.

Formation of the native state is a global property of a protein. In most cases, the entire protein (or at least a large part) is necessary for stability. This is because many of the stabilizing interactions involve parts of the protein that are very distant along the polypeptide chain, but brought into spatial proximity by the folding process.

Proteins are only marginally stable, and achieve stability only within narrow ranges of conditions of solvent and temperature. Outside of these regions, proteins lose their definite compact structure, and even their helices and sheets, and take up states with disorder in the backbone conformation and interactions among residues (8, 9).

Protein structures are stabilized by a variety of chemical interactions for their stability and for their affinity and specificity for ligands:

1. *Covalent and coordinate chemical bonds*. Some proteins contain covalent chemical bonds between side chains. These covalent bonds such as disulfide bridges between cystine residues are quite common and these sets of cystine residues "lock" the polypeptide chain together.
2. *Hydrogen bonding*. Certain groups in proteins can form hydrogen bonds with water or other protein groups. The main chain has one H-bond donor (N–H) and H-bond acceptor (C=O) for each amino acid. In addition, some polar side chains can form hydrogen bonds. The main chain, containing peptide groups, must pass through the interior, and some polar side chains are also buried. They, thereby, lose their interactions with water. To recover the energy, buried polar atoms form protein–protein hydrogen bonds. The standard secondary structures, helices and sheets, are achieved by the formation of hydrogen bonds by the main chain atoms.
3. *Hydrophobic effect*. For proteins to take their native states in the aqueous environment, hydrophobic residues bury themselves in the interior and charged residues come on the surface. The accessible surface area of the protein, calculated from a set of atomic coordinates, measures the thermodynamic interaction between the protein and water.
4. *van der Waals forces and dense packing of protein interiors*. The packing of atoms in protein interiors contributes in two ways to the stability of structure. One is the exclusion of hydrophobic atoms from contact with

water. The other is the dispersive attraction between the protein atoms. The cohesion of ordinary substances shows the existence of attractive forces between atoms and molecules. At matter does not collapse, there must be limits to how far it can be compressed. This observation leads to the presence of repulsive forces at short range. The most general type of interatomic force, the van der Waals force, reflects this principle: the nearer the atoms, the stronger the attractive force, until the atoms are in contact, at which the forces become repulsive and strong. To maximize the total cohesive force, therefore, as many atoms as possible must be brought as close together as possible. It is the requirement for a dense packing that imposes a requirement for structure in the interior of a protein. It produces a fit of the elements of secondary structure packed together in protein interiors.

D. The Protein-Folding Problem

In his pioneering work, C. B. Anfinsen showed that the necessary information for the polypeptide chain to fold into its native structure is contained in its sequence of amino acids. Protein refolding especially demonstrated that the native conformation of many proteins is reproducibly formed even when the proteins are in isolation. This observation can be explained, if the native state is lower in free energy than all other conformations. This observation led to the thermodynamic hypothesis (10) that the native state is the global minimum in the free energy. The stability of each possible conformation of a polypeptide chain depends on the free-energy change between native and unfolded states given by equation

$$\Delta G = \Delta H - T\Delta S, \tag{1}$$

where ΔG, ΔH, and ΔS are the differences in free energy, enthalpy, and entropy, respectively, of the native and unfolded conformation. The enthalpic difference is the difference associated with atomic interactions (electrostatic interactions, van der Waals potentials, hydrogen bonding) whereas the entropy term describes hydrophobic interactions, thereby including the dominant interactions in protein folding, namely, the hydrophobic effect, hydrogen bonding, and configurational entropy. The free energy of stabilization of proteins under ordinary conditions is typically only a few kcal mol^{-1} (5, 11, 12) and slight changes in the surrounding conditions can force a protein to adopt a completely different conformation.

In an unfolded protein, the polypeptide chain can adopt different rotameric positions around ϕ and ψ torsional angels, and side chain can adopt different rotamers around their dihedral angels. When folded, the ϕ and ψ dihedral angles of the polypeptide chain are nearly restricted to a narrow range of values, as are majority of χ angles. This loss of freedom translates into a loss of configurational entropy. This loss of configurational entropy must be overcome by favorable interactions, such as hydrogen bonding, increase in solvent entropy, etc., in order to fold a polypeptide chain into a stable conformation (13–16).

While the experiments by C. B. Anfinsen and coworkers demonstrated that many proteins can adopt their native conformation spontaneously, it immediately raised a fundamental problem known as Levinthal's paradox (17). Anfinsen's experiments suggested that the native state of a protein is thermodynamically the most stable state under biological conditions. But a polypeptide chain has enormous number of possible conformations (at least 2^{100} for a 100 amino acid protein considering are only two possible conformations per amino acid). If one estimates that each state is reached in 1 ps from a related conformation, such a chain would take $\sim 2^{100}$ ps (considering 1 ps per conformation) or $\sim 10^{10}$ years (even more than the estimated age of universe) to sample all possible conformations and to find the lowest energy state. Levinthal, thus, concluded that a specific folding pathway must exist and that protein folding is under kinetic control rather than thermodynamic control.

This issue can be resolved by considering a balance between kinetics and thermodynamics in an energy landscape perspective. According to the energy landscape paradigm, the free-energy landscape has a small gradient in all conformations toward the native state. Even in the absence of a unique folding pathway, the protein dynamics is guided toward the native state. Projected to low dimension, the free-energy surface, thus, has a funnel like slope. The landscape perspective explains the process of reaching a global minimum in free energy (satisfying Anfinsen's experiments) and doing so quickly (satisfying Levinthal's concerns) by multiple folding routes on funnel-like energy landscapes (18) because the new view recognizes that "folding pathways" are not the correct solution to the kinetic problem Levinthal posed. The funnel theory includes ruggedness on the funnel surface (see Fig. 8). The main idea is that while the folding landscape resembles a funnel globally but is to some extent rugged locally, that is, with traps in which the protein can be trapped along the folding pathway. The funnel guides the protein through many different sequences of traps toward the low-energy folded (native) structure. Here, there is no pathway but a multiplicity of folding routes. For small proteins, discrete pathways emerge only late in the folding process when much of the protein has almost reached the native ensemble. The simple parts of the folding process, where most of the real molecular organization is going on, occur in the early events of folding and can be described using a few parameters statistically characterizing the protein-folding funnel (19, 20).

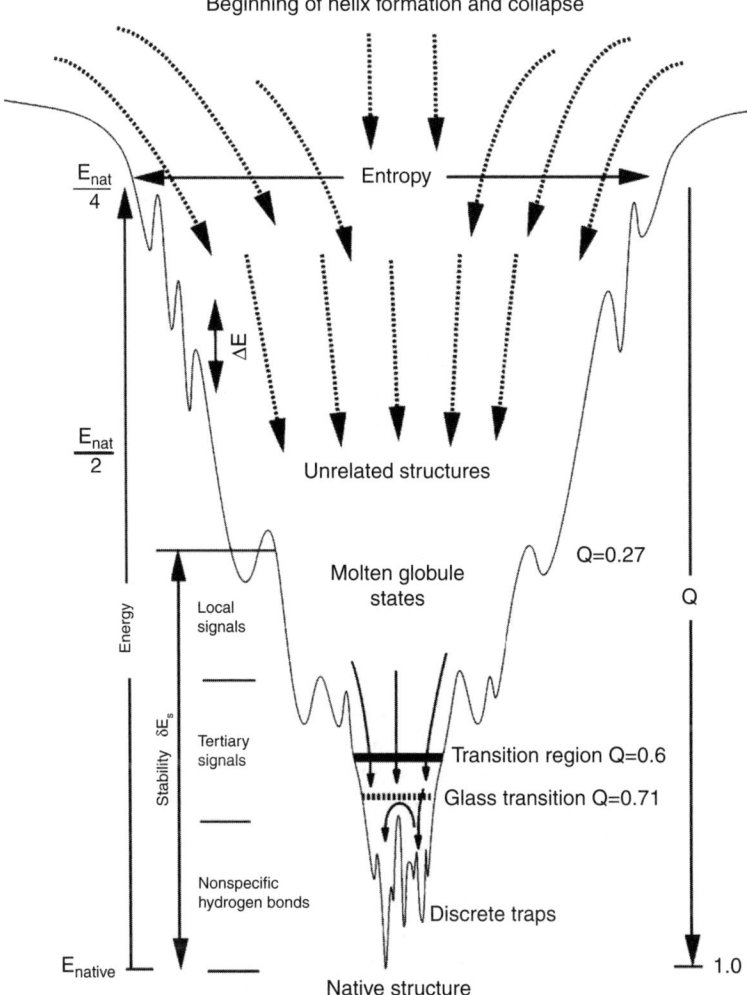

FIG. 8. Schematic representation of the protein-folding landscape (taken from (19)).

E. Early Models for Protein Folding

1. Gō Models

The Gō model (21) is one of the earlier models developed for the study of protein folding. The model employs a minimalistic protein representation (like the c_α trace). The Gō models b minimize the roughness of the free-energy surface by biasing the energy landscape to fold into a given three-dimensional structure. This native interaction biased free-energy landscape avoids the problem of actual physical interactions responsible for the biased surface. The Gō interactions are generally sequence dependent. Advanced Gō models have included sequence independent terms and explicit treatment of solvent interactions (22). Since a protein is foldable by definition in the Gō model, the folding characteristics can be extracted more comfortably compared to MD simulations. They reproduce qualitatively differences between folding kinetics of small and large proteins and explain folding rate and mechanism of folding (23).

2. COARSE-GRAINED MODELS

Protein structure modeling can be performed at various levels of structural detail ranging from simplified two-dimensional or three-dimensional lattice models, continuous representations, and united residue models or with all-atom models. The reduced representations offer several advantages over all-atom models. The number of degrees of freedom in the system is reduced enabling more extensive sampling of the energy landscape. Folding studies of larger proteins, which are not accessible to all-atom MD simulations can be realized with these models. The important drawbacks of coarse-grained models are the lack of accurate reduced representations and the difficulty to design adequate potentials to represent real proteins.

One prototypical example for a coarse-grained model is the UNRES model (24–26) is one of the forcefields for coarse grain protein modeling. It represents a polypeptide chain as sequence of c_α atoms linked by virtual bonds of length 3.8 Å. The interaction sites representing the peptide groups (PG) are located in middle of c_α virtual bonds. The side chains (SC) are represented as a single interaction site attached to the c_α and the c_α...SC bond lengths are fixed. The UNRES forcefield has been derived by averaging the system consisting of protein plus solvent with implementation of Kubo's theory of cluster cumulants. The energy function is parameterized to achieve a hierarchical structure of protein energy landscape. The forcefield contains terms which account for the interactions between PG–PG, SC–PG, SC–SC, local terms that account for rotation about c_α...c_α virtual bond axis, the bending of virtual valence angles, and different rotameric states of SC and multibody correlation terms.

The global minimum is searched by using a hybrid optimization method, where a genetic algorithm generates the populations by a set of crossover operations, followed by a local minimization (27). The method is successful for small proteins (up to 80 residues). For larger proteins, a complicated crossover operators involving nonlocal pattern exchanges (e.g., nonlocal β-sheets), dynamic formation/breaking of disulfide bonds is employed. The conformational space is searched by either starting with a random conformation or with conformation generated using secondary structure predictions.

The UNRES forcefield has been used for blind protein structure predictions (27). Other applications include protein-folding studies using the molecular dynamics and Monte Carlo methods (28).

III. Kinetic Folding Methods

Biomolecular systems are characterized by a large degree of flexibility and the movement of their atoms is correlated and essential for biological function. Protein biological function frequently involves large amplitude motions, called conformational change, resulting in change in the geometry of the molecule. As these changes occur on many different timescales, different strategies are required to answer various different questions. For example, to determine dynamical properties and to understand the processes at nanoseconds to microsecond timescale, kinetic simulation methods can be used. In contrast, stochastic methods are better suited to treat problems of long timescales, such as protein tertiary structure prediction and folding. There are two main approaches in performing molecular simulations: the stochastic approach described below and the kinetic MD method. In this section, we summarize the kinetic approach, in particular the MD method, which has been used for over 30 years to simulate a wide array of systems.

A. Molecular Dynamics

MD was first introduced to study very simple systems, such as gases or fluids of hard spheres. However the methods was extended to almost all conceivable molecular systems, including Lennard–Jones models, liquid water, molten salts and glasses, biomolecules such as proteins, lipids, nucleic acids, etc. (29). MD methods are intuitively appealing and involve mostly classical regime physics.

The MD simulations solve Newton's equations of motion for a system of interacting particles. The equations of motion are given by

$$m_i \frac{d^2 \vec{r}_i}{dt^2} = \vec{F}_i, \qquad (2)$$

where m_i is the mass of particle i, $\vec{r}_i$ is the position vector of ith particle, and $\vec{F}_i$ is the force acting on i by surrounding particles. Forces exerted by the surrounding particles on the ith particle are given by

$$\vec{F}_i = -\sum_j \nabla \phi(r_{ij}), \qquad (3)$$

where $\phi(r_{ij})$ is the corresponding pairwise potential function between particle i and j.

Generation of a MD trajectory involves three essential stages:

1. *Initialization.* For biomolecular simulations, the initial coordinates are often taken from experimental structures. The initial velocity vector is adjusted such that kinetic energy of the system corresponds to the expected temperature of system. The individual particle velocities are drawn from a Gaussian distribution with zero mean and variance, $k_B T_o/m_i$, where k_B is the Boltzmann constant, m_i is the mass of the ith particle, and T_o is the system temperature.
2. *Equilibration.* Starting with assigned coordinates and velocities, the system is evolved for a certain *equilibration* time. During the equilibration, energy exchange between kinetic and potential components takes place. The equations of motion are solved until the total energy the system is equilibrated.
3. *Production.* The properties of the system are measured after the equilibration phase. The typical properties accessible via MD simulations are thermodynamic quantities such as energy, pressure, entropy, specific heat, etc. Structural and transport properties such as radial distribution functions, diffusion coefficient, dynamic structure factors, etc.

In MD simulations, the atoms of the biomolecular system are represented as a collection of point masses (centered at the nuclei) which subject to intra- and intermolecular forces. Each atom has a partial charge, which reflects its molecular environment. A Lennard–Jones radius defines the spatial extent of each atom. The interactions of the atoms are defined by a classical potential energy function, which is commonly referred to as forcefield. The forcefields can be purely *ab initio* (from first principles), empirical, or knowledge based (derived from a distribution).

Some of the standard forcefields presently used for biomolecular simulations are:

AMBER	Assisted Model Building with Energy Refinement (30)
CHARMM	Chemistry at Harvard Macromolecular Mechanics (31)
ECEPP	Empirical Conformational Energy Program for Peptides (32)
ESFF	Extensible Systematic Forcefield (33)
GROMOS	Groningen Molecular Simulation (34)
OPLS	Optimized Potentials for Liquid Simulations (35)
CFF	Consistent Forcefield (36)

In the following, we summarize the key algorithmic ingredients of the molecular dynamics approach.

1. INTEGRATION

Numerical integrators are used to solve the equation of motion. The simplest one among them is Verlet integrator which is accurate to $O(t^4)$:

$$\vec{r}(t + \delta t) = 2\vec{r}(t) - \vec{r}(t - \delta t) + \delta t^2 \vec{a}(t). \tag{4}$$

A slightly modified version of Verlet algorithm, known as Verlet–leapfrog integrator is generally used to minimize the numerical errors associated with the Verlet algorithm (29):

$$\vec{r}(t + \delta t) = \vec{r}(t) + \vec{v}\left(t + \frac{1}{2}\delta t\right)\delta t,$$

$$\vec{v}\left(t + \frac{1}{2}\delta t\right) = \vec{v}\left(t - \frac{1}{2}\delta t\right) + \vec{a}(t)\delta t, \tag{5}$$

$$\vec{v}(t) = \frac{1}{2}\left(\vec{v}\left(t + \frac{\delta t}{2}\right) + \vec{v}\left(t - \frac{\delta t}{2}\right)\right).$$

The time step δt depends on properties which are to be measured in the system. For example, in the folding/unfolding studies a δt of 2 fs is used, whereas δt is set to 1 fs for calculating the spectral properties. When a large δt is used, the high frequency bond stretching is prevented by constrained dynamics. The algorithms SHAKE and RATTLE are commonly used for the rigid bonds (37).

2. THERMOSTATS

The temperature of the system computed via the Virial theorem fluctuates as the system is evolved. Therefore, a temperature control is generally applied to the system. One of simplest method for this purpose is the temperature rescaling (29). If T_A is the instantaneous temperature of the system and T_o is the required temperature, the rescaling factor $f_i(t)$ for velocities of particles is defined as

$$f_i(t) = \sqrt{\frac{3k_B T_o/m_i}{\sum_j v_i(t)^2/N_i}}, \tag{6}$$

where $\sqrt{3k_B T_o/m_i}$ is the mean velocity of particle i with mass m_i, T_o is the desired temperature, N_i is the number of particle of type i, and $\sqrt{\sum_j u_i(t)^2/N_i}$ is the computed mean velocity of particle i for temperature T_A.

The temperature and pressure of the system can be monitored conveniently by using external weak coupling thermostats/barostats. These thermostats/barostats use a fictitious frictional coefficient γ which controls the relaxation rate of the coupling. The commonly used thermostats for constant temperature or pressure MD include Berendsen, Nose–Hoover, Anderson thermostats (38). An alternative method uses the elements of Langevin's dynamics for the thermostat.

3. Long-Range Interactions

The evaluation of nonbonded interactions is of order $O(N^2)$, where N is the number of atoms. To reduce the computational cost, these interactions are evaluated using three alternative schemes:

1. *Cutoff*. The cutoff scheme involves a predefined bounding distance after which the long-range interactions are no longer evaluated. Switching functions are used to alter original function to smoothly fade to zero at cutoff distance. This scheme is used for the van der Waals interactions.
2. *Ewald's summation*. This is applicable only to systems which use periodical boundary conditions. The $1/r$ term in the electrostatics potential is decomposed to into a short-range interaction in real space and long-range interaction in reciprocal space (38). The electrostatics of the system is evaluated completely with such schemes. The variant of Ewald's summation, particle-mesh Ewald (PME) is used often in MD (39). The computation of overall interactions is reduced to $O(N \log N)$ with best implementation of PME.
3. *Multipole expansions*. These schemes rely on a power-series expansion for the electrostatic interaction. The multipoles are expressed in terms of spherical coordinates (r, θ, α), since the expansion involve spherical harmonics $Y_{lm}(\theta, \alpha)$ (38). The multipole expansion scheme can be applied to both periodic and nonperiodic systems. In MD, this scheme is realized by the fast multipole method (FMM) (39). The computation of overall interactions is reduced to $O(N)$ with the best implementation of FMM.

4. Enhanced Sampling

The sampling of the phase space by conventional MD is usually limited. In addition the system can get trapped in local minima. Several schemes are used to overcome these problems. One convenient scheme is REMD (40). In this scheme, identical noninteracting copies of the system are simulated at different temperatures. A swap of copies at different temperatures is attempted, such that all replica remain in thermal equilibrium:

$$W(m \to n) = \begin{cases} 1 \\ \exp(-\Delta) \end{cases}, \quad \Delta = [\beta_m - \beta_n](U_m - U_n), \beta_m = \frac{1}{K_B T_m}. \quad (7)$$

where $W(m \to n)$ is the probability of swaps between temperature m and n, respectively, U_m and U_n are potential energies. During the swap, the high-temperature replicas escape from local traps or jump from one energy basin to another, whereas the low-temperature replicas explore a single region of the energy landscape like conventional MD. In other methods such as local enhanced sampling (LES) (41), a small fragment of the system (a side chain or a ligand molecule) is duplicated to N noninteracting copies. The remainder of system is made to interact at reduced strength (1/N). This facilitates the significant increase of sampling for these fragments.

B. Forcefields

The potential energy function is the core component of a MD methodology. The empirical forcefield-based MD studies are most commonly used for biomolecules including proteins, nucleic acids, and lipids. The success of the empirical forcefield lies in the fact that they are able to reproduce experimentally accessible information, their simplistic functional form, and the efficient algorithms which have enabled the microsecond/millisecond MD for small systems (about 5000 atoms) and nanosecond simulations for very large systems (about 100,000 atoms).

Equation (8) shows a potential energy function which is the most common form for protein forcefields (42). It is composed of simple functions which represent a minimal set of forces to describe a biomolecular structure:

$$\begin{aligned}
U(\vec{R}) &= \sum_{\text{bonds}} K_b (b - b_0)^2 + \sum_{\text{angles}} K_\theta (\theta - \theta_0)^2 \\
&+ \sum_{\text{dihedrals}} K_\chi [1 + \cos(n\chi - \delta)] \sum_{\text{impropers}} K_{\text{imp}} (\phi - \phi)^2 \\
&+ \sum_{\text{nonbonded}} \left(\varepsilon_{ij} \left[\left(\frac{R_{\text{min},ij}}{r_{ij}} \right)^{12} - \left(\frac{R_{\text{min},ij}}{r_{ij}} \right)^6 \right] + \frac{q_i q_j}{\varepsilon_D r_{ij}} \right),
\end{aligned} \quad (8)$$

where b is the bond length, θ is the valence angle, χ is the dihedral angle, φ is the improper angle, and r_{ij} is the distance between atoms i and j. The parameters which constitute the actual forcefield are bond force constant and equilibrium distance, K_b and b_o, respectively; the valence angle force constant and equilibrium angle, θ_b and θ_o, respectively; the dihedral force constant, multiplicity, and phase angle, K_χ, n, and phase angle δ, respectively; K_φ and φ_o

represent the improper force constant and equilibrium improper angle, respectively. All the above terms constitute the internal/intramolecular parameters. The nonbonded interactions are characterized by Lennard–Jones well depth ε_{ij}, minimum interaction radius $R_{\min,ij}$ for van der Waals interactions; the dielectric constant ε and partial charges of atom i and j, q_i and q_j for the Coulombic interactions. The dielectric constant ε is usually set to 1, which corresponds to the permittivity of vacuum, in the calculations that incorporate explicit solvent representations. The alternative methods which treat the solvent environment implicitly have distance dependent dielectric constants. The terms contributing to the potential energy in Eq. (8) are common to majority of popular empirical forcefields such as CHARMM (31), AMBER (43), GROMOS (44), and OPLS (45). Some extended forcefield include higher order terms to treat the bond and valence angle terms, cross terms between bonds and valence angles or valence angles and dihedrals. They facilitate more accurate treatment of vibrational/rotational spectra. Other alternatives terms in the extended forcefield include Morse function for bonds, cosine-based angle terms, and grid-based dihedral energy correction maps.

Among the nonbonded interactions, the van der Waals interaction is sufficiently represented by Lennard–Jones 6–12 potential. On the other hand, most current electrostatic implementations do not treat explicitly electronic polarizability. The polarizability is implicitly taken into account by choosing proper partial charges which overestimate the molecular dipoles. This overestimation mimics the condensed phase environment which occurs in biomolecules. There are cases where an explicit inclusion of polarizability is being tested. The most common ways to include the electronic polarizability are induced dipole models, fluctuating charge models, or their combinations (46).

The solvation treatment is another difficult aspect of an empirical forcefield. Both explicit and implicit solvent treatments are being used. The popular water models used in the biomolecular simulations include the TIP3P, TIP4P, and SPC models (47). Most of the water models yield proper characteristics for bulk water at room temperature. The water models differ in the way the solvent molecule interaction sites are represented. Another important aspect is the fact that each of above water model is linked with a particular empirical forcefield, since forcefields are developed in conjunction with a specific water model. Thus AMBER, OPLS, and CHARMM go with TIP3P, OPLS mostly uses TIP4P and GROMOS uses SPC.

The implicit solvation (48) models offer several advantages over the explicit treatment. The solvation treatment is approximately same as that of explicit models, while the computational costs are considerably lower. So these models are popular with simulations involving extensive conformational sampling. The different methods of implicit solvation treatment involve (1) using a distant dependent dielectric constant, (2) Poisson–Boltzmann (PB) models, and (3) generalized-Born (GB) models. The more accurate treatments involve coupling PB/GB methods with solvent accessibility, which treats the hydrophobic effect.

C. Protein-Folding Simulations

A significant advantage with the MD method is the fact that it tries to capture properties of real proteins which are dynamical structures interacting continuously with their environment. These observables include the molecular geometries and energies, the mean atomic fluctuations, conformational changes which are easily accessible by MD. Since MD deals with the time evolution of the system, kinetic aspects are accessible through the trajectories. It is not surprising that MD is one of most widely used techniques for studying topics related to protein folding and dynamics.

One of the first MD simulations of a complete protein was a simulation of the small bovine pancreatic trypsin inhibitor (BPTI, about 800 atoms) in vacuum for 9.2 ps (49). Recent MD simulations are several orders of magnitude more involved (both in time and system size). The first microsecond simulation was performed on the 36-residue villin headpiece in explicit water and succeeded to fold the protein to a native-like structure (50). Recently the same system was studied for 500 μs by distributed computing project, Folding@HOME (51). Other widely studied proteins on microsecond scale are the trp-cage protein (20 residues) (52), the C-terminal β-hairpin of protein G (16 amino acids) (53), protein A (54), and the designed protein BBA5 (23 residues) (55).

Another application of MD is the study of transition states for two-state protein folding. The transition state is the highest energy point in the reaction pathway from which a protein can fold or unfold with equal probability. Proteins such as 63-residue chymotrypsin inhibitor 2 and 110-residue bacterial protein barnase were studied in conjunction with experiment (56). Other MD applications involve simulation of membrane channels (57), an enzyme reaction (58), protein aggregation studies (59, 60), protein–ligand docking, and protein design (37) and structure refinement.

IV. Free-Energy Forcefields and Simulation Methods

A. Anfinsen's Hypothesis

The free-energy model for protein structure prediction and folding is based on Anfinsen's (10) thermodynamic hypothesis which postulates that native state of protein is the global minimum of the free-energy surface. The principle of this approach is to decouple the sampling method from the free-energy surface. In contrast to forcefields for the internal (or potential energy) used for molecular dynamics, free-energy forcefields assigns to each backbone conformation $\{r_i\}$ an internal free energy $F(\{r_i\})$, such that its occupation probability is given as

$$p(\{r_i\}) = \exp\left(-\beta F(\{r_i\})/Z\right) = \frac{1}{Z}\int d(\{s_i\})\exp\left(-\beta E(\{r_i\},\{s_i\})\right). \quad (9)$$

The variables describing the conformation are only the internal degrees of freedom of the protein $\{r_i\}$, that is, solvent degrees of freedom $\{s_i\}$ have been integrated out. For proteins for which Anfinsen's hypothesis holds, $p(\{r_i\})$ for the native state is nearly unity and nearly zero for all others.

We can, therefore, use any sampling technique, including nonequilibrium methods operating at fictitious temperatures to determine this particular conformation. In the process, we typically sample the entire low-energy region of the free-energy surface, but the occupation probabilities of the individual states depend on the sampling method and do not represent the thermodynamic average. The native conformation is selected on the basis of its energy and the relevant thermodynamic occupation probabilities can be reconstructed by applying the formula above, which holds independently of the method that we have used to sample the surface.

This approach is much more efficient in exploring the low-energy ensemble of the protein than existing kinetic or thermodynamic sampling methods and permits successful unrestrained *in silico* folding of small proteins using all-atom representations. Its enormous advantage stems from the fact that relative free-energy differences between two different conformations can be computed directly from the forcefield without recourse to occupation probabilities that require thermodynamic sampling. Its drawback is that the trajectory generated in the sampling process does not reflect the kinetic barriers of the folding process correctly. As a result folding times, for example, cannot be directly extracted from our trajectories.

B. A Biophysical All-Atom Free-Energy Forcefield

PFF02 is an all-atom (with exception of apolar CH_N groups) free-energy forcefield which identifies the native state of protein as its global minimum. The forcefield models the physical interactions of a protein in an implicit solvent (water) environment at a fixed temperature of 300 K. The bond angles and bond lengths are set to standard values. Rotation about the peptide bond is forbidden. The degrees of freedom are the dihedral angles of backbone (ϕ,ψ) and side chain dihedrals ($\chi_1, \chi_2,\ldots$).

The forcefield PFF02 consists of six nonbonded interactions, including two interactions which were added to original forcefield PFF01 (61):

- *Lennard–Jones.* The van der Waals interactions are included in the forcefield as a Lennard–Jones 6–12 potential:

$$V_{lj}(\vec{r}) = V_0 \sum_{ij} \left[\left(\frac{R_{ij}}{r_{ij}}\right)^{12} - \left(\frac{2R_{ij}}{r_{ij}}\right)^{6} \right],$$

where i, j represent the atoms included in the forcefield, r_{ij} is the distance between these atoms, and R_{ij} are the Lennard–Jones radii ($R_{ij} = \sqrt{R_{ii}R_{jj}}$). The parameters for the Lennard–Jones potential were derived as a potential of mean force from experimental data by fitting short-range (2–5 Å) radial distributions of a set of 138 different proteins which are believed span wide range of different folds (62).

- *Electrostatics.* The electrostatic interaction is modeled using the standard columbic potential. The contribution is split into main chain and side chain contributions:

$$V_{\text{ele}}(\vec{r}) = V_{\text{main}}(\vec{r}) + V_{\text{side}}(\vec{r}) = \sum_{ij} \frac{q_i q_j}{\varepsilon_{g(i)g(j)} r_{ij}},$$

where i, j represent the atoms included in the forcefield, q_i and q_j are the corresponding partial charges, r_{ij} is the distance between these atoms, and $\varepsilon_{g(i)g(j)}$ are group-specific dielectric constants.

- *Hydrogen bonding.* Hydrogen bonds play a vital role in the protein folding (63). The experimental estimate of the hydrogen bonding interaction ranges between -2.8 and $+1.9$ kcal mol^{-1} (64, 65), which is much smaller than covalent bond interactions. Generally, hydrogen bonding is not explicitly modeled, but its contributions are embedded partly in the electrostatics and Lennard–Jones. Since hydrogen bonding and solvent interaction are the two major contributions to protein folding, these interactions are specially emphasized and modeled by two contributions in PFF01/02:

  – Electrostatic interactions considering only the dipole–dipole interaction of the amino and carboxyl groups of the main chain. The long-range interactions are overemphasized due to the cooperative effects:

$$V_{\text{hbdipole}} = \frac{0.1064 e^2}{4\pi\varepsilon\varepsilon_0} \left(\frac{1}{r_{C_i H_j}} - \frac{1}{r_{C_i N_j}} - \frac{1}{r_{O_i H_j}} + \frac{1}{r_{O_i N_j}} \right),$$

where i, j count the amino acids with i belonging to the carboxyl and j the amino group, e equals one elementary charge, and $r_{X_i Y_i}$ gives the distance of the atoms X from amino acid i and Y from amino acid j.

- An additional short-ranged term which corrects the hydrogen bonding by considering the alignment of the hydrogen bond with respect to the donor and acceptor groups (66):

$$V_{hbcorr} = V_0 \sum_{ij} R(r_{H_iO_j})\Lambda(\alpha_{ij}, \beta_{ij}),$$

where $V_0 = -2.12$ kcal (mol Å)$^{-1}$, α is the NHO angle, β is the angle between the CO and NH dipoles, $R(r)$ gives the radial and $\Lambda(\alpha)$ the angular dependence to the correction potential. $R(r)$ and $\Lambda(\alpha, \beta)$ are defined as

$$R(r) = s_{2.4, 0.075}(r),$$

$$\Lambda(\alpha, \beta) = s_{45,5}(\alpha) s_{40,5}(\beta) s_{1.5, 0.05}\left(\sqrt{\frac{\alpha^2}{30} + \frac{\beta^2}{24}}\right)^2,$$

where

$$s_{A,B}(x) = \frac{1}{2}\left(1 - \tanh\left(\frac{x-A}{B}\right)\right).$$

The hydrogen bonding term is interpolates these contributions:

$$V_{hb} = \lambda V_{hbdipole} + (1 - \lambda) V_{hbcorr},$$

where λ gives the strength of correction between $[0, 1]$ with $\lambda = 1$ meaning that the hydrogen bonding is modeled by pure dipole–dipole interaction. In PFF01/02, the value of λ is 0.75.

- *Solvation.* The solvent energy and entropy influences the folding of a protein and contributes to the free energy of the system. On the surface of a protein there are important solvent interactions: hydrophobicity, that is, the entropy of water molecules, the conformational entropy of the protein side chains,[1] and the modulation of the solvation of charged side groups.

The solvation effects are modeled in PFF02 via an implicit solvent model based on the solvent accessible surface area (SASA) of the protein (67). The SASA is calculated by rolling a water sphere of radius 1.4 Å over the protein surface which is defined by the Lennard–Jones radii. The solvation term is given by the relation

$$V_{sol} = \sum_i \sigma_{PT(i)} A(i),$$

---

[1] The main chain is somewhat rigid and its entropic contribution is not significant.

where PT(i) is the potential type of atom i, $\sigma_{PT(i)}$ describes the atomic solvent parameter (ASP) according to the potential type, and $A(i)$ is the SASA of the atom i. The parameters are derived by fitting first a SASA model (68) to reproduce the enthalpies of solvation of tripeptide Gly–X–Gly (69) and then adjusting these parameters to stabilize the native structure of villin headpiece (61).

- *Local electrostatics correction.* Amino acids have a preference for secondary structure elements. For example, tryptophan and threonine occur mostly in β-sheet regions, whereas alanine prefers α-helical region. These preferences are influenced by different electrostatic interactions of the main chain dipoles in their local environment (62). The interaction E_{local} is defined as the electrostatic energy of the main chain CO and NH groups of a residue arising from interactions with the main chain CO and NH groups within that residue and with the adjoining peptide groups. So NH_i interacts with CO_{i-2}, NH_{i-1}, CO_i; NH_{i+1} and CO_{i-1}, NH_i, CO_{i+1}; and NH_{i+2} interacts with CO_i:

$$V_{local} = \lambda_{local} \frac{332.150625 \times \zeta}{2} \sum_j \sum_i \frac{q_i q_j}{r_{ij}},$$

where q_i is the charge on the atom and r_{ij} is the distance between the atoms. The parameter ζ is the amino acid-specific parameter.

- *Torsional energy.* A weak dihedral angle dependent energy term is introduced to stabilize the residues in the β-sheet regions of Ramachandran plot. The interaction has a favorable energy contribution of 0.8 kcal mol^{-1} (maximum) for the residues forming β-sheet:

$$V_{tor} = \lambda_{tor} \sum_i \exp\{\gamma_\phi (\phi_i - \phi_0)^2 + \gamma_\psi (\psi_i - \psi_0)^2\}$$

or all amino acids except proline and glycine. For proline and glycine $E_{tor} = 0$. ϕ_i and ψ_i are the backbone dihedral angles of amino acid i. We used $\phi_i = -110°$, $\psi_i = 130°$, $\gamma_\phi = 5 \times 10^{-3}$ deg^{-2}, and $\gamma_\psi = 1.25 \times 10^{-3}$ deg^{-2}.

C. Stochastic Simulation Methods

Proteins assume unique three-dimensional structures after being synthesized into a linear chain of amino acids following the thermodynamic hypothesis. The thermodynamic hypothesis states that the native state of a protein corresponds to the global minimum of its free-energy surface. The low-energy region of the free-energy landscape of proteins is extremely rugged due to the close packing of the atoms in the native conformation.

Efficient sampling techniques need to reliably locate the global minima of the free-energy surface. This is the central computational bottleneck to fold proteins starting from sequence information alone. As the complexity of the free-energy landscape increase with the size of the protein this task becomes more challenging.

The free-energy surfaces of many proteins have metastable conformations with energies only a few kcal mol^{-1} above the global optimum. In *de novo* protein structure prediction with free-energy models the predicted structure is selected solely on the basis of its energy in comparison to all other conformations. It is therefore important to develop techniques that can identify the global optimum of the forcefield to such accuracy. Because all-atom protein folding requires substantial computational resources it is important to investigate various computational strategies and this section summarizes a few key concepts in their derivation. The basin hopping technique (BHT) (70–72), which has been used to fold the conserved 40 amino acid headpiece of the HIV accessory protein (73, 74), emerges as one suitable approach to perform such simulations. We also discuss an evolutionary algorithm as a multiprocessor generalization of this method.

1. MONTE CARLO

All stochastic methods originate from the Monte Carlo method that explores the energy landscape by random changes in the geometry of the molecule. In this way large regions of the configurational space can be searched in finite time, without regard of the kinetics of the process. A Monte Carlo simulation is composed of the following steps:

1. Specify the initial coordinates (R_0).
2. Generate new coordinates by random change to initial coordinates (R').
3. Compute transition probability $T(R_0, R')$.
4. Generate a uniform random number RAN in range [0, 1].
5. If $T(R_0, R') <$ RAN, then discard the new coordinates and goto step 2.
6. Otherwise accept the new conformation and goto step 2.

The most popular realization of the Monte Carlo method for molecular systems is the Metropolis method (see flowchart in Fig. 9):

1. Specify the initial atom coordinates.
2. Select some atom *i* randomly and move it by a random displacement.
3. Calculate the change of potential energy ΔV corresponding to this displacement.
4. If $\Delta V < 0$, accept the new coordinates and goto step 2.

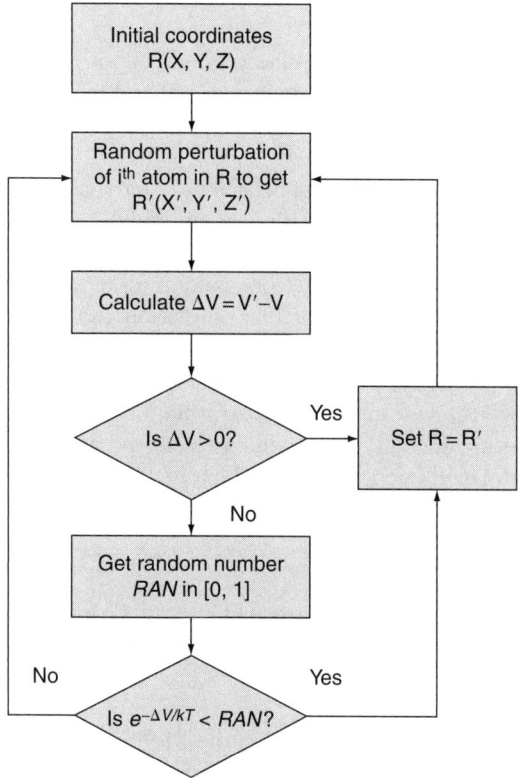

FIG. 9. Schematic representation of Metropolis method.

5. Otherwise, if $\Delta V > 0$, select a random number RAN in the range [0, 1] and:

   (a) If $e^{-\Delta V/kT} <$ RAN, accept the new conformation and goto step 2.

   (b) If $e^{-\Delta V/kT} \geq$ RAN, keep the original coordinates and goto step 2.

In Monte Carlo simulations, the system has no "memory" between two steps, that is, the probability that the system might revert to its previous state is as probable as choosing any other state. As a result of the stochastic simulation, a large number of configurations is accumulated that can be used to calculate thermodynamic properties of the system. Monte Carlo is not a deterministic method (as molecular dynamics), but often gives rapid convergence of the thermodynamic properties for small molecules (75).

2. Improved Sampling Techniques

Due to its popularity a large number of modifications and improvements of the Monte Carlo technique have been suggested and many of them have been used in the context of protein simulations:

- *Simulated annealing.* In this approach (76), barriers in the simulation are avoided by starting the simulation at some high temperature and slowly lower the temperature of the simulation until the target temperature is reached. At high temperature, the exploration of the phase space is very rapid, while near the end of the simulation the true thermodynamic probabilities of the system are sampled.
- *Stochastic tunneling.* Here, a potential energy surface (PES) is transformed by using a nonlinear transformation to suppress the barriers which are significantly above the present best energy estimate (77). The transformed energy surface which is used for exploration of global minimum is given by

$$E_{\text{STUN}} = \ln(x + \sqrt{x^2 + 1}), \qquad (11)$$

with $x = \gamma(E - E_0)$, where E is the present energy, E_0 is the best estimation so far, and γ is the transformation parameter, which controls the rate of rise for the transformation.

- *Parallel tempering.* This method is the Monte Carlo counterpart of the REMD method described. A modified version of this method, which uses an adaptive temperature control and replication step, has been employed for exploration of protein energy surfaces (78).
- *Basin hopping technique.* In this scheme, the original PES is simplified by replacing the energy of each conformation with the energy of a nearby local minimum (79). The minimization is carried out on the simplified potential (see Section IV.C.3).
- *Evolutionary strategy.* This scheme is a multiprocess extension of the BHT. Several concurrent simulations are carried out in parallel on a population. The population is evolved toward a global optimum of energy with a set of rules which enforce energy improvement and population diversity (see Section IV.C.4).

3. Basin Hopping Technique

BHT (70) employs a relatively straightforward approach to eliminate high-energy transition states of the free-energy surface: The original free-energy surface is simplified by replacing the energy of each conformation with the energy of a nearby local minimum. In many applications, the additional effort for the minimization step is more than compensated by the improved efficiency of the stochastic search. This process leads to a simplified potential on which the simulations search for the global minimum. This replacement eliminates high-energy barriers in the stochastic search that are responsible for the freezing problem in simulated annealing. A one-dimensional schematic representation of BHT is shown in Fig. 10. Every basin hopping cycle (minimization step) tries to locate a local minima and thus it simplifies the original PES (black curve) into an effective PES (blue curve) which is then searched for the global minima.

The BHT and derivatives have been used previously to study the PES of model proteins and polyalanines using all-atom models (71, 80–82). Here, we replace the gradient-based minimization step used in many prior studies with a simulated annealing run (76), because local minimization generates only very small steps on the free-energy surface of proteins. In addition, the computation of gradients for the SASA model is computationally prohibitive. Within each

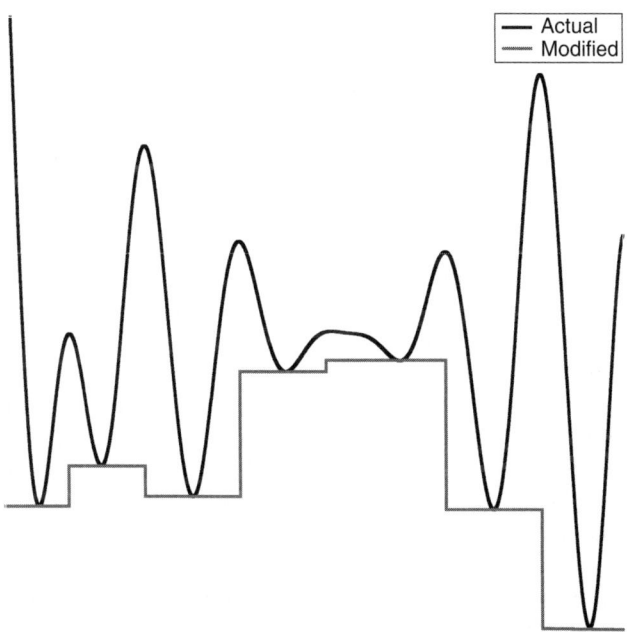

FIG. 10. Schematic representation of basin hopping technique. The modified potential is obtained by replacing every point on the curve to its neared local minimum.

simulated annealing simulation, new configurations are accepted according to the Metropolis criterion, while the temperature is decreased geometrically from its starting to the final value.

The starting temperature and cycle length determine how far the annealing step can deviate from its starting conformation. The final temperature must be chosen small compared to typical energy differences between competing metastable conformations, to ensure convergence to a local minimum. The annealing protocol is, thus, parameterized by the starting temperature T_S, the final temperature T_F, and the number of steps. We investigated various choices for the numerical parameters of the method but have always used a geometric cooling schedule. At the end of one annealing cycle, the new conformation is accepted if its energy difference to the current configuration was no higher than a given threshold energy ε_T, an approach recently proven optimal for certain optimization problems (83). We typically used a threshold acceptance criterion of 1–3 kcal mol^{-1}.

4. Evolutionary Algorithms

The popular BHT method (79, 84) for global optimization eliminates high-energy PES by replacing the energy of each conformation with the energy of a nearby local minimum. For protein folding, we have replaced the original local minimization by simulated annealing (SA). In the course of our folding studies, we find that independent BHT simulations often find identical structures corresponding to the same local (global) minimum. As a result, each independent simulation reconstructs the full folding path independently. It would be very desirable to develop methods, where several concurrent simulations exchange information to *learn* from each other. For a PES having many local minima, independent simulations limit the efficient exploration of the PES. Also, occasionally BHT simulations go astray, ending the search in a wrong energy basin of the PES. We have developed a *greedy* version of BHT (85) which overcomes these problems to a certain extent.

We have, therefore, generalized the BHT approach to a population of size N which is iteratively improved by P concurrent dynamical processes (78). The population is evolved toward an optimum of the free-energy surface with an ES that balances the energy improvement with population diversity. In the ES, conformations are drawn from the *active* population and subjected to an annealing cycle. At the end of each cycle, the resulting conformation is either integrated into the active population or discarded. The algorithm was implemented as a master–client model in which idle clients request a task from the master. The master maintains the *active* conformation of the population and distributes the work to the clients. Each step in the algorithm has three phases:

1. *Selection.* A conformation is drawn randomly from the *active* population. We have used a uniform probability distribution with population of 20 conformers.
2. *Annealing cycle.* We use a simulated annealing schedule with T_{start} drawn from an exponential distribution and T_{end} fixed at 2 K. The number of steps per cycle is increased as $10^5 \times \sqrt{cycle}$.
3. *Population update.* We have adjusted the acceptance criterion for newly generated conformations to balance the population diversity and energy enrichment. We define the two structures as *similar* if they have bRMSD (backbone root mean square deviation) less than 3 Å to each other. We define an *active* population as the pool containing mutually different lowest energy conformers. The master finds a number of similar structures (nc) and then performs one of the following operations on complete population.

   (a) Add: If the new conformation is not *similar* to any structure ($nc = 0$) in the population, we add to the population, provided its energy is less than the energy of conformation with highest energy (E_{worst}).

   (b) Replace: If the new conformation (with energy E_{new}) is *similar to one* existing structure in the population (with energy E_{old}), it replaces that structure provided $E_{new} < E_{old} + \Delta$ (see below).

   (c) Merge: If the new conformation has *several similar* structures, it replaces this group of structures provided its energy is less than the best one of the group E_{best} plus an acceptance threshold Δ.

A flowchart illustrating the population update tasks of the master is shown in Fig. 11. In our first BHT/ES simulations, we have used a fixed energy threshold (Δ) acceptance criterion. Here we have implemented a *variable* energy threshold which we define as $\Delta = A \times \tanh D$, where

$$D = \frac{E_{new} - E_{best}}{A},$$

where A is the energy threshold (3 kcal mol^{-1}), E_{new} is the energy of the new structure, and E_{best} is the lowest energy structure in the population. This choice of the energy criterion ensures that the conformation with the best energy is never replaced, while conformations higher in energy are more easily replaced in the secure knowledge that they are far from optimal. The rules for the *replace* and *merge* operations ensure the structural diversity of the population and its continued energetic improvement (on average).

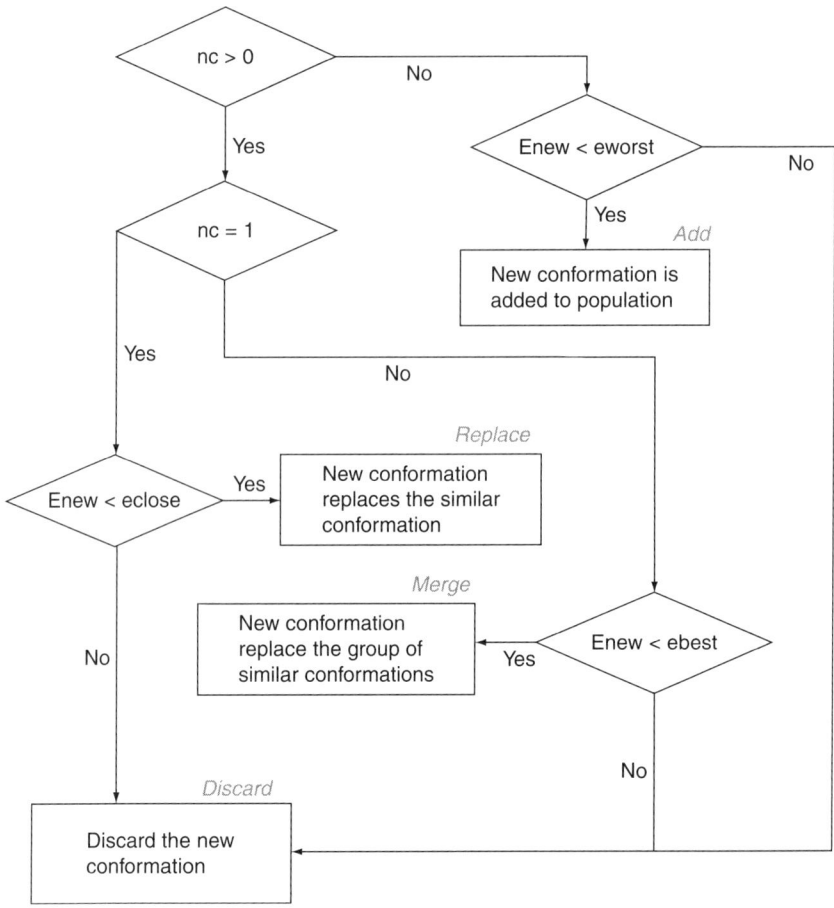

Fig. 11. A flowchart illustrating the population update. See the text for an explanation.

V. Free-Energy Protein Folding

In this section, we review several studies, using a variety of methods described in the previous section, to fold medium-sized proteins.

A. Helical Proteins

1. Tryptophan Cage: 1L2Y

Tryptophan cage or trp-cage protein (86) has been the subject of various theoretical studies and it has been of great scientific interest. It had been reported to fold using REMD and a variety of other simulations (52, 87–92).

We performed 20 independent basin hopping simulations starting with the completely extended conformations in PFF02 with 100 cycles. The starting conformation had a RMS of 12.94 Å to the native conformation and was completely extended manually (by setting all backbone dihedral angles except for proline to 180°). The starting temperatures were chosen from a distribution of exponentially distributed temperatures and the number of steps increased with the BHT cooling cycle by $10^4 \sqrt{n_m}$ where n_m is the number of minimization cycles.

The lowest energy structure converges to a native like conformation with RMSD of 3.11 Å to the native conformation. For the sake of uniformity in case of NMR resolved experimental structures, we compare the RMSD to the first model in the protein data bank file. The lowest energy structure had an energy of -23.4 kcal mol^{-1}. Figure 12C shows the scatter plot of the conformations visited by the basin hopping simulations on the free-energy surface. The overlay of native conformation (green) with the lowest energy conformation (red) is shown in Fig. 12A and the corresponding C_β–C_β overlay matrix is shown in Fig. 12B. The C_β–C_β overlay matrix quantifies the tertiary alignment

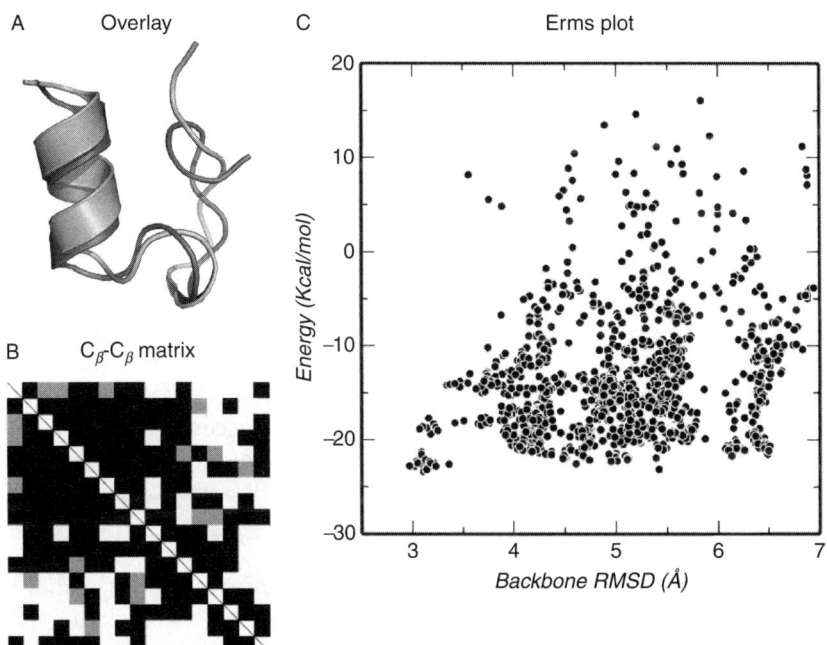

Fig. 12. 1L2Y: overlay of predicted (red) structure to experimental (green) structure. The overlay of the C_β–C_β distance matrix and energy versus RMSD plot. (See Color Insert.)

along with secondary structure formation by taking the difference between all C_β distances of predicted and native conformation. Black regions indicate excellent agreement in the formation of native contacts while white regions indicate larger deviations.

2. POTASSIUM CHANNEL BLOCKER: 1WQE

Potassium channel blockers, 1WQC, 1WQD, 1WQE (93), are of specific interest due to their unusual fold for ion channel blockers. They are toxic venom peptides involved in blocking of potassium channel in cells. They have two helices which are "locked" together by a disulfide bond. Here, we study the folding of 1WQE, a two helical protein which had earlier been folded using PFF01 (94). The starting conformations for this study were completely extended conformations with RMSD of 20.6 Å to the native conformation.

We did 10 independent basin hopping simulations from the extended conformation in PFF02. Nine out of 10 independent BHT simulations converge to conformations that differ by less than 3 Å RMSD to the native conformation. The lowest energy structure found in the simulations has a RMSD of only 2.33 Å to the native conformation with energy of -44.0 kcal mol^{-1}. This is very encouraging; the contribution to the formation of disulfide bridges is yet to be incorporated in PFF02.

Figure 13c shows the scatter plot of conformations visited during the simulations. There are many conformations visited around the native state and the next metastable state can be seen at around 5 Å and is about 4 kcal mol^{-1} higher in energy. This metastable conformation also correctly predicts the two helices but arranges itself in an orthogonal packing instead of up–down arrangement. Independently, the two helices PRO3–THR12 and VAL12–CYS22 have an RMSD of 0.7 and 0.44 Å only. Figure 13a shows the overlay of the native conformation (green) to the lowest energy conformation (red) encountered in the simulations. The overlay shows the perfect agreement of the lowest energy conformation to the experimental structure and the C_β–C_β matrix (Fig. 13B) illustrates the tertiary alignment of the overlay with many black regions.

As nine out of 10 simulations converge to native like conformation and the metastable conformation is 4 kcal mol^{-1} higher in energy, we conclude the folding of 1WQE as predictive and reproducible.

3. HIV ACCESSORY PROTEIN: 1F4I

HIV accessory protein destroys the host cell's ability to survive by binding to a host receptor and restricting an important enzyme to activate the cell's immune system. The 40 amino acid HIV accessory protein 1F4I (73) was earlier folded using PFF01 starting from random starting conformations (74, 95).

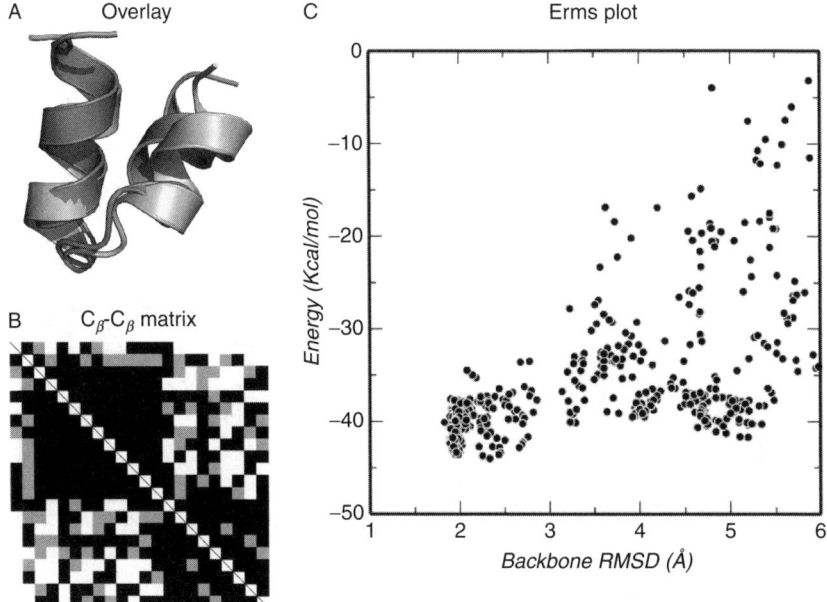

FIG. 13. 1WQE: overlay of predicted (red) structure to experimental (green) structure. The overlay of the C_β–C_β distance matrix and energy versus RMSD plot. (See Color Insert.)

We studied the folding of 1F4I in PFF02 with BHT. We did 20 independent runs with completely extended conformations in PFF02 for 150 basin hopping cycles. The number of cycles is larger than 1L2Y and 1WQE as the complexity of the search space increases with the size of protein. The temperatures were chosen from an exponential distribution and the cooling cycle length was increased as described above. The lowest energy structure encountered in the simulations had an RMSD of 3.29 Å to the native conformation and had an energy of -93.7 kcal mol^{-1}.

The scatter plot of the conformations visited during the simulations is shown in Fig. 14c. Apart from the native-like conformations, there are clusters of low-energy conformations around 6 and 8 Å. The first nonnative conformation is 2 kcal mol^{-1} higher in energy than the lowest native-like conformation. While this misfolded structure differs significantly, it has the same secondary structure. The misfolded conformation is shown in Fig. 15. The corresponding C_β–C_β overlay matrix also shows the secondary structure formation with a different tertiary arrangement. Independently, helix-1 (LYS3–LEU12), helix-2 (GLU16–PHE24), and helix-3 (ASN31–SER39) in this misfolded conformation are nearly perfectly predicted and have RMSDs of only 0.53, 2.0, and 0.52 Å, respectively.

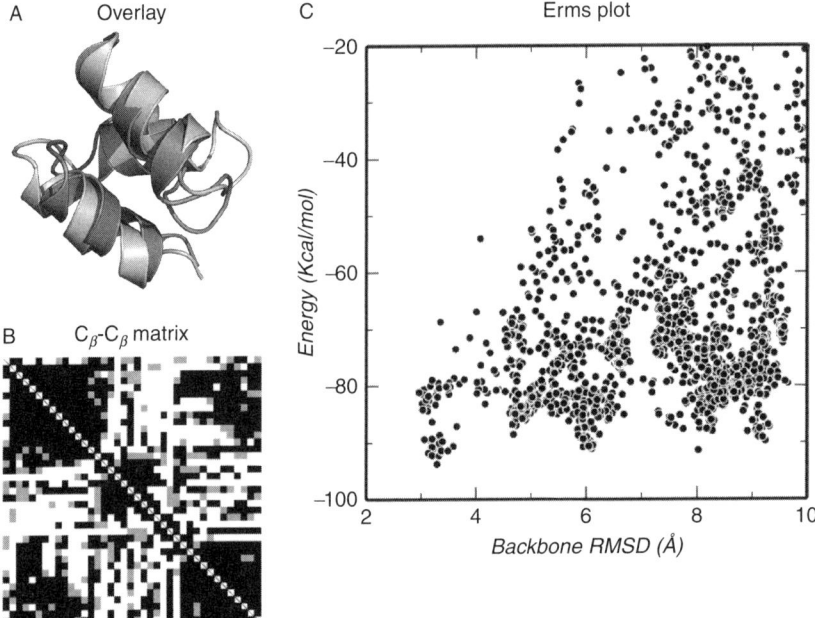

Fig. 14. 1F4I: overlay of predicted (red) structure to experimental (green) structure. The overlay of the C_β–C_β distance matrix and energy versus RMSD plot. (See Color Insert.)

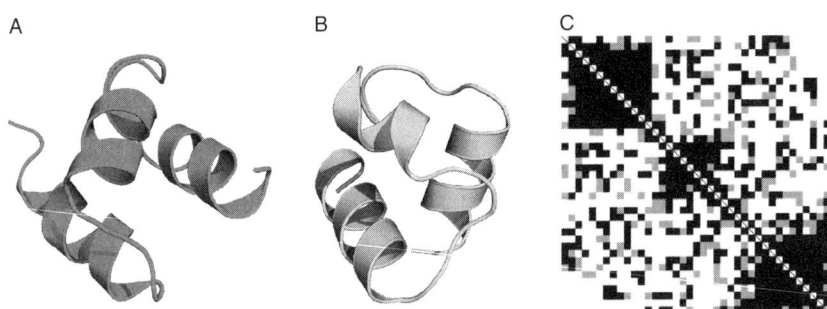

Fig. 15. 1F4I: overlay of misfolded (orange) structure to experimental (green) structure and the overlay of the C_β–C_β distance matrix.

Figure 14A shows the overlay of the native conformation (green) to the lowest energy conformation encountered in the simulations. The overlay shows the agreement of the predicted conformation to the native structure. The C_β–C_β matrix (Fig. 14B) shows the tertiary alignment of the overlay with dark regions. Both the starting (LYS3–LEU12) and the end (ASN31–SER39) helix were correctly predicted, but the middle helix (GLU16–PHE24) had a different tertiary arrangement because of a wrongly predicted turn region. This can also be observed from the C_β–C_β overlay matrix.

Only one of the 20 basin hopping simulations converged to a native-like conformation, but the energy of native-like conformation was significantly lower (less than 2 kcal mol^{-1}) than any other conformation and thus, we conclude the folding study to be predictive but not reproducible.

4. Engrailed Homeodomain: 1ENH

The 54 amino acid engrailed homeodomain protein (96) is a three helical orthogonal bundle protein which has been subjected to detailed MD simulations (97, 98). It was not possible to fold this protein using basin hopping technique and the simulations never reached the energies of the native-like conformations.

Here we studied the folding of engrailed homeodomain in PFF02 using evolutionary algorithm with a maximum population of 64 conformations and 512 processors (99). The lowest energy structure converges to 4.28 Å to the native conformation with the energy of −170.95 kcal mol^{-1}. 1ENH has a unstructured tail at the N-terminus, excluding this seven amino acid region, the RMSD reduces to only 3.4 Å.

The scatter plot of conformations visited during the simulation is shown in Fig. 16C. Seven out of the total population of 64 structures are less than 4.5 Å RMSD to the native conformation. The overlay of the lowest energy conformation (red) with the native conformation (green) is shown in Fig. 16A and the corresponding C_β–C_β overlay matrix is shown in Fig. 16B. There are also competing conformations (within 2 kcal mol^{-1}) with large RMS deviations encountered in the simulations. One such conformation is shown in Fig. 17. These conformations have the same secondary structure, but a different tertiary structure alignment. The C_β–C_β overlay matrix for the misfolded conformation also confirms that all the three helices are properly predicted but their tertiary arrangement is completely different.

No two helices in the misfolded conformation are in agreement with the respective helices in the native state. Independently, helix-1 (E8–E20), helix-2 (E26–L36), and helix-3 (A40–K43) are nearly perfectly predicted and have RMS of only 0.56, 0.42, and 0.47 Å, respectively.

As about 10% of the population is native-like and the misfolded conformations, we can conclude that the folding is reproducible.

5. E-Domain of Staphylococcal Protein A: 1EDK

The E-domain of staphylococcal protein A is one of five homologous Immunoglobulin G-binding domains designated E, D, A, B, and C that comprise the extracellular portion of protein A (100). Its architecture is classified as

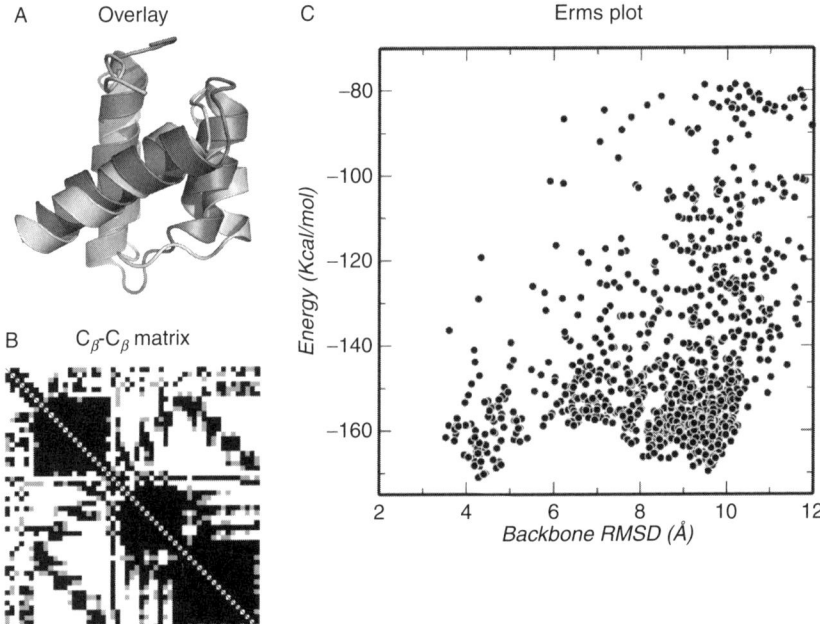

FIG. 16. 1ENH: overlay of predicted (red) structure to experimental (green) structure. The overlay of the C_β–C_β distance matrix and energy versus RMSD plot. (See Color Insert.)

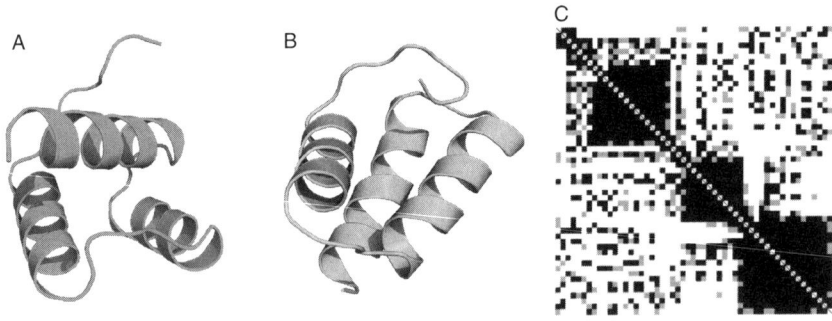

FIG. 17. 1ENH: overlay of misfolded (orange) structure to experimental (green) structure and the overlay of the C_β–C_β distance matrix.

an up–down bundle in CATH (*101*), which makes the topology of the three helical bundles different from the one in engrailed homeodomain protein or HIV accessory protein.

As the final example of helical proteins, we studied the folding of protein A in PFF02 using evolutionary algorithm with a maximum population of 64 conformations and 256 processors for 50 cycles. The lowest energy structure converges to 4.05 Å to the native conformation with the energy of -154.78 kcal mol^{-1}. Excluding the unstructured regions from both the N-terminus and C-terminus, the RMSD of the structured region (GLU5–SER52) reduces to only 2.99 Å.

The scatter plot of conformations visited during the simulation is shown in Fig. 18C. It shows two funnel like regions. Seven out of the total population of 64 structures are less than 4.05 Å RMSD to the native conformation. The overlay of the lowest energy conformation (red) with the native conformation (green) is shown in Fig. 18A and the corresponding C_β–C_β overlay matrix is shown in Fig. 18B.

There are also competing conformations (within 2 kcal mol^{-1}) with large RMS deviations ($\sim$10 Å) encountered in the simulations. One such conformation is shown in Fig. 19. This conformation is the mirror image of the native conformation as helix-1 (GLU5–LEU15) and helix-2 (ALA22–ASP34) align

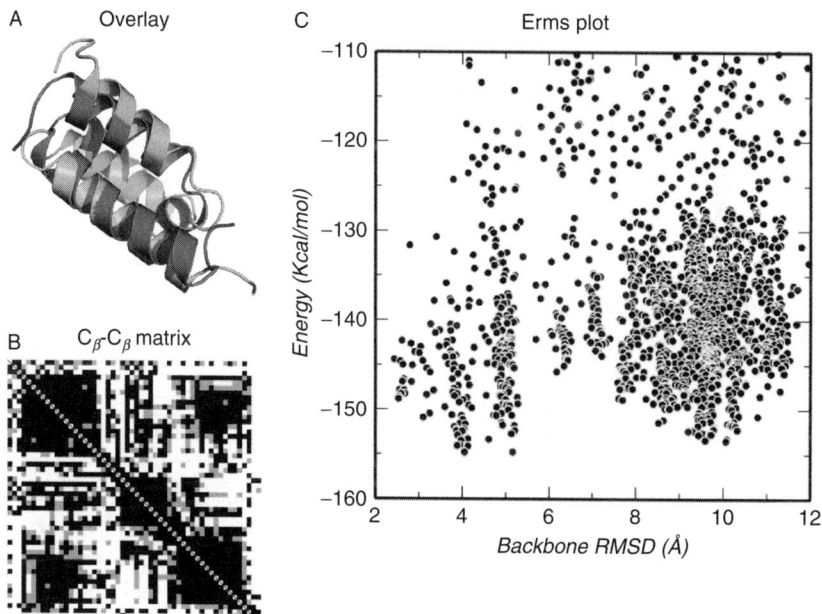

Fig. 18. 1EDK: overlay of predicted (red) structure to experimental (green) structure. The overlay of the C_β–C_β distance matrix and energy versus RMSD plot. (See Color Insert.)

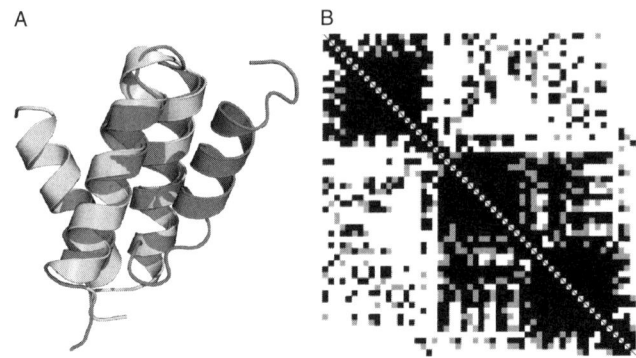

FIG. 19. 1EDK: overlay of misfolded (orange) structure to experimental (green) structure and the overlay of the C_β–C_β distance matrix. (See Color Insert.)

perfectly but helix-3 (ALA40–SER52) is in the opposite direction because of the wrong turn. Independently helix-1, helix-2, and helix-3 have RMS deviations of only 0.40, 0.49, and 0.46 Å, respectively.

Again, about 10% of the population included native-like structures indicating reproducible folding of protein A.

For all proteins which had competing metastable states, the secondary structure was always correctly predicated. This indicates that for proteins in the low-energy region, the secondary structure is almost always formed correctly and what lacks is the tertiary arrangement of these secondary structure elements (102).

B. Hairpins

Hairpins are the simplest β-sheet structures with only two strands in antiparallel directions that are connected together with a turn. Hydrogen bonding and the packing of the protein itself plays a crucial role here in the folding of such small polypeptides. There are not many hairpin proteins that are not stabilized by external interaction with ions or with the formation of disulfide bridges.

In this section, we report the folding studies of various polypeptide chains which are stable in physiological conditions and have no other stabilizing contributions like disulfide bonds arising from cystine side chains. Folding of such small polypeptides is the next step toward the more universal forcefield, as the forcefield selectivity is tested between a helical conformation and a sheet conformation. The helical conformation gives greater contribution with hydrogen bonding energy as it has more number of hydrogen bonds as compared to the sheet (every hydrogen bond in PFF02 gives a contribution of about 2 kcal mol^{-1}). This hydrogen bonding energy should be compensated by the inclusion of new terms in PFF02 and change in other interactions.

1. TRYPTOPHAN ZIPPER: 1LE0

Tryptophan zippers are one of the smallest monomeric, stable β-hairpins that adopt a unique tertiary fold without requiring metal binding, unusual amino acids, or disulfide crosslinks (*103*). We were able to fold various tryptophan zippers using PFF02 and BHT (not shown here).

We studied the folding of 1LE0 using 128 processors on Marenostrum cluster at the Barcelona supercomputer center starting from completely extended conformations. We performed 20 cycles of evolutionary algorithm. The lowest energy conformation reached in the simulation had a RMS of only 1.5 Å to the native conformation with the energy of -29.97 kcal mol^{-1}.

The scatter plot of the conformations visited during the simulations is shown in Fig. 20C. The scatter plot shows that the native-like conformations lie significantly below any other conformation. Twelve out of the 64 conformations from the final population are less than 3.0 Å to the native conformation.

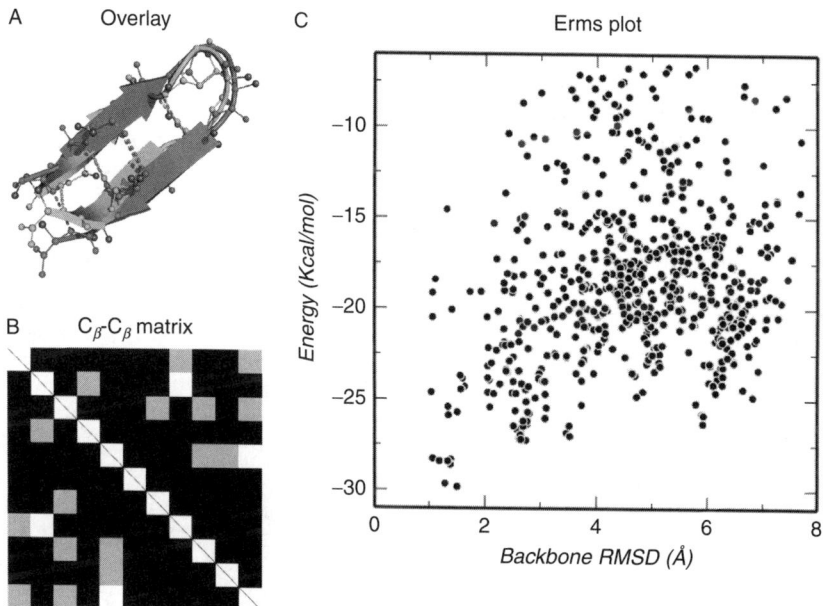

FIG. 20. 1LE0: overlay of predicted (red) structure to experimental (green) structure. The overlay of the C_β–C_β distance matrix and energy versus RMSD plot. (See Color Insert.)

The protein folds in less than 90 min using 128 processors in parallel using the 20 cycles of evolutionary algorithm amounting to 77×10^6 function evaluations or about 9 CPU days.

The overlay of the predicted conformation (red) with the native conformation (green) is shown in Fig. 20A and the corresponding C_β–C_β overlay matrix is shown in Fig. 20B. Large black regions in the C_β–C_β overlay matrix indicate the agreement of native contacts between the two conformations.

As hydrogen bonding plays an important role in the formation and topology of β-sheet structures, it is important to compare the hydrogen bonding pattern in the lowest energy conformations as two β-sheet conformations might look very similar to the eye, but they might have completely different topology resulting from shifting of backbone hydrogen bonds.

The pattern of backbone hydrogen bonds are shown in Table II for the native and the predicted conformation. These were calculated using standard definitions with MOLMOL (distance = 2.4 Å and angle = 35°). Four out of the five backbone hydrogen bonds of the native structure are predicted correctly in the lowest energy structure found in the simulations.

As about 20% of the population converged to native-like conformations with much lower energies, we conclude the folding of tryptophan zipper as reproducible and predictive.

2. HIV-1 V3 Loops

Here we study the folding of two HIV-1 V3 loops. The V3 loop of the HIV-1 envelope glycoprotein gp120 is involved in binding to the CCR5 and CXCR4 coreceptors (104). The structures of an HIV-1 V3 peptides bound to the

TABLE II
1LE0: Backbone Hydrogen Bond Pattern Between Native and Predicted Conformations and Secondary Structure Information

Hydrogen bond	Native	Predicted
03 THR HN → 10 THR 0	X	X
05 GLU HN → 08 LYS 0	X	X
07 ASN HN → 03 THR 0	X	
10 THR HN → 03 THR 0	X	X
12 LYS HN → 01 SER 0	X	X
	Secondary structure	RMSD (Å)
Native	CEEECSSSEEEC	–
Predicted	CEEEETTEEEEC	1.52

respective antibody were found to be a β-hairpin. The hairpin structure with specific side chains on the side is responsible for the binding of viral protein to its receptor. Only when the protein is properly bound, the virus can enter the cell. We studied the folding of two such loops, 1NIZ and 1U6U, which differ themselves by only an insertion of two amino acids in the 1NIZ sequence. The insertion changes the loop structure resulting in different side chains getting exposed to the receptor. This change in the structure of V3 loops is considered to be responsible for coreceptor selectivity by the virus protein.

The $V3_{MN}$ Loop: 1NIZ We studied the folding of 14 amino acid HIV-1 $V3_{MN}$ loop 1NIZ (*104*) in PFF02 using a greedy version of BHT (*105*).

In basin hopping simulations, there is a threshold energy acceptance criterion at the end of every basin hopping cycle. In our simulations, we have used this threshold acceptance criterion of 1–3 kcal mol^{-1} depending upon this size of the protein. In the greedy version of basin hopping, the threshold energy is varied depending upon the best energy found so far in the simulation. Here we calculated the threshold criteria as $(\varepsilon_S - \varepsilon_B)/4$, where ε_S is the starting energy and ε_B is the best energy found so far in the simulation. This choice implies that the conformation with the best energy is never replaced with a conformation that is higher in energy and thus introduces a "memory effect" in the simulation. For the simulations that are higher in energy, the increased threshold value implies a higher acceptance probability of conformations with higher energy.

We did 200 cycles of greedy basin hopping simulations in PFF02. The simulations were started with completely extended conformation which had a RMS of 12 Å to the native state. The lowest energy structure found in the simulation had a RMSD of only 2.04 Å to the native state.

The scatter plot of the conformations visited during the simulations is shown in Fig. 21C. The scatter plot shows a single downhill folding funnel for this hairpin. Eight out of the 10 independent simulations converged to less than 3.5 Å RMSD to the native conformation.

The overlay of the lowest energy conformation (red) with the native conformation (green) is shown in Fig. 21A and the corresponding C_β–C_β distance matrix is shown in Fig. 21B. Large black regions in the C_β–C_β overlay matrix indicate the agreement of native contacts between the two conformations.

Again, we did the backbone hydrogen bond analysis and four out of the five backbone hydrogen bonds of the native structure are correctly predicted in the lowest energy structure found in the simulations. The pattern of backbone hydrogen bonds is shown in Table III. The secondary structure of the predicted and native conformation is also shown in Table III. The letters in the secondary structure correspond to DSSP definitions.

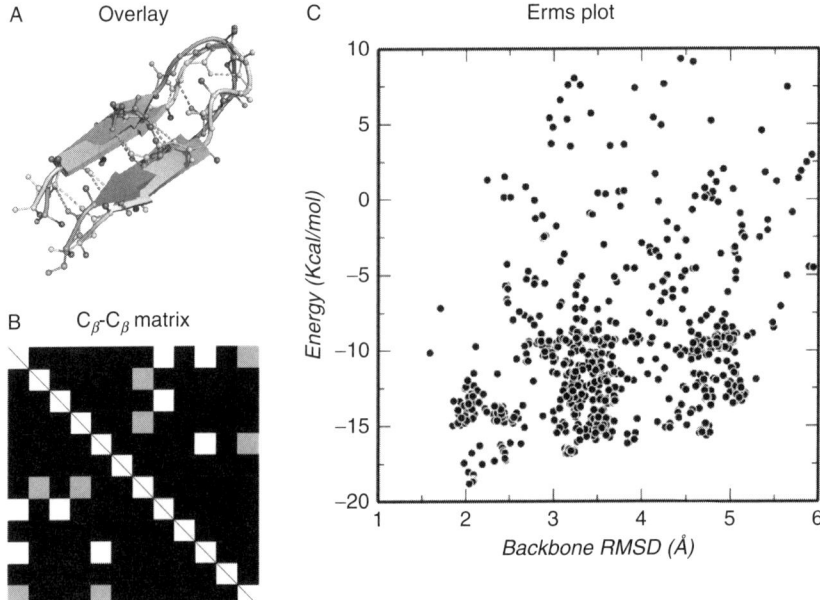

FIG. 21. 1NIZ: overlay of predicted (red) structure to experimental (green) structure. The overlay of the C_β–C_β distance matrix and energy versus RMSD plot. (See Color Insert.)

As eight of the 10 simulations converged to native-like conformation without any competing metastable conformations, the folding is concluded as reproducible and predictive.

The HIV-1 $V3_{IIIB}$ Loop: 1U6U. Comparison of the known V3 structures leads to a model in which a 180° change in the orientation of the side chains and the resulting one residue shift in backbone hydrogen bonding patterns in the N-terminal strand of the β-hairpins markedly alters the topology of the surface that interacts with antibodies and that can potentially interact with the HIV-1 coreceptors (*106*).

We studied the folding of 17 amino acid HIV-1 $V3_{IIIB}$ loop-1U6U in PFF02 using a greedy version of BHT for same 200 cycles as for 1NIZ. The simulations were started with completely extended conformation which had a RMS of 15 Å to the native state.

All the 10 independent simulations after 200 cycles of greedy basin hopping found the β-sheet like conformations. The lowest energy conformation (-32.9 kcal mol^{-1}) found in the simulation had a RMS of 4.57 Å to the native state, which is relatively higher for a β-hairpin. This happens because of an overall bend in the loop resulting from solvent interactions, which can be expected as the peptide is a fragment of a larger protein.

TABLE III
1NIZ: Backbone Hydrogen Bond Pattern Between Native and Predicted Conformations and Secondary Structure Information

Hydrogen bond	Native	Predicted
02 ARG HN → 13 THR 0	X	X
04 HIS HN → 11 PHE 0	X	X
06 GLY HN → 09 ARG 0		X
08 GLY HN → 06 GLY 0	X	
11 PHE HN → 03 HIS 0	X	X
13 THR HN → 01 ARG 0	X	X
	Secondary structure	RMSD (Å)
Native	CEEEECSSCEEEEC	–
Predicted	CEEEECSSCEEEEC	2.04

The scatter plot of all the conformations visited during the simulations is shown in Fig. 22C which shows the single funnel-like landscape. The overlay of the lowest energy conformation (red) with the native conformation (green) is shown in Fig. 22A and the corresponding C_β–C_β distance matrix (Fig. 22B) shows that the two strands are correctly predicted and has the correct tertiary arrangement. The lowest energy structure still correctly predicts four out of five native backbone hydrogen bonds, thus indicating the correct pattern found in PFF02.

The hydrogen bond analysis helps us understand the topology better as the lowest energy conformation had larger deviations from the native structure. The backbone hydrogen bonds of both these conformations are shown in Table IV. As all four backbone hydrogen bonds are predicted correctly in the lowest energy conformation, it is evident that this conformation has correct topology regardless of its high RMS deviation which occurs due to dislocated turn region.

We had predictively and reproducibly folded two very similar (sequence) proteins with different topologies in PFF02. PFF02 can thereby differentiate between these two HIV-1 V3 loops.

3. HP7, A 12-Residue β-Hairpin: 2EVQ

HP7 is a 12 amino acid designed β-hairpin (107). Here, we studied the folding of this protein with the greedy version of basin hopping simulations (108). We performed 10 independent simulations of greedy basin hopping

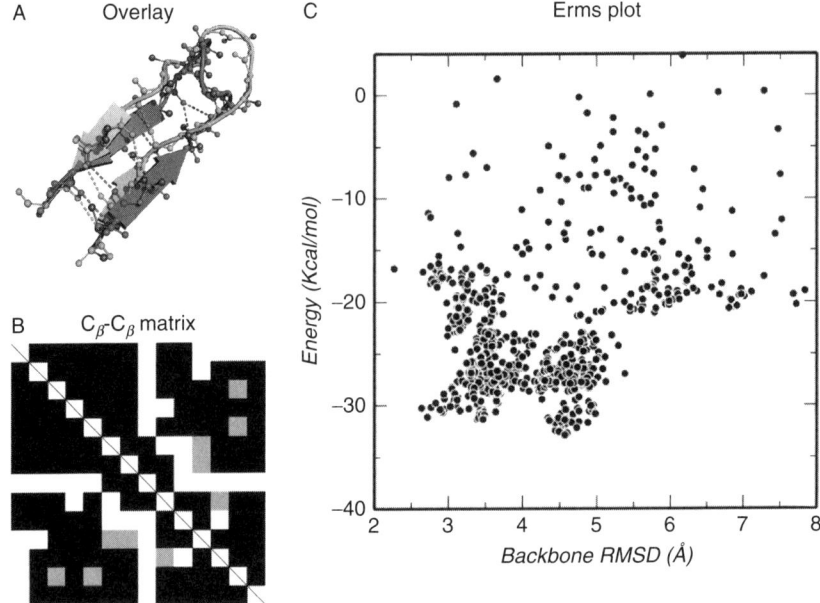

FIG. 22. 1U6U: overlay of predicted (red) structure to experimental (green) structure. The overlay of the C_β–C_β distance matrix and energy versus RMSD plot. (See Color Insert.)

TABLE IV
1U6U: Backbone Hydrogen Bond Pattern Between Native and Predicted Conformations and Secondary Structure Information

Hydrogen bond	Native	Predicted
02 SER HN → 16 ILE 0	X	X
04 ARG HN → 14 VAL 0	X	X
14 VAL HN → 04 ARG 0	X	X
16 ILE HN → 02 SER 0	X	X

	Secondary structure	RMSD (Å)
Native	CEEEECCSSTTCCEEEEC	–
Predicted	CEEEEEEETTTEEEEEC	4.57

method with 100 cycles in PFF02. The simulations were started from completely extended conformation of the protein which had a RMSD of 10.5 Å to the native state.

Eight out of the 10 independent simulations after 100 cycles of greedy basin hopping find the β-sheet like conformations and converge to less than 3.0 Å RMSD to the native conformation. The lowest energy conformation has an RMSD of 2.62 Å to the native conformation and had energy of -26.0 kcal mol^{-1}.

The scatter plot of the conformations visited during the simulations is shown in Fig. 23C. The scatter plot shows two funnels on the free-energy surface for this hairpin. The metastable conformations corresponding to the funnel at around 5.5 Å populated helical conformations, but is 7 kcal mol^{-1} higher than the lowest energy conformation. This shows that the lowest energy conformation is native-like and significantly lower than other metastable conformations.

The overlay of the lowest energy conformation (red) with the native conformation (green) is shown in Fig. 23A and the corresponding $C_\beta-C_\beta$ distance matrix is shown in Fig. 23B. Large black regions in the $C_\beta-C_\beta$ overlay matrix indicate the agreement of native contacts between the two conformations.

Again, we did the hydrogen bond analysis and four out of the five backbone hydrogen bonds of the native structure are predicted in the lowest energy structure found in the simulations. The pattern of backbone hydrogen bonds is shown in Table V. The secondary structure of the predicted and native conformation is also shown in Table V.

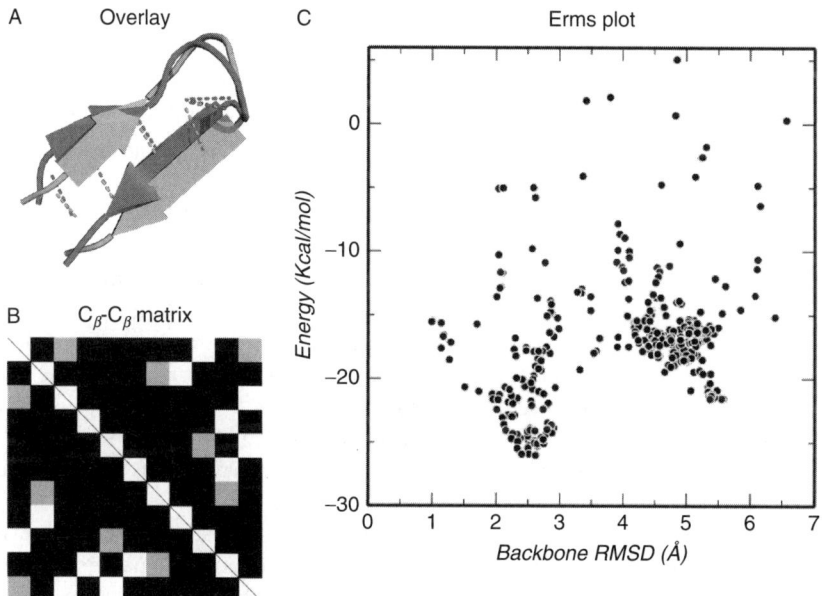

FIG. 23. 2EVQ: overlay of predicted (red) structure to experimental (green) structure. The overlay of the $C_\beta-C_\beta$ distance matrix and energy versus RMSD plot. (See Color Insert.)

TABLE V
2EVQ: Backbone Hydrogen Bond Pattern Between Native and Predicted Conformations and Secondary Structure Information

Hydrogen bond	Native	Predicted
02 THR HN → 11 THR 0	X	X
04 ASN HN → 09 LYS 0	X	X
07 THR HN → 04 ASN 0	X	
08 GLY HN → 04 ASN 0	X	X
11 THR HN → 02 THR 0	X	X

	Secondary structure	RMSD (Å)
Native	CEEETTTTEEC	–
Predicted	CEEETTTTEEEC	2.62

As eight of 10 simulations converged to native-like conformation without any competing metastable conformations, the folding is concluded as reproducible and predictive.

4. C-Terminal Hairpin of the Protein G

The C-terminal hairpin of protein G has been subjected to various scientific studies on β-sheet formations (*109–112*) and is considered stable in isolation from the rest of the protein.

Here, we study the folding of this hairpin domain in PFF02 with basin hopping simulations. We started 10 independent basin hopping simulations for 100 cycles in PFF02. The staring conformation was completely extended and had RMSD of 15.8 Å to the native conformation.

We found that only one of 10 simulations converged to a sheet-like conformation, while the remaining nine simulations are always stuck at the helical conformations. The lowest energy conformation is a β-hairpin and has only 1.27 Å RMSD to the native conformation with energy of -27.3 kcal mol^{-1}. The energy of the lowest helical conformation is -26.9 and is thus only 0.4 kcal mol^{-1} away.

The scatter plot of all conformation visited during the simulations is shown in Fig. 24C. It can be easily seen that very few conformations are native like and there are many conformations at about 6 Å RMSD. There are almost no conformations in the region between 6 and 1 Å indicating the presence of a huge barrier between the helical conformation and the native conformation,

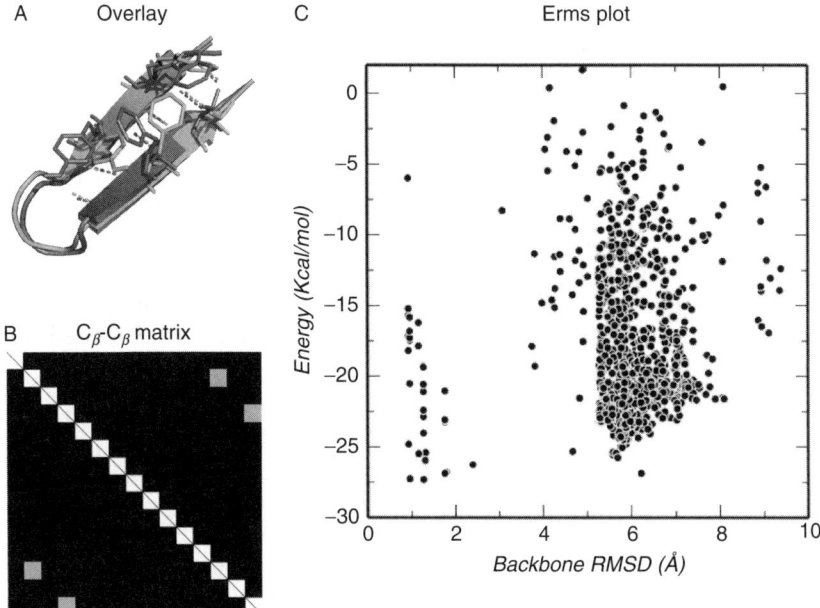

FIG. 24. C-terminal hairpin of protein G: overlay of predicted (red) structure to experimental (green) structure. The overlay of the C_β–C_β distance matrix and energy versus RMSD plot. (See Color Insert.)

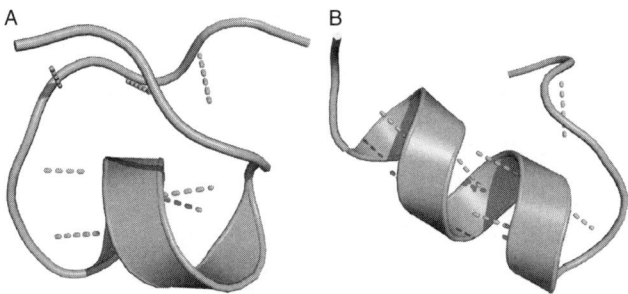

FIG. 25. C-terminal hairpin of protein G: misfolded structures with more backbone hydrogen bonds and more helical content.

which is not crossed by most of the simulations. The landscape for this hairpin appears to be very complex. Some of the misfolded helical conformations are shown in Fig. 25.

The overlay of the predicted (red) and native (green) conformation is shown in Fig. 24A and the corresponding C_β–C_β distance matrix is shown in Fig. 24B. The C_β–C_β overlay matrix is completely black indicating complete agreement of native contacts between the two conformations.

Again, we did the hydrogen bond analysis and all six backbone hydrogen bonds of the native conformation are predicted in the lowest energy conformation found in the simulations. The pattern of backbone hydrogen bonds is shown in Table VI. The secondary structure of the predicted, native and misfolded conformation is also shown in Table VI.

Although the lowest energy conformation has near perfect native contacts and backbone hydrogen bonding pattern, the simulation is neither predictive nor reproducible.

5. Designed Stable β-Hairpin: 1J4M

The hairpin 1J4M is a designed stable β-hairpin (113). It is designed to be extremely stable in the β-sheet conformation.

Here, we studied the folding of 1J4M with basin hopping simulations. We performed 10 independent simulations with 100 basin hopping cycles in PFF02. The simulations were started from completely extended conformation of the protein which had a RMSD of 13.3 Å to the native state.

TABLE VI
C-Terminal Hairpin of Protein G: Backbone Hydrogen Bond Pattern Between Native and Predicted Conformations and Secondary Structure Information

Hydrogen bond	Native	Predicted
02 GLU HN → 15 THR 0	X	X
04 THR HN → 13 THR 0	X	X
06 ASP HN → 11 THR 0	X	X
11 THR HN → 06 ASP 0	X	X
13 THR HN → 04 THR 0	X	X
11 THR HN → 02 GLU 0	X	X
	Secondary structure	RMSD (Å)
Native	CEEEEETTTTEEEEEC	–
Predicted	CEEEEETTTTEEEEEC	1.27
Misfolded	CEECHHHHHHHSEECC	6.22

Nine out of 10 independent simulations after 100 cycles of greedy basin hopping found the β-sheet like conformations and converge to less than 3.0 Å RMSD to the native conformation. The lowest energy conformation has an RMSD of 2.46 Å to the native conformation and had energy of 29.9 kcal mol^{-1}.

The energies reported here are positive as the native conformation has some covalently bound atoms which are clashing in PFF02. As the bond distances are kept fixed in PFF02 from the starting conformation for all simulations, thus introducing a constant bias and keep the energies comparable.

The scatter plot of all conformations visited during the simulation is shown in Fig. 26C. The overlay of the predicted (red) and native (green) conformation is shown in Fig. 26A and the corresponding C_β–C_β distance matrix is shown in Fig. 26B. The C_β–C_β overlay matrix is completely black indicating complete agreement of native contacts between the two conformations.

Again we did the hydrogen bond analysis and four out of the five backbone hydrogen bonds of the native conformation are predicted in the lowest energy conformation found in the simulations. The pattern of backbone hydrogen bonds is shown in Table VII. The secondary structure of the predicted, native and misfolded conformation is also shown in Table VII.

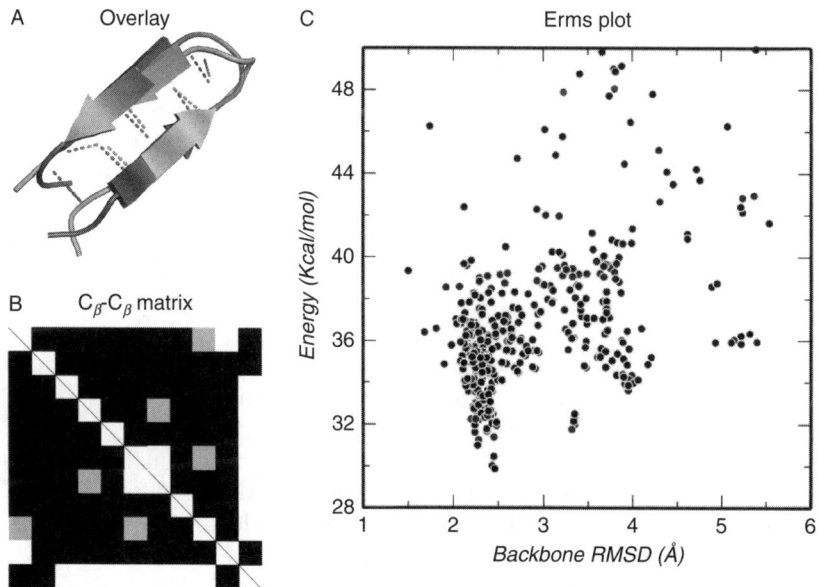

FIG. 26. 1J4M: overlay of predicted (red) structure to experimental (green) structure. The overlay of the C_β–C_β distance matrix and energy versus RMSD plot. (See Color Insert.)

TABLE VII
1J4M: Backbone Hydrogen Bond Pattern Between Native and Predicted Conformations and Secondary Structure Information

Hydrogen bond	Native	Predicted
04 TRP HN → 11 TYR 0	X	X
06 TYR HN → 09 ILE 0	X	X
09 ILE HN → 06 TYR 0	X	X
11 TYR HN → 04 TRP 0	X	X
13 GLY HN → 11 TYR 0	X	
	Secondary structure	RMSD (Å)
Native	CCCEEETTEEECCC	–
Predicted	CCEEEETTEEEECC	2.46

As nine of 10 simulations converged to native-like conformation without any competing metastable conformations, the folding is concluded as reproducible and predictive.

C. Three-Stranded Sheet (GSGS Peptide)

In this section, we move our folding studies from simple two-stranded β-hairpins to slightly more complicated β-sheet structures. The GSGS peptide is an antiparallel β-sheet with three strands (114) which was extensively investigated with phenomenological and all-atom MD studies (115–117).

We studied the folding of this three-stranded peptide with BHT in PFF02. We performed 200 cycles of basin hopping simulations for 20 independent simulations. The starting conformations were chosen randomly and had no secondary structure information.

We found that three of four lowest energy trajectories converge to near-native conformations with a bRMSD to the native conformation of 2.19, 2.26, and 2.67 Å, respectively.

The scatter plot of all conformations visited during the simulation is shown in Fig. 27C. There are metastable conformations around 4.5 Å and have a random coil conformation. The overlay of the predicted (red) and native (green) conformation is shown in Fig. 27A and the corresponding C_β–C_β distance matrix is shown in Fig. 27B. Many blocks in the C_β–C_β overlay matrix are black indicating good agreement of native contacts between the two conformations.

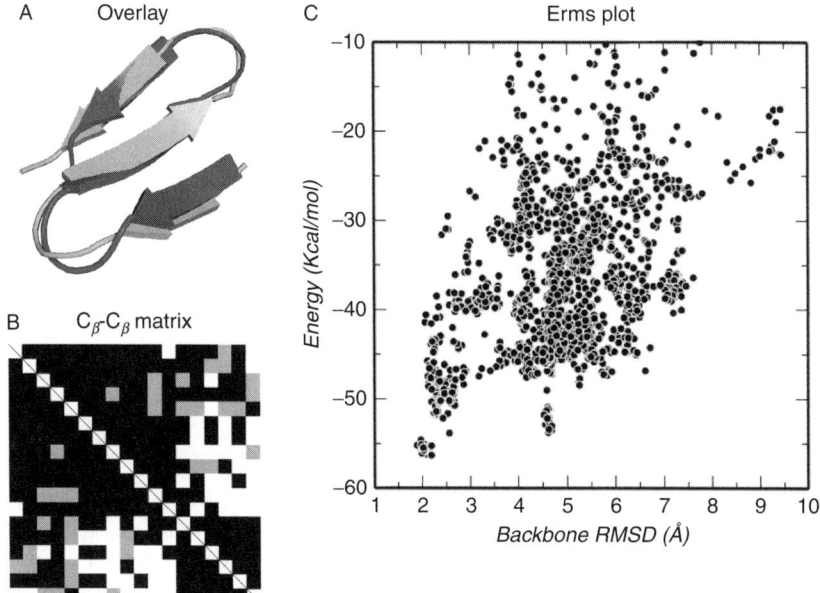

FIG. 27. GSGS peptide: overlay of predicted (red) structure to experimental (green) structure. The overlay of the C_β–C_β distance matrix and energy versus RMSD plot. (See Color Insert.)

The folded conformation of the GSGS shows a perfect alignment of the three secondary structure elements and only small deviations in the loops connecting the defined secondary structure elements. We have performed 20 independent basin hopping simulations on the 20 amino acid GSGS peptide. Predictive reproducible folding of the mini-protein is thereby achieved.

Lined up independently the β-sheet the regions from (2 to 5, 8 to 13, 16 to 19) agree to within 0.50, 0.55, 0.55 Å with the native conformation. The C_β–C_β distance difference matrix for the GSGS peptide indicates perfect alignment to within experimental resolution.

D. Folding of a DNA-Binding Zinc Finger Motif

Zinc fingers are among the most abundant proteins in eukaryotic genomes and occur in many DNA-binding domains and transcription factors (118). They function in DNA recognition, RNA packaging, transcriptional activation protein folding and assembly and apoptosis. Many zinc fingers contain a Cys_2His_2-binding motif that coordinates the Zn-ion in $\alpha\beta\beta$-framework (119–121) and much effort is toward the engineering of novel zinc fingers (122). A classical zinc finger motif-binding DNA is illustrated in Fig. 28. The reproducible

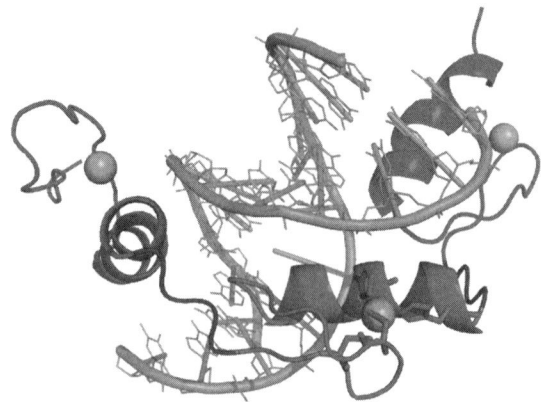

FIG. 28. A classical Cys$_2$His$_2$ zinc finger motif with Zn-ion (orange) and DNA (magenta). (See Color Insert.)

folding of such proteins with mixed secondary structure, however, remains a significant challenge to the accuracy of the all-atom forcefield and the simulation method (123). We use the all-atom free-energy forcefield PFF02 to predictively fold the 23–51 amino acid segment of the N-terminal subdomain of ATF-2 (PDB ID 1BHI) (124), a 29 amino acid peptide that contains the basic leucine zipper motif. 1BHI folds into the classical TFIIIa conformation found in many zinc finger-like subdomains. The fragment contains all the conserved hydrophobic residues (PHE25, PHE36, LEU42) of the classical zinc finger motif and the CYS27, CYS32, HIS45, HIS49 zinc-binding pattern.

Starting from a completely unfolded conformation with no secondary structure (16 Å bRMSD to native), we performed 200 cycles of the evolutionary algorithm. The distribution of bRMSD versus energy of all accepted conformations during the simulation (Fig. 29) demonstrates that the simulation explores a wide variety of conformations, with regard to their free energy and their deviation from the native conformation. Among the 10 energetically lowest conformations (see Table VIII) sixfold into near-native conformations with bRMSDs of 3.68–4.28 Å, while fourfold to conformations with a larger bRMSD. The three energetically best conformations are all near-native in character, an overlay with the experimental conformation (left panel of Fig. 15) illustrates that the helix, β-sheet, and both turns are correctly formed. The hydrophobic residues, which determine the packing of the β-sheet against the helix, are illustrated in blue in the figure. The helical section (GLU39–GLU50) and the β-sheet (PHE25–LEU26 and ARG35–PHE36) deviate individually by 1.6 and 2.4 Å bRMSD from their experimental counterparts, respectively. The overall deviation between the experimental and the folded conformations stems from the relative arrangement of the β-sheet with respect

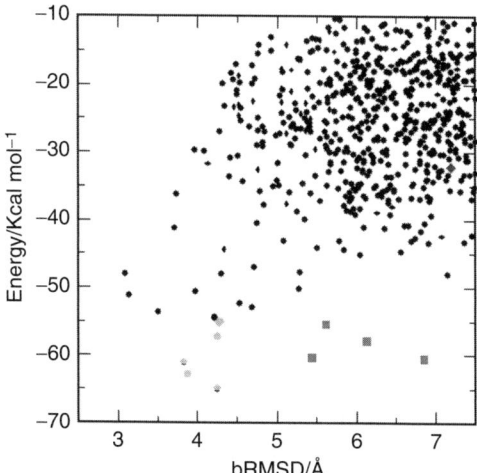

FIG. 29. Free energy versus bRMSD of all accepted conformations in the simulation. The best 10 structures are highlighted as red circles (native-like), green squares (nonnative). The folding intermediate is denoted by blue diamond. (See Color Insert.)

TABLE VIII
ENERGY, BRMSD, AND SECONDARY STRUCTURES OF BEST 10 LOWEST ENERGY STRUCTURES

No.	Energy (kcal mol^{-1})	bRMSD (Å)	Secondary structure
E01	−64.94	4.25	CCEECTTTTSCCEESSCHHHHHHHHHHHC
E02	−62.84	3.88	CCEECTTTTSCCEESSCHHHHHHHHSTTC
E03	−61.05	3.83	CCEECTTTTCCCEESSCHHHHHHHHSTTC
E04	−60.51	6.85	CCEECTTTTSCCEECSCHHHHHHSCCCCC
E05	−60.40	5.44	CCBBCTTTTCCCBCCSCHHHHHHHCCCBC
E06	−57.93	6.12	CCEECTTTTSCCEECSCHHHHHHSCCCCC
E07	−56.21	4.25	CCEEEECSSSSCEEEESCHHHHHHHHHHC
E08	−55.44	5.61	CCSSSCSSCCSSCCCSCHHHHHHHHTTTC
E09	−55.18	4.27	CCCCEECTTSSCEECSHHHHHHHHHCSCC
E10	−55.02	−4.29	CCCCBTTTTBTTCCCSSHHHHHHHHHHHC

to the helix, which are dominated by unspecific hydrophobic interactions. All conserved hydrophobic side chains are also buried in the folded structure. The zinc-coordinating cysteine residues (CYS27, CYS32) are within 2 Å of their native positions and available association with the Zn-ion.

Figure 31 (top panel) shows the convergence of the energy. After about 120 attempted updates per population member (3.5×10^8 function evaluations), the population converged to the native ensemble. According to the funnel paradigm for protein folding (19), tertiary structure forms as the protein slides downhill on the free-energy surface from the unfolded ensemble toward the native conformation. Each annealing cycle generates a small perturbation on the existing conformation, which averages to a 0.5 Å bRMSD change (max 3 Å initially). As new low-energy conformations replace old conformations, the population slides as a whole down the funnel of the free-energy landscape.

Ensemble averages as a function of time over the moving population are thus associated with different stages of the structure formation process. In the lower panels of Fig. 31, we plot the average helical content and the number of β-sheet H-bonds as a function of the cycle number. Following a rapid collapse to a compact conformation, the helix forms first, followed by the formation of the β-sheet. An analysis of the folding funnel upward in energy illustrates that the lowest energy metastable conformations correspond to a partial unzipping of amino acids PHE25–ARG35, while the conserved cysteine residues are still buried. Even much higher on the free-energy funnel (blue diamond in Fig. 29), we find many structures that have much residual structure, but essentially not long-range native contacts. The preformed sheet region is stabilized by H-bonds (LEU26–CYS27, ARG35) and packs at a right angle to the helix, the hydrophobic residues are only partially buried. This conformational freedom may be relevant in DNA binding, where the helical part of the zinc finger packs into the major groove of the DNA.

De novo folding of the zinc finger domain permits a direct sampling of the relevant low-energy portion of the free-energy surface of the molecule as a first step toward the elucidation of the structural mechanisms involved in DNA binding (125). We find that much of the structure of the zinc finger is formed even in the absence of the metal ion that is ultimately required for the stabilization of the native conformation. Because the algorithm tracks the development of the population, it is possible to reconstruct a folding pathway by reconstructing the sequence of events starting with converged conformation and moving backward to the completely unfolded conformation.

Crucial steps along the continuous folding pathway are illustrated in Fig. 32 (note that there is no quantitative mapping onto the time axis). The early folding process is characterized by helix nucleation and concurrent collapse into a globular conformation with a radius of gyration that is comparable to that of the native conformation. The simulation then explores conformations of the same spatial extent with increasing helical, but no β-sheet content. Lower in free energy the simulation samples conformations in which partially formed β-sheets pack against the helix. On the basis of the free-energy estimate to conformations without the helix (8 kcal mol^{-1}) such conformations can be

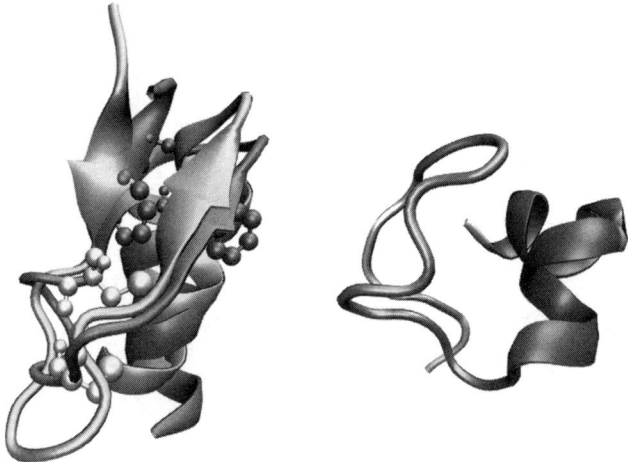

FIG. 30. Left: overlay of the native (green) and folded (magenta) conformations. The conserved hydrophobic residues are shown in blue and Zn-binding cysteines are shown in yellow. Right: the intermediate conformation with partially formed helix and β-sheet. (See Color Insert.)

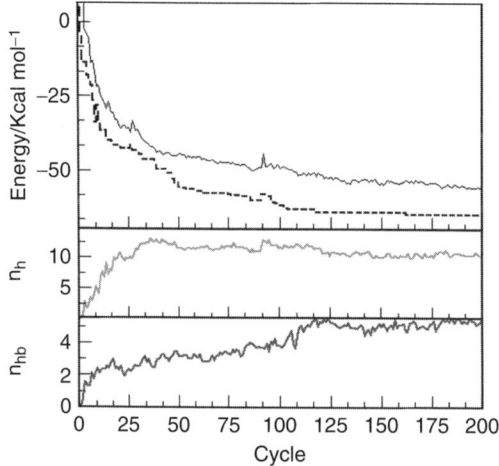

FIG. 31. Top: average (solid line) and best (dashed line) energies; middle: number of amino acids (n_h) in a helical conformation (as computed by DSSP); and bottom: number of hydrogen bones (n_{hb}) as function of the ES cycle number.

explored in DNA binding and transcription. Our simulation approach permits a rapid exploration of this free-energy region and thus characterizes the biologically active ensemble. MET51 packs in all low-energy conformations against

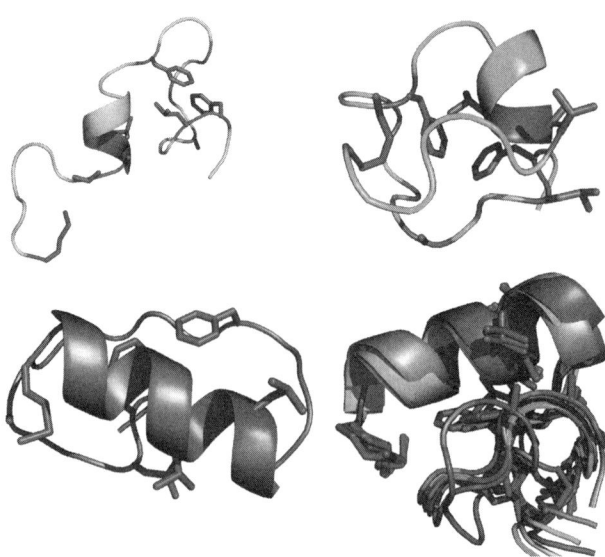

FIG. 32. Key events in the folding: helix nucleation (top left), collapsed globular conformation (top right), fully formed helix (bottom left), partially formed β-sheets using the helix as a template (bottom right). (See Color Insert.)

the combined scaffold and acts as a closure of the DNA-binding motif. It may thus provide an enthalpic contribution to a nonstandard helix-capping motif that differs from the TGEKP linker sequence observed in multifinger domains (118) in several zinc fingers.

We have thus demonstrated predictive all-atom folding of the DNA-binding zinc finger motif in a free-energy forcefield PFF02. This investigation offers the first unbiased characterization of the low-energy free-energy surface of the zinc finger motif, which is unattainable in coarse-grained, knowledge-based models.

VI. Summary

Using the free-energy approach, we have folded more than 13 proteins to near-native conformations in simulations that started from extended conformations. We started with three helical proteins using the optimized BHT that were earlier folded in PFF01. The forcefield correctly predicts the native-like states at lowest energies for these proteins. We then studied folding of larger helical proteins (50–60 amino acids) using evolutionary algorithm. In certain cases, we observe many metastable conformations which have same secondary structures with different arrangement. Mirror Abhinav/images also seem to have competing energies and make selection difficult.

Next, we studied six hairpins in PFF02 to study to folding of proteins β-sheet secondary structure. All of the six hairpins fold into native-like conformations with correct pattern of backbone hydrogen bonds. The correct hydrogen bond pattern ensures that the hairpin has the correct bend and the side chains are also projected in the same directions. There are not many helical metastable conformations for these hairpins except the C-terminal domain of protein G. This hairpin did not fold predictively and most simulations were stuck in higher energy helical like conformations.

We finally studied a three-stranded β-sheet and a mixed protein which constitutes both helix and β-sheets with optimized BHT. Both these proteins were reproducibly and predictively folded in PFF02.

The overview of these folding simulations is given in Table IX.

We have, therefore, studied proteins spanning both helix and sheet secondary structural elements. Five helical, six hairpins, one three-stranded β-sheet and one mixed protein were folded in PFF02 using stochastic optimization methods. The average RMSD for the lowest energy structures to their respective native conformation for these 13 proteins is only 2.87 Å. The study

TABLE IX
OVERVIEW OF FOLDING STUDIES IN PFF02

PDB ID	N	Topology	RMSD (Å)
1L2Y	20	α	3.11
1WQE	23	$\alpha\alpha$	2.33
1F4I	40	$\alpha\alpha\alpha$	3.29
1ENH	54	$\alpha\alpha\alpha$	3.40
1EDK	56	$\alpha\alpha\alpha$	4.05
1LE0	12	$\beta\beta$	1.50
1NIZ	14	$\beta\beta$	2.04
1U6U	17	$\beta\beta$	4.57
2E4Q	12	$\beta\beta$	2.62
G Cterm	16	$\beta\beta$	1.67
1J4M	14	$\beta\beta$	2.46
GSGS	20	$\beta\beta\beta$	2.19
IBHI	29	$\alpha\beta\beta$	4.28

G Cterm is the C-terminal hairpin of protein G, GSGS is the synthetic three-stranded β-peptide, and N indicates the number of amino acids in the protein.

included both helical and sheet like proteins along with a mixed system varying from 12 to 56 amino acids. PFF02 is thus able to predict the native state of a wide range of proteins at the global minimum of their free-energy surface and the basin hopping technique and evolutionary algorithm were able to locate this free-energy surface.

Protein folding with free-energy methods is much faster than the direct simulation of the folding pathway by kinetic methods such as MD. Using just standard PCs we can fold a simple hairpin with fifteen to 20 amino acids in a matter of hours, at most in a day. Unfortunately even for free-energy methods the computational cost rises steeply with the system size and for this reason it is impossible to test the full range of applicability of PFF02 for large family of proteins in a direct folding study. There is, however, an indirect way to test the viability of the free-energy forcefield using a large database (decoy set) of possible conformations for a given protein, including some near-native conformations.

The second ingredient in protein-folding studies, aside from the forcefield, is the simulation protocols, which ultimately determine whether the global optimum of the forcefield is determined accurately and reliably. We have therefore attempted to develop and adopt such methods, for example, the stochastic tunneling or the basin hopping technique, which had proven successful in early folding studies for small proteins, in order to find a particularly efficient algorithm. We experimented with all parameters of these methods that included the number of steps and starting and final temperatures.

BHT was modified by increasing the number of steps with every basin hopping cycle and the starting temperatures for annealing were taken from an exponential distribution. This protocol increased the convergence of the basin hopping simulations. This protocol was further modified to a "greedy" version, which always retains the best energy conformation found so far. These improvements together increased the speed and reliability of the simulations and resulted in lower final energies, which is the goal for these optimization problems.

One of the key limitations of these methods is that they map the global optimization problem onto a single fictitious dynamical process. In this type of simulation protocol, the molecule constructs one trajectory starting somewhere in the unfolded ensemble, which hopefully converges toward the native conformation. Even with standard basin hopping simulations, several simulations are necessary to obtain a predictive and reproducible result. In the standard protocol the simulations are completely independent of one another. This raises the obvious question, whether an improved convergence can be obtained by coupling a number of concurrent dynamical processes. The second, related question concerns the largest number of concurrent processes that can be coupled together to speed the overall search. In this respect, optimization-based methods have a significant advantage over traditional kinetic methods, because the latter must ultimately strive to construct one single consecutive trajectory. The only option to speed the simulation for a single trajectory is the parallelization of the energy and force evaluation, which requires a large amount of data transfer. The optimization methods using a large number of concurrent dynamical processes, on the other hand, are able to use coarse-grained strategies in which a single processor performs one of many largely independent simulations.

We have implemented an evolutionary algorithm on massively parallel architectures such as the BlueGene computer. The algorithm is implemented in a master–client model which keeps a diverse population on the master and the clients sample the protein landscape simultaneously and return to the master. The algorithm scales very well with the number of processors used (up to 4096 tested on the IBM BlueGene). Using this algorithm, we folded various proteins such as 40 amino acid HIV accessory protein (1F4I) and 54 amino acid engrailed homeodomain protein (1ENH) in a single day. The folding of the engrailed homeodomain protein was carried out in a single day using 512 processors on the Barcelona Mare Nostrum Supercomputer, the current largest supercomputer in Europe. This is a great achievement as the folding of a protein of comparable size required about 4 months using 50 processors in earlier studies. The folding of the tryptophan zipper proteins (1LE0) was possible in only 14 min using 128 processors.

Using PFF02 along with modified versions of the BHT, we could fold several protein structures starting from completely extended conformations. These include various helical proteins, the tryptophan cage protein (1L2Y), the HIV accessory protein (1F4I), and a potassium channel blocker protein (1WQC) which were earlier folded in PFF01. The tryptophan cage protein is a widely studied model for protein folding both theoretically and experimentally. The HIV accessory protein and potassium channel blockers are biologically important proteins. The HIV accessory protein destroys the host cell's ability to survive by binding to a host receptor and restricting an important enzyme to activate the cell's immune system. Potassium channel blockers are toxic venom peptides involved in blocking of potassium channel in cells. We also folded much larger and widely studied model proteins (both experimentally and theoretically) like the engrailed homeodomain protein (1ENH) and E-domain of the staphylococcal protein A (1EDK), which were folded with the evolutionary algorithm.

We then investigated the folding of various β-hairpins in PFF02. These hairpins included tryptophan zipper protein (1LE0), HIV-1 V3 loops (1NIZ, 1U6U), designed stable β-proteins (2EDK, 1J4M), and the C-terminal hairpin of G protein to experimental resolution. The tryptophan zipper protein and C-terminal hairpin of protein G have been subjected to many theoretical and experimental studies. The HIV-1 V3 loops are highly homologous loops which have a different hydrogen bonding pattern responsible for coreceptor selectivity by the virus. The loop conformation is responsible for selecting infection of T cells or macrophages. The folding of these loops (1NIZ and 1U6U) is particularly encouraging because PFF02 can distinguish these very similar sequences and correctly predicts a one residue shift in backbone hydrogen bonding pattern resulting in different side chains orientation responsible for coreceptor selectivity of the virus protein. The experimentally stable hairpins serve as good model systems for studies on β-sheet formation and folding.

Apart from two-stranded β-hairpins, we also studied the folding of the three-stranded GSGS peptide. The GSGS peptide is a designed stable three-stranded β-sheet with glycine–serine (GS) bends and has been a model system for three-stranded β-sheet formation. We finally studied the folding of $\alpha\beta\beta$-zinc finger domain protein 1BHI. Zinc fingers are among the most abundant proteins in eukaryotic genomes and occur in many DNA-binding domains and transcription factors.

VII. Outlook

To date, we have succeeded to develop methods to find the native state of various proteins by locating the global minimum of the free-energy surface. There are, however, a large number of questions that remain to be addressed. Fortunately, there are complementary methods, which in combination with the free-energy methodology developed here, can address these problems. For example, we have neglected the details of the kinetics of protein folding in our approach. As stated earlier, it is important to study kinetics of folding to understand protein-folding mechanism and to predict folding rates. Because free-energy methods sample exhaustively the low-energy conformations of the protein that are accessible under physiological conditions it may be possible to reconstruct the folding kinetics on the basis of that ensemble of conformations. This can be achieved by a dynamical analysis of the low-energy region by using master equations assuming diffusive processes between similar conformations.

A related interesting aspect of protein folding is the study of transition states. Transition states are the saddle points of the free-energy surface that connect the unfolded state and the folded state. Computationally transition states can be determined by a so-called p-fold analysis, that is, searching for protein conformations that fold or unfold with the equal probability at some finite temperature. Experimentally, transition-state analysis is carried out by mutating the sequence of the protein and measuring the changes in kinetics and equilibria of protein folding (ψ-value analysis). This raises the question of protein stability under mutations. The latter question can be addressed by computing the free-energy difference between the folded and unfolded ensemble for a variety of mutations.

Also, further developments could be made in the direction of protein–protein interactions. These studies can help understand protein aggregation, which are responsible for various diseases, such as Alzeimer's or Parkinson's disease. We have already implemented modules in our simulation package that can treat protein–protein interactions and the first studies regarding protein–protein docking are presently under way.

Finally, we must address the question how we can fold even larger proteins, with more than 100 amino acids. We have encountered the problem of freezing when studying large proteins. Once the protein collapses, it is difficult to generate no clashing Monte Carlo moves, which leads to poor acceptance ratio. As a result, the protein cannot explore the conformational space. Further development of methods which are faster in locating the global minimum is still required to study all-atom folding of proteins over hundred amino acids. One possible solution to this is by splitting the protein into fragments and later joining them to obtain tertiary structure. This method can generate native like conformations for the protein which can be further relaxed and identified. Such methods have been very useful in the field of protein structure prediction.

With the development of the all-atom protein forcefield (PFF02), we have made a significant step toward a universal free-energy approach to protein folding and structure prediction. The massively parallel simulation methods developed in the last few years now permit the protein folding of medium-size proteins from random initial conformations. This work thus lays the foundations to further explore the mechanism of protein folding, to understand protein stability and ultimately develop methods for *de novo* protein structure prediction.

REFERENCES

1. Branden, C., and Tooze, J. (1999). "Introduction to Protein Structure." Routledge, New York.
2. Berg, J., Tymoczko, J., and Stryer, L. (2001). "Biochemistry." Michelle Julet.
3. Weatherford, D. W., and Salemme, F. R. (1979). Conformations of twisted parallel β-sheets and the origin of chirality in protein structures. *Proc. Natl Acad. Sci. USA* **76,** 19–23.
4. Mason, S. A. (1984). Origins of biomolecular handedness. *Nature* **311,** 19–23.
5. Lesk, A. M. (2001). "Introduction to Protein Architecture." Oxford University Press, Oxford.
6. Ramachandran, G. N., and Mitra, A. K. (1976). An explanation for the rare occurrence of *cis* peptide units in proteins and polypeptides. *J. Mol. Biol.* **107,** 85–92.
7. Ramachandran, G. N., Ramakrishnan, C., and Sasisekharan, V. (1963). Stereochemistry of polypeptide chain configurations. *J. Mol. Biol.* **7,** 95–99.
8. Hollecker, M., and Creighton, T. E. (1982). Effect on protein stability of reversing the charge on amino groups. *J. Mol. Biol.* **701,** 395–404.
9. Matthews, B. W. (1987). Genetic and structural analysis of the protein stability problem. *Biochemistry* **26,** 6885–6888.
10. Anfinsen, C. B. (1973). Principles that govern the folding of protein chains. *Science* **181,** 223–230.

11. Privalov, P. L. (1979). Stability of proteins: Small globular proteins. *Adv. Protein Chem.* **33**, 167–241.
12. Privalov, P. L., and Gill, S. J. (1988). Stability of protein structure and hydrophobic interaction. *Adv. Protein Chem.* **39**, 191–234.
13. Baldwin, R. (1986). Temperature dependence of the hydrophobic interaction in protein folding. *Proc. Natl Acad. Sci. USA* **83**, 8069–8072.
14. Dill, K. A. (1990). Dominant forces in protein folding. *Biochemistry* **29**, 7155–8133.
15. Dill, K. A., Bromberg, S., Yue, K., Fiebig, K. M., Yee, D. P., Thomas, P. D., and Chan, H. S. (1995). Principles of protein folding—A perspective from simple exact models. *Protein Sci.* **4**, 561–602.
16. Makhatadze, G. I., and Privalov, P. L. (1996). On the entropy of protein folding. *Protein Sci.* **5**, 501–510.
17. Levinthal, C. (1968). Are there pathways for protein folding? *J. Chem. Phys.* **65**, 44–45.
18. Leopold, P. E., Montal, M., and Onuchic, J. N. (1992). Protein folding funnels: A kinetic approach to the sequence–structure relationship. *Proc. Natl Acad. Sci. USA* **89**, 8721–8725.
19. Onuchic, J. N., Luthey-Schulten, Z., and Wolynes, P. G. (1997). Theory of protein folding: The energy landscape perspective. *Annu. Rev. Phys. Chem.* **48**, 545–600.
20. Chan, H. S., and Dill, K. A. (1998). Protein folding in the landscape perspective: Chevron plots and non-Arrhenius kinetics. *Proteins Struct. Funct. Genet.* **30**, 2–33.
21. Go, N. (1983). Theoretical studies of protein folding. *Annu. Rev. Biophys. Bioeng.* **12**(1), 183–210.
22. Cheung, M. S., Garcia, A. E., and Onuchic, J. N. (2002). Protein folding mediated by solvation: Water expulsion and formation of the hydrophobic core occur after the structure collapse. *Proc. Natl Acad. Sci. USA* **11**, 2351–2361.
23. Head-Gordon, T., and Brown, S. (2003). Minimalist models for protein folding and design. *Curr. Opin. Struct. Biol.* **13**, 160–167.
24. Liwo, A., Oldziej, S., Pincus, M. R., Wawak, R. J., Rackovsky, S., and Scheraga, H. A. (1997). A united-residue force field for off-lattice protein-structure simulations. I. Functional forms and parameters of long-range side-chain interaction potentials from protein crystal data. *J. Comp. Chem.* **18**, 849–873.
25. Liwo, A., Pincus, M. R., Wawak, R. J., Rackovsky, S., Oldziej, S., and Scheraga, H. A. (1997). A united-residue force field for off-lattice protein-structure simulations. 2. Parameterization of short-range interactions and determination of weights of energy terms by z-score optimization. *J. Comp. Chem.* **18**(7), 874–887.
26. Liwo, A., Kazmierkiewicz, R., Czaplewski, C., Groth, M., Oldziej, S., Wawak, R. J., Rackovsky, S., Pincus, S., and Scheraga, H. A. (1998). United-residue force field for off-lattice protein-structure simulations. III. Origin of backbone hydrogen-bonding cooperativity in united-residue potentials. *J. Comp. Chem.* **19**(3), 259–276.
27. Oldziej, S., Czaplewski, C., Liwo, A., Chinchio, M., Nanias, M., Vila, J. A., Khalili, M., Arnautova, Y. A., Jagielska, A., Makowski, M., Schafroth, H. D., Kazmierkiewicz, R. *et al.* (2005). Physics-based protein-structure prediction using a hierarchical protocol based on the UNRES force field: Assessment in two blind tests. *Proc. Natl Acad. Sci. USA* **102**(21), 7547–7552.
28. Nanias, M., Czaplewski, C., and Scheraga, H. A. (2006). Replica exchange and multicanonical algorithms with the coarse-grained united-residue (unres) force field. *J. Chem. Theory Comput.* **2**(3), 513–528.
29. Allen, M. P., and Tildesley, D. J. (1987). "Computer Simulation of Liquids." Clarendon Press, Oxford.
30. Ponder, J. W., and Case, D. A. (2003). Force fields for protein simulations. *Adv. Protein Chem.* **66**, 27–85.

31. MacKerell, A. D., Jr., Bashford, D., Bellott, M., Dunbrack, R. L.,, Jr., Evanseck, J. D., Field, M. J., Fischer, S., Gao, J., Guo, H., Ha, S., Joseph-McCarthy, D., Kuchnir, L. et al. (1998). All-atom empirical potential for molecular modeling and dynamics studies of proteins. *J. Phys. Chem. B* **102**, 3586.
32. Momany, F. A., McGuire, R. F., Burgess, A. W., and Scheraga, H. A. (1975). Energy parameters in polypeptides. VII. Geometric parameters, partial atomic charges, nonbonded interactions, hydrogen bond interactions, and intrinsic torsional potentials for the naturally occurring amino acids. *J. Phys. Chem.* **79**, 2361.
33. Biosym/MSI, San Diego. (1995). Discover 2.9.7/95.0/3.0.0 User Guide.
34. van Gunsteren, W. F., and Berendsen, H. J. C. (1987). "Groningen Molecular Simulation (GROMOS) Library Manual (Biomos)." AG Groningen, The Netherlands.
35. Jorgeson, W. L. (1981). Quantum and statistical mechanical studies of liquids. 11. Transferable intermolecular potential functions. Application to liquid methanol including internal rotation. *J. Am. Chem. Soc.* **103**, 341.
36. Hagler, A. T., and Ewig, C. S. (1994). On the use of quantum energy surfaces in the derivation of molecular force fields. *Comp. Phys. Commun.* **84**, 131–155.
37. Adcock, S. A., and McCammon, J. A. (2007). Molecular dynamics: Survey of methods for simulating the activity of proteins. *Chem. Rev.* **106**, 1589–1615.
38. Frenkel, D., and Smit, B. (2001). "Understanding Molecular Simulations: From Algorithms to Applications." Academic Press, San Diego.
39. Sagui, C., and Daren, T. A. (1999). Molecular dynamics simulations of biomolecules: Long-range electrostatic effects. *Annu. Rev. Biophys. Biomol. Struct.* **28**, 155–179.
40. Sugita, Y., and Okamoto, Y. (1999). Ab initio replica-exchange Monte Carlo method for cluster studies. *Chem. Phys. Lett.* **314**, 141–151.
41. Czerminski, R., and Elber, R. (1991). Computational studies of ligand diffusion in globins. I. Leghemoglobin. *Proteins Struct. Funct. Bioinf.* **10**, 70.
42. Mackerell, A. D. (2004). Empirical force fields for biological macro-molecules: Overview and issues. *J. Comp. Chem.* **25**, 1584–1604.
43. Pearlman, D. A., Case, D. A., Caldwell, J. W., Ross, W. R., Cheatham, T. E., DeBolt, S., Ferguson, D., Seibel, G., and Kollman, P. (1995). Amber, a computer program for applying molecular mechanics, normal mode analysis, molecular dynamics and free energy calculations to elucidate the structures and energies of molecules. *Comp. Phys. Commun.* **91**, 1–41.
44. Scott, W. R. P., Hunenberger, P. H., Tironi, I. G., Mark, A. E., Billeter, S. R., Fennen, J., Torda, A. E., Huber, T., Kruger, P., and van Gunsteren, W. F. (1999). The GROMOS biomolecular simulation program package. *J. Phys. Chem. A* **103**, 3596–3607.
45. Jorgensen, W. L., Maxwell, D. S., and Tiradorives, J. J. (1996). Development and testing of the OPLS all-atom force field on conformational energetics and properties of organic liquids. *J. Am. Chem. Soc.* **118**, 11225–11236.
46. Halgren, T. A., and Damm, W. (2001). Polarizable force fields. *Curr. Opin. Struct. Biol.* **11**, 236–242.
47. Jorgensen, W. L., Chandrasekhar, J., Madura, J. D., Impey, R. W., and Klein, M. L. (1983). Comparison of simple potential functions for simulating liquid water. *J. Chem. Phys.* **79**(2), 926–935.
48. Orozco, M., and Luque, F. J. (2000). Theoretical methods for the description of the solvent effect in biomolecular systems. *Chem. Rev.* **100**, 4187.
49. McCammon, J. A., Gelin, B. R., and Karplus, M. (1977). Dynamics of folded proteins. *Nature* **267**, 585–590.
50. Duan, Y., and Kollman, P. A. (1998). Pathways to a protein folding intermediate observed in a 1-microsecond simulation in aqueous solution. *Science* **282**, 740–744.

51. Jayachandran, G., Vishal, V., and Pande, V. (2007). Using massively parallel simulation and markovian models to study protein folding: Examining the dynamics of the villin headpiece. *J. Chem. Phys.* **124**, 164902.
52. Snow, C. D., Zagrovic, B., and Pande, V. S. (2002). Folding kinetics and unfolded state topology via molecular dynamics simulations. *J. Am. Chem. Soc.* **124**, 14548–14549.
53. Zagrovic, B., Sorin, E. J., and Pande, V. (2001). Beta-hairpin folding simulations in atomistic detail using an implicit solvent model. *J. Mol. Biol.* **313**, 151–169.
54. Garcia, A. E., and Onuchic, N. (2003). Folding a protein in a computer: An atomic description of the folding/unfolding of protein A. *Proc. Natl Acad. Sci. USA* **100**, 13898–13903.
55. Rhee, Y. M., Sorin, E. J., Jayachandran, G., Lindahl, E., and Pande, V. S. (2004). Simulations of the role of water in the protein-folding mechanism. *Proc. Natl Acad. Sci. USA* **101**, 6456–6461.
56. Fersht, A. R., and Daggett, V. (2002). Protein folding and unfolding at atomic resolution. *Cell* **108**, 573–582.
57. Roux, B., and Schulten, K. (2004). Computational studies of membrane channels. *Structure* **12**, 1343.
58. Neria, E., and Karplus, M. (1997). Molecular dynamics of an enzyme reaction: Proton transfer in TIM. *Chem. Phys. Lett.* **267**, 23.
59. Wei, G., Mousseau, N., and Derreumaux, P. (2007). Simulations of the early steps of protein aggregation. *Prion J.* **1**, e1–e6.
60. Klimov, D. K., and Thirumalai, D. (2003). Dissecting the assemble of $a\beta_{16-22}$ amyloid peptides into antiparallel β sheets. *Structure* **11**, 295–307.
61. Herges, T., and Wenzel, W. (2004). An all-atom force field for tertiary structure prediction of helical proteins. *Biophys. J.* **87**(5), 3100–3109.
62. Avbelj, F., and Moult, J. (1995). Role of electrostatic screening in determining protein main chain conformational preferences. *Biochemistry* **34**, 755–764.
63. Berg, J. M., Tymoczky, J. L., and Stryer, L. (2002). "Biochemistry," 5th edn. W. H. Freeman and Company, New York.
64. Avbelj, F. (1992). Use of a potential of mean force to analyze free energy contributions in protein folding. *Biochemistry* **31**, 6290–6297.
65. McDonald, I. K., and Thornton, J. M. (1994). Satisfying hydrogen bonding potentials in protein folding. *J. Mol. Biol.* **238**, 777–793.
66. Sippl, M. J., Nemethy, G., and Scheraga, H. A. (1984). Intermolecular potentials from crystal data. 6. Determination of empirical potentials for O–HO=C hydrogen bonds from packing configurations. *J. Phys. Chem.* **88**, 6231–6233.
67. Lee, B., and Richards, F. M. (1971). The interpretation of protein structures: Estimation of static accessibility. *J. Mol. Biol.* **55**(3), 379–380.
68. Eisenberg, D., and McLachlan, A. D. (1986). Solvation energy in protein folding and binding. *Nature* **319**, 199–203.
69. Sharp, K. A., Nicholls, A., Friedman, R., and Honig, B. (1991). Extracting hydrophobic free energies from experimental data: Relationship to protein folding and theoretical models. *Biochemistry* **30**, 9686–9697.
70. Nayeem, A., Vila, J., and Scheraga, H. A. (1991). A comparative study of the simulated-annealing and Monte Carlo-with-minimization approaches to the minimum-energy structures of polypeptides: [Met]-enkephalin. *J. Comp. Chem.* **12**, 594–605.
71. Abagyan, R. A., and Totrov, M. (1994). Biased probability Monte Carlo conformational searches and electrostatic calculations for peptides and proteins. *J. Mol. Biol.* **235**, 983–1002.
72. Wales, D. J., and Doye, J. P. K. (1997). Global optimization by basin-hopping and the lowest energy structures of Lennard–Jones clusters containing up to 110 atoms. *J. Phys. Chem. A* **101**, 5111–5116.

73. Withers-Ward, E. S., Mueller, T. D., Chen, I. S., and Feigon, J. (2000). Biochemical and structural analysis of the interaction between the UBA(2) domain of the DNA repair protein HHR23A and HIV-1 Vpr. *Biochemistry* **39**, 14103–14112.
74. Herges, T., and Wenzel, W. (2004). An all-atom force field for tertiary structure prediction of helical proteins. *Biophys. J.* **87**, 3100–3109.
75. Leach, A. R. (2001). "Molecular Modelling: Principles and Applications." Pearson Education Ltd, Harlow, England.
76. Kirkpatrick, S., Gelatt, C. D., Jr., and Vecchi, M. P. (1983). Optimization by simulated annealing. *Science* **220**, 671–680.
77. Hamacher, K., and Wenzel, W. (1999). A stochastic tunnelling approach for global minimization. *Phys. Rev. E* **59**, 938.
78. Schug, A., Herges, T., Verma, A., and Wenzel, W. (2005). Investigation of the parallel tempering method for protein folding. *Phys. Cond. Matter, Special Issue: Structure and Function of Biomolecules* **17**, 1641–1650.
79. Leitner, D. M., Chakravarty, C., Hinde, R. J., and Wales, D. J. (1997). Global optimization by basin-hopping and the lowest energy structures of Lennard–Jones clusters containing up to 110 atoms. *Phys. Rev. E* **56**, 363.
80. Wales, D. J., and Dewsbury, P. E. J. (2004). Effect of salt bridges on the energy landscape of a model protein. *J. Chem. Phys.* **121**, 10284–10290.
81. Mortenson, P. N., and Wales, D. J. (2001). Energy landscapes, global optimisation and dynamics of the polyalanine Ac(ala)8 NHMe. *J. Chem. Phys.* **114**, 6443–6454.
82. Mortenson, P. N., Evans, D. A., and Wales, D. J. (2002). Energy landscapes of model polyalanines. *J. Chem. Phys.* **117**, 1363–1376.
83. Schneider, J., Morgenstern, I., and Singer, J. M. (1998). Bouncing towards the optimum: Improving the results of Monte Carlo optimization algorithms. *Phys. Rev. E* **58**, 5085–5095.
84. Nayeem, A., Vila, J., and Scheraga, H. A. (1991). A comparative study of the simulated-annealing and Monte Carlo-with-minimization approaches to the minimum-energy structures of polypeptides: [Met]-enkephalin. *J. Comp. Chem.* **12**(5), 594–605.
85. Wenzel, W. (2006). Predictive folding of a β hairpin in an all-atom free-energy model. *Europhys. Lett.* **76**, 156.
86. Neidigh, J. W., Fesinmeyer, R. M., and Andersen, N. H. (2002). Designing a 20-residue protein. *Nat. Struct. Biol.* **9**, 425–430.
87. Schug, A., Herges, T., and Wenzel, W. (2003). Reproducible protein folding with the stochastic tunneling method. *Phys. Rev. Lett.* **91**, 1581021–1581024.
88. Ding, F., Buldyrev, S. V., and Dokholyan, N. V. (2005). Folding Trp-cage to NMR resolution native structure using a coarse-grained protein model. *Biophys. J.* **88**, 147–155.
89. Linhananta, A., Boer, J., and MacKay, I. (2005). The equilibrium properties and folding kinetics of an all-atom Go model of the Trp-cage. *J. Chem. Phys.* **122**, 1–15.
90. Schug, A., Wenzel, W., and Hansmann, U. H. E. (2005). Energy landscape paving simulations of the trp-cage protein. *J. Chem. Phys.* **122**, 1–7.
91. Schug, A., Herges, T., Verma, A., Lee, K. H., and Wenzel, W. (2006). Comparison of stochastic optimization methods for all-atom folding of the trp-cage protein. *ChemPhysChem* **6**, 2640–2646.
92. Juraszek, J., and Bolhuis, P. G. (2006). Sampling the multiple folding mechanisms of trp-cage in explicit solvent. *Proc. Natl Acad. Sci. USA* **103**, 15859–15864.
93. Chagot, B., Pimentel, C., Dai, L., Pil, J., Tytgat, J., Nakajima, T., Corzo, G., Darbon, H., and Ferrat, G. (2005). An unusual fold for potassium channel blockers: NMR structure of three toxins from the scorpion Opisthacanthus madagascariensis. *Biochem. J.* **388**, 263–271.
94. Wenzel, W. (2006). Predictive folding of a β-hairpin in an all-atom free-energy model. *Europhys. Lett.* **76**, 156–162.

95. Schug, A., Herges, T., and Wenzel, W. (2004). All-atom folding of the three-helix HIV accessory protein with an adaptive parallel tempering method. *Proteins* **57**, 792–798.
96. Clarke, N. D., Kissinger, C. R., Desjarlais, J., Gilliland, G. L., and Pabo, C. O. (1994). Structural studies of the engrailed homeodomain. *Protein Sci.* **3**, 1779–1787.
97. Mayor, U., Guydosh, N. R., Johnson, C. M., Grossmann, J. G., Sato, S., Jas, G. S., Freund, S. M., Alonso, D. O., Daggett, V., and Fersht, A. R. (2003). The complete folding pathway of a protein from nanoseconds to microseconds. *Nature* **421**, 863–867.
98. Daggett, V., and Fersht, A. (2003). The present view of the mechanism of protein folding. *Nat. Rev. Mol. Cell Biol.* **4**, 497–502.
99. Verma, A., and Wenzel, W. (2006) All-atom protein folding in a single day (submitted for publication).
100. Starovasnik, M. A., Skelton, N. J., O'Connell, M. P., Kelley, R. F., Reilly, D., and Fairbrother, W. J. (1996). Solution structure of the E-domain of staphylococcal protein A. *Biochemistry* **35**, 15558–15569.
101. Orengo, C. A., Michie, A. D., Jones, S., Jones, D. T., Swindells, M. B., and Thornton, J. M. (1997). CATH—A hierarchic classification of protein domain structures. *Structure* **5**, 1093–1108.
102. Herges, T., and Wenzel, W. (2005). Characterization of the free energy landscape of the villin headpiece in an all-atom force field. *Structure* **13**, 661–668.
103. Cochran, A. G., Skelton, N. J., and Starovasnik, M. A. (2001). Tryptophan zippers: Stable, monomeric β-hairpins. *Proc. Natl Acad. Sci. USA* **98**, 5578–5583.
104. Sharon, M., Kessler, N., Levy, R., Zolla-Pazner, S., Gorlach, M., and Anglister, J. (2003). Alternative conformations of HIV-1 V3 loops mimic β-hairpins in chemokines, suggesting a mechanism for coreceptor selectivity. *Structure* **11**, 225–236.
105. Verma, A., and Wenzel, W. (2008). Conformational landscape of the HIV-V3 hairpin loop from all-atom free-energy simulations. *J. Chem. Phys.* **128**, 105103.
106. Rosen, O., Chill, J., Sharon, M., Kessler, N., Mester, B., Zolla-Pazner, S., and Anglister, J. (2005). Induced fit in HIV-neutralizing antibody complexes: Evidence for alternative conformations of the gp120 V3 loop and the molecular basis for broad neutralization. *Biochemistry* **44**, 7250–7258.
107. Andersen, N. H., Olsen, K. A., Fesinmeyer, R. M., Tan, X., Hudson, F. M., Eidenschink, L. A., and Farazi, S. R. (2006). Minimization and optimization of designed β-hairpin folds. *J. Am. Chem. Soc.* **128**, 6101–6110.
108. Verma, A., and Wenzel, W. (2007). Predictive and reproducible de-novo all-atom folding of a β-hairpin loop in an improved free energy force field. *J. Phys. Cond. Matt.* **19**, 285213.
109. Zhou, R., Berne, B. J., and Germain, R. (2001). The free energy landscape for β-hairpin folding in explicit water. *Proc. Natl Acad. Sci. USA* **98**, 14931–14936.
110. Islam, S. A., Karplus, M., and Weaver, D. L. (2004). The role of sequence and structure in protein folding kinetics: The diffusion–collision model applied to proteins L and G. *Structure* **12**, 1833–1845.
111. Nguyen, P. H., Stock, G., Mittag, E., Hu, C. K., and Li, M. S. (2005). Free energy landscape and folding mechanism of a β-hairpin in explicit water: A replica exchange molecular dynamics study. *Proteins Struct. Funct. Genet.* **61**, 705–808.
112. Nguyen, P. H. (2006). Complexity of free energy landscapes of peptides revealed by nonlinear principal component analysis. *Proteins Struct. Funct. Genet.* **65**, 893–913.
113. Pastor, M. T., Lopez de la Paz, M., Lacroix, E., Serrano, L., and Perez-Paya, E. (2002). Combinatorial approaches: A new tool to search for highly structured β-hairpin peptides. *Proc. Natl Acad. Sci. USA* **99**, 614–619.
114. De Alba, E., Santoro, J., Rico, M., and Jimenez, M. A. (1999). De novo design of a monomeric three-stranded antiparallel β-sheet. *Protein Sci.* **8**, 854–865.

115. Wang, H., and S-Sung, S. (2000). Molecular dynamics simulations of three-strand β-sheet folding. *J. Am. Chem. Soc.* **122,** 1999–2009.
116. Ferrara, P., and Caflisch, A. (2000). Folding simulations of a three-stranded antiparallel sheet peptide. *Proc. Natl Acad. Sci. USA* **97**(20), 10780–10785. Cited by (since 1996): 89.
117. Caflisch, A. (2006). Network and graph analyses of folding free energy surfaces. *Curr. Opin. Struct. Biol.* **16,** 71–78.
118. Laity, J. H., Lee, B. M., and Wright, P. E. (2001). Zinc finger proteins: New insights into structural and functional diversity. *Curr. Opin. Struct. Biol.* **11,** 39–46.
119. Lee, M. S., Gippert, G. P., Soman, K. V., Case, D. A., and Wright, P. E. (1989). Three-dimensional solution structure of a single zinc finger–DNA-binding domain. *Science* **245** (4918), 635–637.
120. Pavletich, N. P., and Pabo, C. O. (1991). Zinc finger–DNA recognition: Crystal structure of a Zif268–DNA complex at 2.1 A. *Science* **252**(5007), 809–817.
121. Wolfe, S. A., Nekludova, L., and Pabo, C. O. (2000). DNA recognition by Cys2His2 zinc finger proteins. *Annu. Rev. Biophys. Biomol. Struct.* **29**(1), 183–212.
122. Urnov, F. D., Miller, J. C., Lee, Y. L., Beausejour, C. M., Rock, J. M., Augustus, S., Jamieson, A. C., Porteus, M. H., Gregory, P. D., and Holmes, M. C. (2005). Highly efficient endogenous human gene correction using designed zinc-finger nucleases. *Nature* **435**(7042), 646–651.
123. Abagyan, A., and Totrov, M. (1999). Ab initio folding of peptides by the optimal-bias Monte Carlo minimization procedure. *J. Comput. Phys.* **151,** 402–412.
124. Nagadoi, A., Nakazawa, K., Uda, H., Okuno, K., Maekawa, T., Ishii, S., and Nishimura, Y. (1999). Solution structure of the transactivation domain of ATF-2 comprising a zinc finger-like subdomain and a flexible subdomain. *J. Mol. Biol.* **287,** 593–607.
125. Laity, J. H., Dyson, H. J., and Wright, P. E. (2000). DNA-induced alpha-helix capping in conserved linker sequences is a determinant of binding affinity in cys_2–his_2 zinc fingers. *J. Mol. Biol.* **295,** 719–727.

Folding Considerations for Therapeutic Protein Formulations

LIOUBOV G. KOROTCHKINA*,
KARTHIK RAMANI[†],
AND SATHY V. BALU-IYER[*,1]

*Department of Pharmaceutical Sciences, State University of New York at Buffalo, Amherst, New York 14260
[†]Biocon Ltd., Bangalore, India

I. Introduction.. 256
II. Folding and Stability of Therapeutic Proteins................... 257
   A. Protein Instability 257
   B. Impact of Folding on Safety and Efficacy of Protein Therapeutics 257
   C. Folding Characteristics of Protein Therapeutics 258
   D. Protein Folding Considerations During Product Development............. 260
   E. Analytical Methods for Characterization and Understanding Structure–Function Relationship 263
   F. Folding Consideration in the Development of Second Generation Protein Therapeutics and Biosimilars... 265
III. Case Study: Rational Formulation Development of Human Recombinant Factor VIII 266
References.. 268

Proteins as therapeutic agents are highly specific and efficient; however, development of therapeutic proteins as drugs presents difficulties. The difficulties are determined by the complexity of the protein structure, folding, and its sensitivity to the environmental conditions. Proteins are subjected to chemical and physical instability, which greatly affects their efficacy as therapeutic agents. Changes in protein conformation resulting in protein unfolding, aggregation, and denaturation can abolish the activity of the protein and additionally increase its immunogenicity and toxicity. Different factors during therapeutic protein development affect its stability and folding, for example, pH shift, temperature and pressure changes, presence of salts, metal ions and surfactants, pressure, shaking and shearing, absorption to surfaces, and protein concentration. Monitoring of stability and folding of therapeutic proteins and

[1]Formerly Sathyamangalam V. Balasubramanian

biosimilars by analytical techniques is extremely important for their development and confirmation of their structure and function. The case study of recombinant factor VIII demonstrates how investigation of folding by a variety of biochemical and biophysical techniques led to the rational development of a more stable formulation of factor VIII.

I. Introduction

Proteins play important functions in the body, that is, catalysis of intracellular and extracellular reactions with high speed and specificity (enzymes); regulation of growth, development, and metabolism (receptors, proteins of signal transduction pathways, growth factors, hormones); motion and support (e.g., myosin in muscle, collagen in skin and bones); immune protection (antibodies); and transport and storage (e.g., ion channels, hemoglobin, ferritin). Proteins are highly specific and often potent, providing advantage over small organic molecule as therapeutic agents. It is not surprising that the number of protein-therapeutic drugs is increasing. Recombinant-DNA technology has allowed producing proteins through genetic engineering rather than isolation from animal and human material resulting in increased production capacity and safety. Another advantage of recombinant protein drugs is that proteins can be modified to improve their function. Currently, more than 130 therapeutic proteins are available and many more are in development (1). This list includes proteins that are used for treatment of different diseases and conditions (e.g., diabetes, hemophilia, cancer, myocardial infarction, autoimmune disease, infectious disease, HIV). Therapeutic proteins are used for replacement therapy (when specific protein is deficient, e.g., insulin, factor VIII), to increase activity of certain protein (e.g., erythropoietin is used in anaemia); to interfere with certain molecules' or organisms' action (monoclonal antibodies in cancer treatment); to protect body from infectious or other diseases (protein vaccines); and as diagnostic tools (tuberculosis detection) (1).

Because of their complex structure and folding, the pharmaceutical development of proteins, from production to formulation to bedside, poses challenges. The safety and therapeutic efficacy of proteins depends on their folding and conformation that is sensitive to environmental conditions that are encountered during various stages of product development. In this review, we provide our perspective on stability and folding considerations in the pharmaceutical development of protein products. This article is not meant to be exhaustive review on several seminal scientific contributions made in protein folding but is focused on folding considerations in the development of proteins as therapeutic entities.

II. Folding and Stability of Therapeutic Proteins

A. Protein Instability

Because of its complex structure, proteins undergo physical and chemical instability (2). Chemical instability of proteins refers to any process that involves modification of the protein via bond formation or cleavage. There are different degradation reactions that can damage therapeutic proteins depending on protein structure and environmental conditions. These reactions are: deamidation, oxidation, hydrolysis, disulfide bond formation or breakage, isomerization, succinimidation, racemization, β-elimination, and Maillard reaction. The rate of these chemical reactions is greatly reduced in solid state and thus development of freeze-dried powder is attempted to increase shelf-life.

Physical stability refers to changes in the higher-order structure (secondary and above). This includes denaturation, adsorption to surfaces, aggregation, and precipitation. In the global fold, the exposure of the hydrophobic groups is minimized and this conformation is required for optimal biological activity of the protein. The denaturation is a molecular process in which the protein unfolds and as a result, the global fold is lost. Once the protein unfolded, polypeptide chain undergoes further inactivation by association with surfaces, aggregation, and precipitation. Physical aggregation can be followed by chemical aggregation, that is, formation of covalent and noncovalent modifications in the protein leading to irreversible aggregation. Growth of the aggregate can continue till the solubility limit is reached, after which protein precipitates. Exposure of hydrophobic regions in a protein during unfolding can also lead to surface absorption to vials, injection tubes, etc., and also can accumulate at the surfaces (air–liquid, liquid–container).

B. Impact of Folding on Safety and Efficacy of Protein Therapeutics

Protein folding related molecular events, aggregation, precipitation, and surface absorption are significant issues in the development of therapeutic protein because these conditions result in reduction or complete elimination of protein activity. Additionally, protein aggregates can have increased immunogenicity and toxicity (3). It has been shown that presence of aggregates in the formulation enhances the immune response to monomeric protein that can have significant impact on the safety of the therapy. The protein antigens presented in a highly ordered, multimeric form as observed in aggregates, can elicit antibody response against the therapeutic protein. Further, immune response elicited by protein aggregates can have significant adverse clinical effects. Antibodies generated against the exogenously administered protein can either neutralize its activity thereby reducing the efficacy (*4, 5*) or can cross-react with the endogenous protein with life-threatening consequences. For example, presence of aggregates in the formulation of erythropoietin has been shown to break the tolerance and these antibodies cross react with endogenous counter part resulting in red cell aplasia (*6*). It has also been reported that presence of aggregates can elicit severe immediate anaphylaxis reactions.

C. Folding Characteristics of Protein Therapeutics

The folding and mechanisms by which proteins aggregate have significant pharmaceutical implications. The simplest model used to describe the reversible protein folding is a two-state model in which only two species are observed at any time, native (N) and denatured (D), and are in equilibrium (N ⟷ D). This simple assumption allows the estimation of free energy changes between folded and unfolded states. The free energy change for the folded and unfolded state is small, typically in the range of 5–20 kcal/mol. The major protein-destabilizing force is conformational entropy. Entropy of the unfolded protein is high as a result of loss in structure and unrestricted rotation of the side chains. The enthalpy of the unfolding/denaturation can be either negative or positive. This simple two-state model, however, is inadequate to define unfolding for several therapeutic proteins.

The mechanism of aggregation of proteins is complex. It has been reported that aggregates are not necessarily formed from completely unfolded proteins but rather from the partially folded/unfolded intermediate states (Scheme 1) (*7*).

N (native) ⇆ I (intermediate) ⇆ U (unfolded/denatured)

⇅   (Scheme 1)

A (aggregated)

Proteins can unfold to form near-native (partially unfolded) states. Different factors (temperature, pressure, presence of chaotropes) can shift the equilibrium resulting in protein unfolding/denaturation (*8, 9*). It has been shown that the partially unfolded states are aggregation prone states that display substantial native-like features with only minor tertiary structural changes and can promote aggregation. For proteins, such as rhIFN-γ and rhGCSF, aggregation can occur even under solution conditions that favor native-like physiological conditions (*10*). In many such cases, the rate-limiting step appears not to be the generation of extensively unfolded states. In the case of rhIFN-γ, aggregation has been shown to occur through the formation of a transiently expanded conformational species, whose surface area is greater than the native state by only 9% (*11*). Similarly, native rhGCSF has been shown to aggregate through the formation of a monomeric transition state in which the surface area increase was only 15% (12). In the case of rFVIII, neither complete unfolding of the protein nor the generation of partially unfolded states of the protein appears to be a prerequisite for the protein to aggregate. Rather, aggregation appears to involve subtle conformational changes in the C2 domain encompassing the lipid-binding region (13, 14).

Further, equilibrium analysis of unfolding is complicated by the aggregation kinetics and is observed for some therapeutic proteins. In addition, the aggregation kinetics of the aggregation prone state also determines the stability of protein therapeutics. In the Lumry–Eyring model (8, 15), the intermediate or aggregation-competent state (A) can associate to form multimers (A_{m+1}) (Eq. (1)). A_m has lower free energy than A and N and is thermodynamically favored.

$$A_m + A \rightarrow A_{m+1} \qquad (1)$$

In general, the heating rate dependence of protein thermal transitions can be understood in terms of the Lumry–Eyring model (16) as described above. According to this framework, a reversible unfolding step is followed by an irreversible event. If the conversion of unfolded states to aggregated state occurs at much faster rate, and the process can be represented by the simpler kinetic scheme (17):

$$N \xrightarrow{k} A \qquad \text{(Scheme 2)}$$

Scheme 2 represents the limiting case of the Lumry–Eyring model consisting of only two populated states, the native (N) and the final aggregated state (A). This predicts that k is a first-order kinetic constant that varies with temperature, as given by the Arrhenius equation. Based on this simple kinetic model, T_m should vary with heating rate (v) according to the Eq. (2):

$$\ln(v/T_m^2) = \ln(AR/E_a) - E_a/RT_m \qquad (2)$$

where A is the frequency factor, E_a is the activation energy of the unfolding step, and R is the gas constant. Therefore, a linear plot of $\ln(v/T_m^2)$ versus $1/T_m$ should indicate a first-order reaction with slope of $-E_a/R$.

The unfolding profile of recombinant FVIII and transition temperature (T_m) was found to be dependent on heating rate and based on this data, the process of rFVIII unfolding appears to be kinetically determined. Such heating rate dependency on unfolding transitions has also been observed for other therapeutic proteins, including IFN-γ (18) and Immunoglobulin G (IgG)(19). The activation energy (E_a) for the transition of rFVIII was found to be ~535 kJ/mol (~128 kcal/mol) and compares well with the E_a associated with the two transitions observed for the multidomain protein, IgG (E_a for the two transitions were reported to be 456 kJ/mol (~109 kcal/mol) and 692 kJ/mol (~165 kcal/mol)) (19).

This molecular level understanding of protein folding and stability is critical to the development of protein therapeutics.

D. Protein Folding Considerations During Product Development

1. MANUFACTURING

Proteins are manufactured using recombinant technology in different expression systems (bacterial, yeast, insect, plant, and mammalian). Mammalian system is a system of choice to provide posttranslational modifications of the recombinant protein necessary for its activity. Bacterial system is preferred when posttranslational modification are not important because expression in bacterial system is fast, inexpensive, and provides high yield of protein. However, expression of a foreign protein in bacteria can lead to the loss of its proper fold and formation of inclusion bodies. These folding issues greatly influence the specific activity and stability of a therapeutic protein. There are several strategies used to increase production of the target protein in soluble form, that is, expression at lower temperature, modification of media, usage of specifically designed bacterial strains, coexpression of molecular chaperons. Some therapeutic proteins are purified from inclusion bodies, for example, human growth hormone (20), and human macrophage colony-stimulating factor (21). The advantage of using inclusion bodies is that protein can be easily purified from inclusion bodies with high level of purity. The challenge is protein refolding which is achieved by solubilization of purified inclusion bodies in chaotropic agents such as guanidinium hydrochloride and urea followed by protein refolding by removal of denaturants through dilution and buffer exchange by dialysis, diafiltration, or chromatographic techniques (22). Techniques for refolding have to be carefully chosen to prevent formation of inactive improperly folded intermediates and aggregation.

2. PRODUCTION AND PURIFICATION

The purification and manufacturing conditions involve shear forces such as mixing and passage through pumps. It has been shown that unfolding and aggregation of protein can occur at high shear rates as the shear force can result in the loss of global fold as the free energy change between folded and unfolded conformation is typically in the range of 5–20 kcal/mol. In addition, exposure to organic solvents and surfactants is another stress that leads to unfolding of the protein. The unfolding behavior of protein in water-solvent mixtures and surfactants has been extensively studied and it is generally believed that such exposure leads to denaturation and loss of activity. For monoclonal antibody products purification steps involve pH shock to elute the protein from affinity columns and such pH shift can cause unfolding and aggregation of the protein.

During the purification, it is important to maintain the global fold that defines the native conformation of the protein, as this conformation is critical for the safety and efficacy of the therapy.

3. Bulk Protein Drugs

Pharmaceutical development of proteins involves handling of proteins at significantly higher bulk concentrations. This problem is exemplified for monoclonal-antibody-based products. Since EC50 are higher for Mabs, a relatively higher concentration (typically few hundred milligrams to gram per vial) of Mabs needs to be formulated. The limited dose volume, particularly for s.c. administration, further complicates this but this route of administration is preferred due to ease of administration and patient compliance. The investigating folding behavior and stability of protein at higher concentrations poses several problems. The analytical techniques used as quality control and stability indicating assays do not effectively provide folding and stability information at this high concentration. For example, folding of proteins in solution conditions are often studied by circular dichroism (CD) measurements, but use of such spectroscopic methods with high concentration and OD is not reliable. One approach to address this issue is to develop analytical methodology to study folding and stability at higher concentration (23).

4. Formulation Development

Following production and purification, the next step in the product development is the formulation of the protein in a suitable dosage form. At this stage, several formulation additives and excipients are added to ensure proper stability. Screening of excipients is one of the development steps that involve clear understanding of folding behavior of protein product under development. It is a common practice to use thermal stress and pH changes to unfold the protein to mimic manufacturing and processing cycles and T_m and ΔG are determined, assuming a simple two-state model for unfolding. Thermal methods such as differential scanning calorimetry and spectroscopic techniques are used as analytical methods to follow unfolding of the protein. Any excipient that increases the Tm is considered as a stabilizer (24).

However, several proteins display complex folding behavior and in such cases formulation development is a challenge. For example, protein aggregation involves minor conformational changes that resemble the native state (13). Several physical techniques could not distinguish between these states from that of native states. The aggregation kinetics involving such states further complicates simple excipient screening procedure and use of higher bulk protein concentrations can accelerate this kinetically controlled aggregation process. Thus, formulation development requires the characterization of minor conformational changes or aggregation prone states and their kinetics. These considerations are critical for multidomain proteins and monoclonal-based therapeutics. Previous findings on the molecular details of the aggregation prone state of rFVIII and their kinetics (Please refer to the case study at the end of this chapter) have contributed towards the rational design of a rFVIII–phosphoserine (PS) complex with improved physical stability (25, 26). Based on folding behavior, some investigators attempted to use molecular chaperone as formulation excipients (23).

5. Accelerated Stability Testing

The dosage form thus developed should meet FDA required shelf-life of two years and predicting stability and storage condition is complicated because of complex folding behavior of proteins. Accelerated stability testing is a common method for the prediction of the long term storage conditions for small molecule based drugs. This is typically done at elevated temperatures, above normal storage conditions and stability is monitored as a function of time. This reaction normally follows a first-order process, that is

$$A_t = A_0 e^{-kt}$$

where A_t and A_0 are the activities or concentrations at time t and time 0. From these data, the rate of degradation can be calculated at each of the different temperatures. The rates may then be fitted to Arrhenius equation to predict the rates at any temperature. Generally protein products are developed for storage at 4 °C with a minimum target shelf-life of 2 years. However, protein degradation cannot be predicted by Arrhenius equation, as it may involve inactivation by several mechanisms. The rate of disappearance of parent compound will not be identical to rate of accumulation of final product. For example, accelerated stability testing performed at higher temperature also involves unfolding of the protein and aggregation. Thus, the accelerated stability testing is done at lower temperatures such as 25, 30, and 40 °C and formation of native-like folding intermediates in this temperatures range and their kinetics complicate the prediction of stability. Thus, understanding of the folding behavior under isothermal conditions is critical for predicting stability and recommendation of storage conditions.

6. Freeze-Drying

In order to improve shelf-life of protein therapeutics, most of the protein drugs are sold as lyophilized powder. Physical instability can occur during lyophilization by several mechanisms. There may be effects due to cold denaturation, concentration/crystallization of salts, changes in pH, and the creation of solid–liquid interfaces when ice crystals are formed. For example, freezing stage of lyophilization leads to huge pH shift; freeze crystallization of phosphate buffer results in the reduction of pH from 7.4 to 3.5. This pH swing can unfold the protein leading to aggregation of the protein (27). Further, residual moisture in lyophilized protein can have sufficient mobility for noncovalent aggregation as well as chemical reactions.

7. AT THE BED SIDE

At the bedside, the protein is reconstituted using manufacturer supplied reconstitution medium with gentle swirling and patients are advised not to shake the solution vigorously as it can unfold the protein leading to aggregation. Further, surfactants present in the formulation can foam and protein molecules that come in contact with the interface can unfold and aggregate. Healthcare professionals and patients are always advised to visually examine the solution and look for precipitates. A typical advice found in the package insert is provided below:

> "To reconstitute lyophilized Betaseron for injection, use a sterile syringe and needle to inject 1.2 ml of the diluent supplied, Sodium Chloride, 0.54% Solution, into the Betaseron vial. Gently swirl the vial of Betaseron to dissolve the drug completely; do not shake. Foaming may occur during reconstitution or if the vial is swirled or shaken too vigorously. If foaming occurs, allow the vial to sit undisturbed until the foam settles. Inspect the reconstituted product visually and discard the product before use if it contains particulate matter or is discolored. After reconstitution with the accompanying diluent, 1 ml of Betaseron solution contains 0.25 mg Interferon β-1b."
>
> (Taken from Package insert Betaseron)

Overall, protein folding and stability play a critical role in all stages of protein product development.

E. Analytical Methods for Characterization and Understanding Structure–Function Relationship

Unlike small molecule that are chemically well-defined, structural complexities associated with proteins and their inherent heterogeneity requires the use of several techniques to fully understand the structure–function relationship (Table I). Given the close relationship between biological structure, folding and function, incorporating biological potency/bioassays/receptor binding studies (Table II), in addition to chemical and structural analysis, is an essential component of the development program of protein therapeutics.

TABLE I

ANALYTICAL METHODS FOR CHARACTERIZATION OF STRUCTURE–FUNCTION RELATIONSHIP OF PROTEINS

Aspect of characterization	Techniques
Identity	Amino acid analysis, peptide mapping, Chromatography (RP-HPLC, SEC)
Purity	Chromatography (RP-HPLC, SEC), Electrophoresis (SDS-PAGE, IEF)
Charge heterogeneity	IEF, IEXC, CIEF
Hydrophobicity	HIC, RP-HPLC
Post-translational modification (Glycosylation and pattern), especially for complex biologicals such as monoclonal antibodies	LC-ESI, CE, peptide mapping, fluorescence labeling
Structure/conformation/folding	
-Primary	Amino acid analysis, MS, N and C-terminal sequencing, peptide mapping (protease digestion)
- Secondary	Far UV–CD, FTIR, Raman spectroscopy, peptide mapping
- Tertiary	Near-UV CD, fluorescence spectroscopy, NMR, Raman spectroscopy, second derivative UV spectroscopy
Conformational stability	DSC
Size and molecular weight distribution	AUC, SV, DLS, SEC-LS, MALDI-TOF
Molecular interaction	SPR, ITC

AUC, analytical ultracentrifugation; CD, circular dichroism; CE, capillary electrophoresis; CIEF, capillary isoelectric focusing; DLS, dynamic light scattering; DSC, differential scanning calorimety; FTIR, Fourier transform infrared spectroscopy; HIC, hydrophobic ion chromatography; IEF, isoelectric focusing; IEXC, ion exchange chromatography; ITC, isothermal titration calorimetry; LC-ESI, liquid chromatography electrospary ionization; MALDI-TOF, matrix assisted laser desorption ionization-time of flight; MS, mass spectrometry; NMR, nuclear magnetic resonance; RP-HPLC, reverse phase high performance liquid chromatography; SDS-PAGE, sodium dodecyl sulfate polyacrylamide gel electrophoresis; SEC-LS, size exclusion chromatography coupled with light scattering; SPR, surface plasmon resonance; SV, sedimentation velocity; UV, ultraviolet spectroscopy

TABLE II

BIOASSAYS FOR THE CHARACTERIZATION OF MOLECULAR PROPERTIES OF PROTEINS

Aspect of characterization	Techniques
Molecular interactions	ELISA, SPR, ITC, receptor binding, radioligand binding, flow cytometry
Three dimensional conformation	Biological assays (e.g., CDCL for monoclonal antibodies), receptor binding

CDCL, complement dependent cell lysis; ELISA, enzyme linked immunosorbant assay; ITC, isothermal titration calorimetry; SPR, surface plasmon resonance

F. Folding Consideration in the Development of Second Generation Protein Therapeutics and Biosimilars

1. Folding and Delivery

The safety and efficacy of several marketed products could be improved by developing second-generation protein therapeutics that is less immunogenic and long acting. Covalent attachment of poly(ethylene glycol) (PEG) to specific protein residues has been successfully used for few smaller therapeutic proteins (28). Due to the improved *in vivo* behavior, these modified proteins have prolonged circulation in the blood. However, one of the major issues is the loss of global fold and activity of the protein following the attachment of bulky PEG polymer. Another approach to improve circulation half-life is to deliver therapeutic proteins using formulation of proteins in delivery vehicles such as liposomes and PLGA microspheres. Loading of protein in PLGA microspheres involve exposure to organic solvents that has been shown to denature proteins, leading to significant loss of activity. Even though, loading of proteins in liposomes do not involve exposure to organic solvents, spontaneous loading is limited. In order to improve protein loading in liposomes, triggered loading procedure have been developed (29). This involves engineering a hydrophobic interaction between partially folded states of protein and liposomes. Maintenance of the protein activity is critical during the association of the protein to liposomes. To achieve this, a rational structure-based approach was used to generate folding intermediates with exposed hydrophobic domain but which are reversible to native conformation.

2. Protein Folding and Development of Biosimilars (Biogeneric or Follow on Biologics)

Biotechnology-derived protein therapeutics or biologicals have become a critical part of modern healthcare landscape. The exorbitant cost of these therapeutic agents, however, has considerably increased the financial burden on patients requiring the usage of these drugs for their survival, government institutions, and insurers alike. More importantly, many of these life saving products are not affordable and hence not accessible to many patients in developing countries. The expiry of patent protection for some protein therapeutics recently has resulted in the arrival of few follow-on protein products or biosimilars, which are biological medicinal products similar to licensed reference medicinal products.

As discussed earlier, protein therapeutics are structurally very complex molecules that are usually manufactured using biological processes in living cells or organisms. They are much more heterogeneous and contain a mixture of different isoforms, unlike small molecule therapeutics, which are chemically well-defined (30). Production of proteins usually involves complex manufacturing and quality control processes and is very sensitive to variation in process parameters and the type of living system used (31). This can potentially have a significant impact on the folding and chemical properties of the molecule and influence the clinical response of the administered therapeutic.

III. Case Study: Rational Formulation Development of Human Recombinant Factor VIII

Factor VIII (FVIII) is a multidomain protein and is a critical cofactor in the blood coagulation cascade. Its deficiency or dysfunction causes Hemophilia A, a bleeding disorder. Replacement therapy using administration of exogenous recombinant FVIII is the first line of therapy for Hemophilia A. However, this complex protein is susceptible to aggregation that is often associated with loss of activity. In this section, we discuss folding-based rational approach for the formulation development of FVIII.

Thermal unfolding of FVIII was monitored using biophysical and biochemical techniques such as far and near UV CD, fluorescence emission and anisotropy, size exclusion chromatography and antibody binding, and the results were corroborated to develop a folding model. The folding model was used to rationally develop a stable formulation for FVIII (14).

The far UV CD spectrum of FVIII showed a negative band around 215 nm indicating that the protein predominantly existed in β-sheet conformation and is consistent with X-ray and molecular modeling studies (32, 33). There were no changes observed in the spectral characteristics as the temperature was increased to 50 °C, indicating that there are no changes in the secondary structural elements. However, antibody binding to FVIII as studied by sandwich ELISA showed a decrease in binding in the same temperature range. The sandwich ELISA system determines the binding affinity of pair of antibodies. The binding affinity studies performed using ESH4 antibody that recognizes lipid binding region of the protein as probe antibody and ESH8 as capture antibody, showed that the binding of protein to ESH4 decreased significantly as the temperature was increased from 20 to 50 °C indicating that there may be conformational changes as the protein unfold. It is possible that the CD spectroscopy may not be sensitive to such small conformational changes to a multidomain protein. As the temperature is further increased from 50 to 65 °C, the intensity of the negative band observed in CD studies increased, and possibly, suggesting an increase in β-sheet content or loosening of β-sheet clusters. However, SEC and fluorescence polarization studies conducted in the same temperature range suggested formation of aggregates as the temperature was increased. At 55 °C, two peaks were observed in SEC profiles, one at

6.56 min and the other at 5.64 min. The peak at 6.56 min corresponds to monomeric protein and the second peak may be due to small aggregates. Further, it was observed that the peak at 5.64 min is sensitive to the temperature. The peak position progressively shifted to lower elution time and at 75 °C, the protein eluted in the void volume with a retention time of 5.1 min. Thus, the increase in ellipticity observed in melting profile suggested that the transition coincided with the formation of aggregates. This is consistent with fluorescence polarization studies; between 20 and 50 °C, no change in fluorescence polarization was observed but as the protein is heated above 50 °C, an increase in fluorescence polarization was observed. The increase in polarization is due to increase in size of the protein as a result of aggregation. The CD spectrum for FVIII taken above 75 °C displayed a red shift in the negative band and a positive band around 210 nm, typical of intermolecular β strands.

The thermal unfolding of FVIII was also studied at various heating rates and this approach was used to investigate the impact of aggregation kinetics on equilibrium unfolding (34). The transition temperature, T_m, was found to be dependent on heating rate; the T_m increased as the heating rate was increased. Such dependence clearly demonstrates that the thermal unfolding at least in part is under kinetic control.

Based on the unfolding studies and aggregation kinetics, Lumry–Eyring nucleated polymerization model (35) is proposed for the thermal unfolding of FVIII. This model describes different stages in unfolding that captures several experimental observations.

Native ⟷ Intermediate (Conformational change in lipid binding domain) ⟷ Oligomer formation → Nucleation (irreversible) → Growth → Higher order aggregates

In step I, native state undergoes conformational changes in the temperature range of 20–50 °C and the data suggests that this conformational change would involve lipid binding domain. As this domain is hydrophobic in nature, it is possible that this conformational change results in the formation of aggregation prone state that initiates the oligomerization. Once the nucleation occurs, further thermal stress (between 50 and 60 °C) results in the formation of small clusters (step II) that is reflected in CD spectral changes, an increase in the ellipticity. Formation of small aggregates is further confirmed by the SEC and fluorescence polarization studies as discussed above. The next stages, III and IV, would involve the nucleation and growth of these aggregates and would be in the temperature range of 60 °C and above. In this stage, conformational changes and formation of intermolecular β strands are observed in CD studies. The growth phase (stage IV) involves addition and rearrangement of aggregation prone state monomers that results in the formation of soluble high molecular weight aggregates. However, analytical techniques used in this study possibly could not capture this stage very effectively. It is appropriate to mention here that lack of suitable analytical techniques that can capture visible and subvisible particle contribute to our incomplete understanding of this stage of the unfolding (36). In stage V, assembly of aggregates or condensation occur leading to formation of bundles and filaments. This stage requires threshold aggregate size from nucleation and growth phase (stages III and IV) of the model.

Use of this folding model in formulation development: Lumry–Eyring nucleated polymerization model would be useful for the rational formulation development as this model captures several critical steps that determine stability and also provides relationship between formulation variations and stability. The model shows relationship between rate of monomer loss (and rate of aggregation) to formulation factors such as initial protein concentration, unfolding free energy, aggregate size distributions and nucleation, and growth rates.

The aggregation prone state is formed as a result of conformational changes in the lipid binding region of the protein. Since FVIII binds to phosphatidyl serine (PS) with great affinity, PS based excipients that form soluble and particulate carriers such as liposomes were developed (25, 26). The sandwich ELISA studies clearly showed that O-phospho-L-serine, the head group of phosphatidyl serine, competes with ESH 4, an antibody that binds to lipid binding region, confirming that OPLS binds to lipid binding region. Further this binding alters unfolding profile of FVIII. The CD spectral characteristics and melting profile of FVIII was altered in the presence of OPLS. The spectral analysis showed that in the presence of OPLS, no spectral changes assigned to formation of intermolecular β strands was observed, indicating that OPLS interferes with aggregate formation. Such conclusion was also supported by fluorescence polarization and SEC studies. In the presence of OPLS, the polarization values were lower than that observed for free FVIII. In SEC studies, the relative faction of aggregate to monomer is in favor of monomeric species in the presence of OPLS. Further, FVIII–OPLS complex also reduced the formation of antibodies against FVIII in Hemophilia A mice.

References

1. Leader, B., Baca, Q. J., and Golan, D. E. (2008). Protein therapeutics: A summary and pharmacological classification. *Nat. Rev. Drug Discov.* **7**, 21–39.
2. Manning, M. C., Patel, K., and Borchardt, R. T. (1989). Stability of protein pharmaceuticals. *Pharm. Res.* **6**, 903–918.
3. Rosenberg, A. S. (2006). Effect of protein aggregates: An immunologic perspective. *AAPS J.* **8**, E501–E507.
4. Jacquemin, M. G., and Saint-Remy, J. M. (1998). Factor VIII immunogenicity. *Haemophilia* **4**, 552–557.

5. Lollar, P., Healey, J. F., Barrow, R. T., and Parker, E. T. (2001). Factor VIII inhibitors. *Adv. Exp. Med. Biol.* **489**, 65–73.
6. McKoy, J. M., Stonecash, R. E., Cournoyer, D., Rossert, J., Nissenson, A. R., Raisch, D. W., Casadevall, N., and Bennett, C. L. (2008). Epoetin-associated pure red cell aplasia: Past, present, and future considerations. *Transfusion* **48**, 1754–1762.
7. Wang, W. (2005). Protein aggregation and its inhibition in biopharmaceutics. *Intern. J. Pharm.* **289**, 1–30.
8. Krishnamurthy, R., and Manning, M. C. (2002). The stability factor: Importance in formulation development. *Curr. Pharm. Biotechnol.* **3**, 361–371.
9. Wang, W. (1999). Instability, stabilization, and formulation of liquid protein pharmaceuticals. *Intern. J. Pharm.* **185**, 129–188.
10. Chi, E. Y., Krishnan, S., Randolph, T. W., and Carpenter, J. F. (2003). Physical stability of proteins in aqueous solution: Mechanism and driving forces in nonnative protein aggregation. *Pharm. Res.* **20**, 1325–1336.
11. Kendrick, B. S., Carpenter, J. F., Cleland, J. L., and Randolph, T. W. (1998). A transient expansion of the native state precedes aggregation of recombinant human interferon-gamma. *Proc. Natl. Acad. Sci. USA* **95**, 14142–14146.
12. Krishnan, S., Chi, E. Y., Webb, J. N., Chang, B. S., Shan, D., Goldenberg, M., Manning, T. W., Randolph, T. W., and Carpenter, J. F. (2002). Aggregation of granulocyte colony stimulating factor under physiological conditions: Characterization and thermodynamic inhibition. *Biochemistry* **41**, 6422–6431.
13. Grillo, A. O., Edwards, K. L., Kashi, R. S., Shipley, K. M., Hu, L., Besman, M. J., and Middaugh, C. R. (2001). Conformational origin of the aggregation of recombinant human factor VIII. *Biochemistry* **40**, 586–595.
14. Ramani, K., Purohit, V. S., Miclea, R. D., Middaugh, C. R., and Balasubramanian, S. V. (2005). Lipid binding region (2303–2332) is involved in aggregation of recombinant human FVIII (rFVIII). *J. Pharm. Sci.* **94**, 1288–1299.
15. Chi, E. Y., Krishnan, S., Randolph, T. W., and Carpenter, J. F. (2003). Physical stability of proteins in aqueous solution: Mechanism and driving forces in nonnative protein aggregation. *Pharm. Res.* **20**, 1325–1336.
16. Lumry, R., and Eyring, H. (1954). Conformation changes of proteins. *J. Phys. Chem.* **58**, 110–120.
17. Sanchez-Ruiz, J. M., Lopez-Lacomba, J. L., Cortijo, M., and Mateo, P. L. (1988). Differential scanning calorimetry of the irreversible thermal denaturation of thermolysin. *Biochemistry* **27**, 1648–1652.
18. Kendrick, B. S., Cleland, J. L., Lam, X., Nguyen, T., Randolph, T. W., Manning, M. C., and Carpenter, J. F. (1998). Aggregation of recombinant human interferon gamma: Kinetics and structural transitions. *J. Pharm. Sci.* **87**, 1069–1076.
19. Vermeer, A. W., and Norde, W. (2000). The thermal stability of immunoglobulin: Unfolding and aggregation of a multi-domain protein. *Biophys. J.* **78**, 394–404.
20. Patra, A. K., Mukhopadhyay, R., Mukhija, R., Krishnan, A., Garg, L. C., and Panda, A. K. (2000). Optimization of inclusion body solubilization and renaturation of recombinant human growth hormone from *Escherichia coli*. *Protein Expr. Purif.* **18**, 182–192.
21. Tran-Moseman, A., Schauer, N., and De Bernardez Clark, E. (1999). Renaturation of *Escherichia coli*-derived recombinant human macrophage colony-stimulating factor. *Protein Expr. Purif.* **16**, 181–189.
22. Sahdev, S., Khattar, S. K., and Saini, K. S. (2008). Production of active eukaryotic proteins through bacterial expression systems: A review of the existing biotechnology strategies. *Mol. Cell. Biochem.* **307**, 249–264.

23. Harn, N., Allan, C., Oliver, C., and Middaugh, C. R. (2007). Highly concentrated monoclonal antibody solutions: Direct analysis of physical structure and thermal stability. *J. Pharm. Sci.* **96**, 532–546.
24. Tsai, P. K., Volkin, D. B., Dabora, J. M., Thompson, K. C., Bruner, M. W., Gress, J. O., Matuszewska, M., Keogan, M., Bondi, J. V., and Middaugh, C. R. (1993). Formulation design of acidic fibroblast growth factor. *Pharm. Res.* **10**, 649–659.
25. Purohit, V. S., Ramani, K., Sarkar, R., Kazazian, H. H., Jr., and Balasubramanian, S. V. (2005). Lower inhibitor development in hemophilia A mice following administration of recombinant factor VIII-O-phospho-L-serine complex. *J. Biol. Chem* **280**, 17593–17600.
26. Miclea, R. D., Purohit, V. S., and Balu-Iyer, S. V. (2007). O-phospho-L-serine, multi-functional excipient for B domain deleted recombinant factor VIII. *AAPS J* **9**, E251–E259.
27. Pikal, M. J. (1994). Freeze-drying of proteins. In "Formulation and Delivery of Proteins and Peptides" (J. L. Cleland and R. Langer, Eds.), pp. 120–133. ACS Symposium Series.
28. Kozlowski, A., Charles, S. A., and Harris, J. M. (2001). Development of pegylated interferons for the treatment of chronic hepatitis C. *Biodrugs* **15**, 419–429.
29. Balasubramanian, S., Straubinger, R. M., Ramani, K., Besman, M., and Kashi, R. (2006). Method of complexing a protein by the use of a dispersed system and proteins thereof. (WO/2006/002195).
30. Roger, S. D. (2006). Biosimilars: How similar or dissimilar are they. *Nephrology* **11**, 341–346.
31. Kuhlmann, M., and Covic, A. (2006). The protein science of biosimilars. *Nephrol. Dial. Transplant.* **21**(Suppl. 5), v4–v8.
32. Shen, B. W., Spiegel, P. C., Chang, C. H., Huh, J. W., Lee, J. S., Kim, J., Kim, Y. H., and Stoddard, B. L. (2008). The tertiary structure and domain organization of coagulation factor VIII. *Blood* **111**, 1240–1247.
33. Stoilova-McPhie, S., Villoutreix, B. O., Mertens, K., Kemball-Cook, G., and Holzenburg, A. (2002). 3-Dimensional structure of membrane-bound coagulation factor VIII: Modeling of the factor VIII heterodimer within a 3-dimensional density map derived by electron crystallography. *Blood* **99**, 1215–1223.
34. Ramani, K., Purohit, V., Middaugh, C. R., and Balasubramanian, S. V. (2005). Aggregation kinetics of recombinant human FVIII (rFVIII). *J. Pharm. Sci.* **94**, 2023–2029.
35. Andrews, J. M., and Roberts, C. J. (2007). A Lumry–Eyring nucleated polymerization model of protein aggregation kinetics: 1. Aggregation with pre-equilibrated unfolding. *J. Phys. Chem. B Condensed Matter Mater. Surfaces Interfaces Biophys.* **111**, 7897–7913.
36. Carpenter, J. F., Randolph, T. W., Jiskoot, W., Crommelin, D. J., Middaugh, C. R., Winter, G., Fan, Y.-X., Kirshner, S., Verthelyi, D., Kozlowski, S., Clouse, K. A., Swann, P. G. *et al.* (2008). Overlooking subvisible particles in therapeutic protein products: Gaps that may compromise product quality. *J. Pharm. Sci.* DOI: 10.1002/jps.21530.

Index

A

Aberrant aggregation, 81–83
AGADIR, LR-based model, 23–24. *See also* α-Helical peptides
Aggresomes, misfolding defense mechanism, 102
Amino acids
  free energies and helix propagation propensities, 27
  helix N-and C-terminal position, 30–31
Amyloidogenic aggregation, 80–82
Andersen–Tong model, 21
AQ peptides, 11
Atomic solvent parameter (ASP), 209

B

Basin hopping technique (BHT)
BH4. *See* Tetrahydrobiopterin
Biopterins, 112
BiP/PDI chaperone system, ER quality control, 144, 157–158

C

Calnexin chaperone system
  chaperone deletions
    calreticulin and calnexin, 147
    E13–14, 150–151
    ERp57, 147–149
    glucosidases, 146–147
    GT1, 149–150
    private, tissue-specific and conventional, 151
  cycling, 143
    folding-competent and defective glycoproteins, 144
    glycoprotein folding, mechanistic features, 145
    ER quality control, 157
CD. *See* Circular dichroism
CFTR. *See* Cystic fibrosis transmembrane conductance regulator
Chaperones
  chemical, 117–118
  ER-resident
    BIP, BIP cofactors, and GRP94, 138
    protein disulfide insomerase (PDI) superfamily, 139–140
  function of, 100–101
  mutant proteins, 116
  natural ligands
    ion-free apo-protein, 120
    kinetic (K_m) effect, 119
  pharmacological
    specific stabilizers, 118
    thermodynamic stability, 119
  therapeutic applications, 116
Circular dichroism, 98, 261
Coarse-grained models
  blind protein structure predictions, 199
  drawbacks of, 198
Cystic fibrosis transmembrane conductance regulator, 103, 153
Cytosolic coat protein II (COPII), 145

D

Dehydronic force, 56–57
Dehydrons, 58
De-Mannosylation, extensive
  misfolded proteins and ERAD
    cell lines, oligosaccharide, 142
    futile cycles, 155
    MnsI, mammalian ortholog, 156–157
    N-glycans and folding-defective polypeptides, 156
Diet therapy, phenylketonuria (PKU), 109
Differential scanning calorimetry (DSC), 98

Dihydrobiopterin (BH$_2$), 92, 112
Dimethyl sulfoxide (DMSO), 117

E

Endoplasmic reticulum, polypeptide chains
 ERAD, cytosol substrate recognition and translocation
  defective protein folding, 153–154
  folding-competent and, 154–155
  folding-defective polypeptides, fate, 162–163
  macroautophagy, 161
  mammalian orthologs, Yos9p, 160
  misfolded proteins deviation, De-Mannosylation, 155–157
  nonglycosylated proteins, disposal, 157–158
  polypeptides, terminally misfolded, 152–153
  to retro-translocation site, 158–160
  tuning concepts, 161–162
 in mammalian, protein translocation and maturation
  calnexin and calreticulin, recruitment, 142–143
  calnexin chaperone system, cycling, 143–145
  calnexin chaperone system, members deletion, 146–150
  cytosolic coat protein II (COPII), 145
  ER lumen, 136
  ER lumen, emerging, 138–140
  folding-competent polypeptides, 151–152
  oligosaccharides and nascent chains, 140–142
  other chaperones, deletion, 150–151
  to plasma membrane, 136
  polypeptide translocation, 136–138
Enzyme replacement therapy, phenylketonuria (PKU), 109
ER-Golgi intermediate compartment (ERGIC), 145
ERp57, PDI superfamily member
 analysis
  primary B-cells and fibroblasts lacking, 148–149
  protein maturation, 149
 deletion, 147–148

N-glycosylated polypeptides, oxidative maturation, 147
Escherichia coli, phenylketonuria (PKU), 106

F

Fast multipole method (FMM), 202
Flory's isolated-pair hypothesis, 19
Free-energy forcefields, protein folding
 anfinsen's hypothesis
  prediction and folding, 205
  sampling technique, 206
 biomolecular structure, 203
 biophysical
  electrostatics, corrections, 207, 209
  hydrogen bonding, 207–208
  peptide bond, 206
  solvation, 208–209
 generalized-Born (GB) models, 205
 Poisson–Boltzmann (PB) models, 204
 valence angles and dihedrals, 204

G

Generalized-Born (GB) models, 205
Gene therapy, phenylketonuria (PKU), 109
Glycine–serine (GS) bends, 246
GroEL
 chaperones, 100, 116

H

Heat shock proteins (HSP), 100
α-Helical peptides
 $3_{10}\alpha$ and π helix, 6
 capping box and side-chain interactions, 21
 capping motifs, 4
 design of
  acetylation, amidation and side-chain spacings, 10
  concentration determination, 11–12
  dipole, 8–9
  3_{10}-helices, 13–15
  and helix parameters measurement, 12
  history, 7–8
  host–guest studies and helix lengths, 8
  π helices, 15

solubility, 10–11
templates, 13–14
helix–coil theory, 15
  AGADIR, 23–24
  3_{10}-and α helices, 22–23
  Lifson–Roig model, 16–19
  Lomize–Mosberg model, 24–25
  N1, N2, and N3 preferences, 21–22
  N-and C-caps, 20–21
  single-sequence approximation, 20
  tertiary interactions, 25
  unfolded state and polyproline II
    helix, 19–20
  Zimm–Bragg model, 16
metal binding, 4–6
stability, forces affecting
  capping motifs, 33, 35
  caps, 29–32
  coil transition, enthalpy
    change, 36–37
  covalent side-chain interactions, 33
  helix interior, 25–29
  ionic strength, 35–36
  noncovalent side-chain interactions, 32–35
  phosphoserine, 32
  pK_a values, 37–38
  proteins relevance, 38
structure
  conformation, 3
  definition, 2–3
3_{10}-Helix
  design, 13–15
  and α-helices differentitation, 22–23
  stabilization, 6
π–Helix, 6
  hydrogen bond formation, 23
Helix-coil theory
  AGADIR, 23–24
  3_{10}- and α helices, 22–23
  Lifson–Roig model, 16–19
  Lomize–Mosberg model, 24–25
  N1, N2, and N3 preferences, 21–22
  N- and C-caps, 20–21
  single-sequence approximation, 20
  tertiary interactions, 25
  unfolded state and polyproline II
    helix, 19–20
  Zimm–Bragg model, 16
Hemophilia A, 266
Hsp organizing protein (HOP), 102

human Phenylalanine hydroxylase (hPAH), 93
  active site region, 112
  folding and stability, 97
  denaturation, 98
  irreversible denaturation, 99
  transition temperatures, 98
Human recombinant factor VIII, rational
  formulation development
  CD spectrum for, 267
  characteristics and melting profile, 268
  ELISA studies, 268
  folding-based rational approach, 266
  thermal unfolding, 267
  UV CD spectrum of, 266
human tyrosine hydroxylase (hTH), 106
Huntington's disease, 103
Hydrogen-bond dehydration, 55–56
Hydrophobicity, 57. See also Dehydronic force
Hyperphenylalaninemia, 92

I

Intramolecular hydrogen bonds
  dielectric environment, 64, 76
  polar-group hydration, 54–55
  in soluble proteins, 58–59
  strength and stability of, 64
  water-exposed, 57–58

L

Large neutral amino acids (LNAA), 109
Lennard–Jones models, 199
Lifson–Roig model
  AGADIR
    current version, 24
    side-chain interaction energy,
      determination, 23–24
  coil residues, 16–17
  helix–coil equilibrium, 17
  helix dipole effects, 22
  N1 and C1 residues, 21–22
  partition functions equation, 18–19
  residues weights assignment, 20
Local enhanced sampling (LES), 203
Lomize–Mosberg model, 24–25
Lumry–Eyring model, 259
Lysosomal storage diseases (LSDs), 118

M

Macroautophagy, 161
Misfolding diseases, 89–90, 103
Molecular dynamics (MD)
  applications, 205
  biomolecular simulations, 200
  enhanced sampling
    high-temperature replicas, 203
    REMD scheme, 202
  integration, 201
  long-range interactions, 202
  Newton's equations, 199
  thermostats
    external weak coupling, 202
    Virial theorem, 201

N

Nascent polypeptide-associated complex (NAC), 101–102

O

Oligosaccharyl transferase (OST), 137
  core structure, 142
  oligosaccharide transformation, 141
  subunits of, 140–141

P

Particle-mesh Ewald (PME), 202
Pathogenically wrapped proteins
  aberrant aggregation, 81–83
  amyloidogenic aggregation, 80–82
Peptidylprolyl *cis/trans* isomerases (PPI), 140
Pharmacological chaperones
  specific stabilizers, 118
  thermodynamic stability, 119
Phenylalanine hydroxylase (PAH), cytosolic protein
  deamidation, 104
  ligand-binding studies
    binding sites, 113
    cofactor binding, 111–115
    N-terminal sequence, 113
    substrate binding, 115
  mammalian domain structure, 93
Phenylketonuria (PKU)
  BH4 supplementation, 91
  diet and gene therapy, 109
  FoldX analysis, 106
  misfolding disease
    mutational analysis, 105
    mutation-dependent destabilization, 108
  pharmacological chaperones, 123
  tetrahydrobiopterin-responsive
    BH_4 treatment, 120, 122
    data on, 123
    free-energy differences, 121
  therapeutic approaches, 109
Phosphatidyl serine (PS), 268
Potential energy surface (PES)
Protein disulfide insomerase (PDI), 138–139, 147–149
Proteins, low kinetic stability
  amyloids formation, 102
  correct misfolding, 116–117
  cytosol, protein folding and degradation, 101
  folding and unfolding, 90
    energy landscape, 94
    native and intermediate states, 93
    stabilities, 94
  hPAH, folding and stability, 97–99
  ligand binding, protein stabilization
    chaperoning role, 111
    folding/unfolding equilibrium, 110
    ligand concentration, 110–111
  misfolding, intracellular control
    aggregation prone, 105
    consensus concept, 103
    molecular chaperones, 100
    mutations, 103
    oxidative stress, 104–105
    post-translational modifications and translational misincorporation, 104
    temperature, 105
  native structure
    free-energy representations, 95
    kinetic barrier, 96
    Lumry–Eyring scenario, 97
    monotonic unfolding transition, 95
    relevant and residual stabilities, 96
    thermodynamic stability, 94
  and ubiquitination, 102
Proteins folding, free-energy forcefields
  amino acids, composition

INDEX

auto-induced folding reaction, 182
  genetic code in, 186
  helix and sheet, 188
  L and D chiral forms of, 188
  peptide groups and planes, 189
  polypeptide chain, 188–190
  structures and color coded, 187
DNA-binding zinc finger motif, free-energy Cys$_2$His$_2$, 238
De novo folding, 240
  eukaryotic genomes, 237
  free-energy forcefield PFF02, 242
  free energy *vs.* bRMSD, 239
  helix nucleation, key events, 242
  Zn-binding cysteines, 241
dominant forces
  native state and chemical reactions, 194
  packing of atoms, 195
free-energy forcefields
  Anfinsen's hypothesis, 205–206
  biophysical all-atom, 206–209
hairpins, free-energy
  basin hopping cycle, 227
  C-terminal, 232–234
  disulfide bridges, 224
  HIV-1 V3 loops, 226–229
  HP7, a 12-residue β-hairpin, 229–232
  tryptophan zipper, 225–226
helical, free-energy
  basin hopping simulations, 217
  Cβ–Cβ distance matrix and energy *vs* RMSD plot, 217–218
  E-domain of staphylococcal, 221–224
  engrailed homeodomain, 221
  HIV accessory, 218–221
  immunoglobulin g-binding domains, 221
  potassium channel blocker, 218
  PRO3–THR12 and VAL12–CYS22, 218
  tryptophan cage, 216–218
kinetic folding methods
  forcefields, 203–205
  molecular dynamics, 199–203
  simulations, 205
landscape, schematic representation of, 197
models
  Coarse-grained, 197–198
  Go, 197–198
molecular dynamics (MD) simulations of, 184

molecules production, 181
primary structure, 190
problems
  energy landscape perspective, 196
  free-energy change, 195
quaternary structure, physiological conditions, 193
secondary structure
  elements, 191
  α-helices and β-sheets, 191–192
  structural parameters for, 192
stochastic simulation methods
  basin hopping technique, 212–214
  evolutionary algorithms, 214–216
  improved sampling techniques, 212
  Monte Carlo method, 210–211
structure classifications, 193
tertiary structure, 193
three-stranded sheet, free-energy
  all-atom MD studies, 236
  GSGS peptide, 237
transmembrane family, 183
6-Pyruvoyl-tetrahydropterin synthase (PTPS) gene, 120

Q

Quality control systems (QCS), 90
  eukaryotic cytosol, 101
  molecular chaperones, 100

R

Replica exchange molecular dynamics (REMD), 185

S

Soluble proteins
  cooperativity, 54–55
  folded structure wrapping
    dehydrons, 58
    intermolecular wrapping, 61
    intramolecular hydrogen bonds, 58–59
    structural integrity, 57–58
    wrapping defects, 59–60
  folding cooperativity

Soluble proteins (cont.)
  generating wrapping/folding
    trajectories, 66–71
  solvation nanoscale model, 76–80
  wrapping, 62–66
  wrapping patterns and motifs, 71–76
  folding process
    dehydronic force, 56
    hydrogen-bond dehydration, 55–56
    hydrophobicity, 57
Solvation nanoscale model
  effective permittivity, 79
  Fourier transformation, 78–79
  hydrogen bond stabilization, 80
  pivotal components, 76
  Poisson equation, 76, 78
  solvent-structuring effect, 76–78
Solvent accessible surface area
    (SASA), 208
Superoxide dismutase (SOD), 119–120

T

Tetrahydrobiopterin (BH4), 90, 111–115
Therapeutic proteins
  and biosimilars, development of second
      generation
    development of, 265–266
    folding and delivery, 265
  factors, 255
  folding and stability
    accelarated stability testing, 262
    aggregation mechanism, 258–259
    characteristics of, 258–260
    chemical instability of, 257
    first-order reaction, 259
    folding considerations, product
        development, 260
    formulation development, 261–262
    free energy changes, 258
    freeze-drying, 261–262
    genetic engineering and treatments, 256
    manufacturing, 260
    monitoring, 255
    pharmaceutical development, 261
    and purification, 260–261
    reconstitution medium, 263
    safety and efficacy, 257–258
  molecular properties of, 264
  structure–function relationship, analytical
      methods for, 263–264

Translocating chain-associating membrane
    protein (TRAM), 136–137
Translocon-associated complex (TRAP), 137
Trifluoroethanol (TFE), 37
Tryptophan hydroxylase 1 (TPH1), 111

U

Ubiquitin proteasome pathway (UPP), 90

V

Verlet algorithm, molecular dynamics, 201

W

Wrapping, soluble proteins
  epigenetic polymorphism
    consequences, 84
    prion-like aggregation, 83–84
  folded structure
    dehydrons, 58
    intermolecular wrapping, 61
    intramolecular hydrogen bonds, 58–59
    structural integrity, 57–58
    wrapping defects, 59–60
  folding cooperativity
    generating wrapping/folding
        trajectories, 66–71
    solvation nanoscale model, 76–80
    wrapping, 62–66
    wrapping patterns and motifs, 71–76
  folding process
    dehydronic force, 56
    hydrogen-bond dehydration, 55–56
    hydrophobicity, 57
  pathogenically wrapped proteins
    aberrant aggregation, 81–83
    amyloidogenic aggregation, 80–82

Y

Yos9p, mammalian orthologs, 160

Z

Zimm–Bragg model, 16, 25

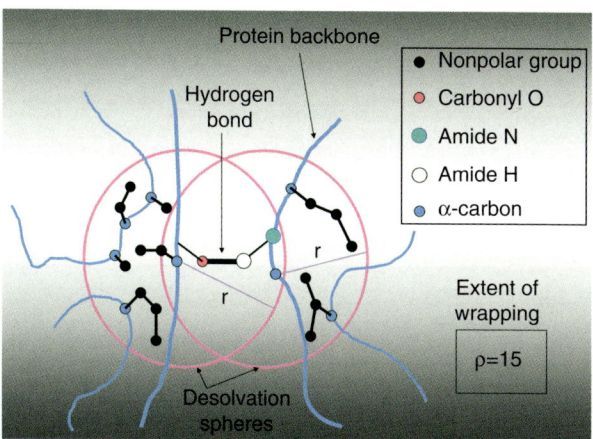

FERNÁNDEZ ET AL., FIG. 1. Wrapping of a backbone hydrogen bond in a soluble protein. Intramolecular hydrogen bonds in soluble proteins prevail only if they are protected from water attack. Thus, their extent of intramolecular wrapping by nonpolar groups (black balls) becomes central to define their stability and strength (12). The extent of hydrogen-bond wrapping indicates the number of nonpolar groups contained within a desolvation domain defined as two intersecting balls of fixed radius centered at the α-carbons of the residues paired by the hydrogen bond. While the wrapping statistics on hydrogen bonds vary with this value, the tails of the distribution remain invariant, thus enabling a unique identification of under-wrapped hydrogen bonds (dehydrons). Dehydrons are packing defects that become stabilized upon removal of surrounding water through protein associations. Thus, dehydrons may be regarded as structural features defining the gene sensitivity to its interactive context and its reliance on binding partnerships to maintain structural integrity.

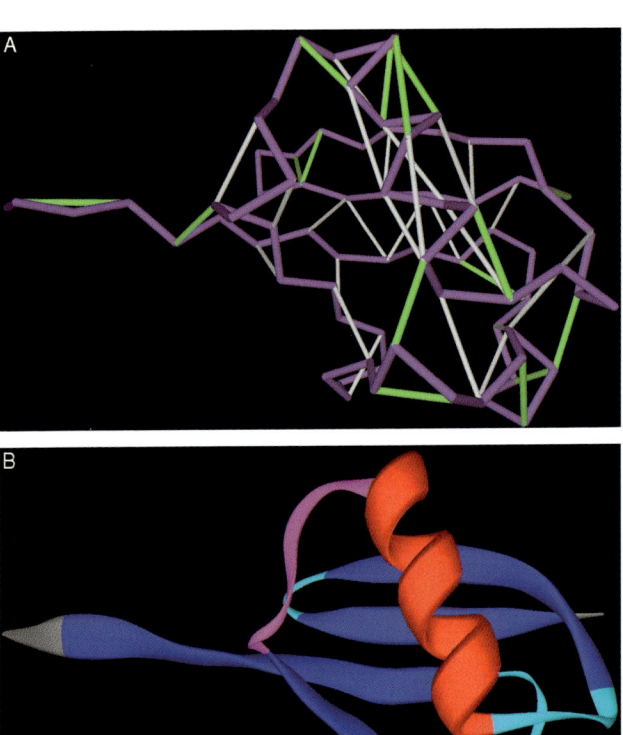

FERNÁNDEZ ET AL., FIG. 2. Illustration of the under-wrapping of protein structure. (A) Dehydron pattern for human ubiquitin (PDB accession code 1UBI). Dehydrons are indicated as green segments joining the α-carbons of the paired units, well-wrapped hydrogen bonds ($\rho > 19$) are shown in light grey, and the protein backbone is conventionally shown as blue virtual bonds joining the α-carbons of consecutive amino acid units. The displayed structure has 64 backbone hydrogen bonds, out of which 16 are dehydrons. Thus, the extent of under-wrapping for this protein is 16/64 = 25%. (B) The ubiquitin structure is also displayed in ribbon representation for easy visualization.

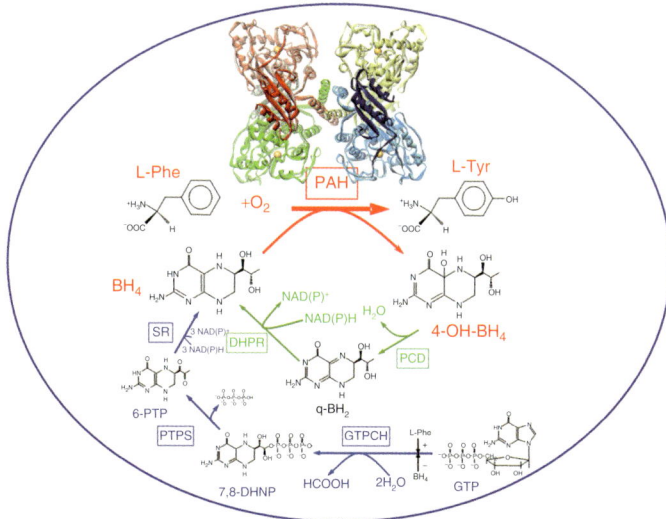

MARTINEZ ET AL., FIG. 1. Hydroxylation of L-Phe by mammalian PAH (red pathway) including the BH_4 regenerating (green) and the *de novo* biosynthetic pathways of BH_4 *in vivo* (blue). The enzymes implicated in these processes are enclosed by frames and the regulatory effects of L-Phe (+) and BH_4 (−) on the rate-limiting step for BH_4 synthesis are also shown. Abbreviations used: BH_4, (6R)-L-*erythro*-5,6,7,8-tetrahydrobiopterin; 4-OH-BH_4, pterin-4a-carbinolamine; q-BH_2, quinoinoid 7,8-dihydrobiopterin; 6-PTP, 6-pyruvoyl-5,6,7,8-tetrahydropterin; 7,8-DHNP, 7,8-dihydroneopterin triphosphate; PCD, pterin 4a-carbinolamine dehydratase; DHPR, dihydropteridine reductase; SR, sepiapterin reductase; PTPS, 6-pyruvoyl-5,6,7,8-tetrahydropterin synthase; GTPCH, GTP cyclohydrolase I. Adapted from *(25)* with permission from SPS Verlagsgesellschaft MBH. The structural model of tetrameric PAH is also shown; composite model created using the structures PDB 2PHM (rat) and PDB 2PAH (human).

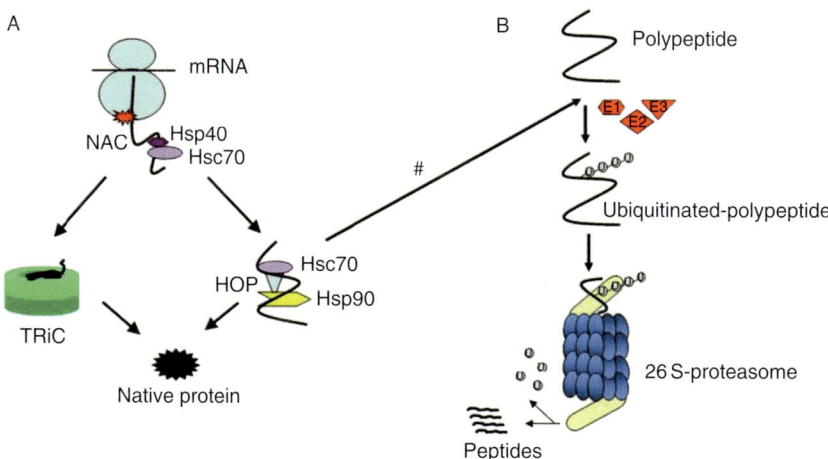

MARTINEZ ET AL., FIG. 4. Quality control system (QCS) in the eukaryotic cytosol. (A) Protein assisted folding for newly synthesized polypeptides, involving the interactions with different molecular chaperones; NAC, nascent polypeptide-associated complex; HOP, Hsp organizing protein; for the names of other components see text (Section IIIA and IIIB). (B) Ubiquitin–proteasome pathway (UPP) for the degradation of defective proteins. U, ubiquitin; E1, ubiquitin activating enzyme; E2, ubiquitin-conjugating enzyme; E3, ubiquitin ligase. The Hsc70–HSP90 complex interacts with the UPP system through cochaperones (#), for example, CHIP *(7, 80)*.

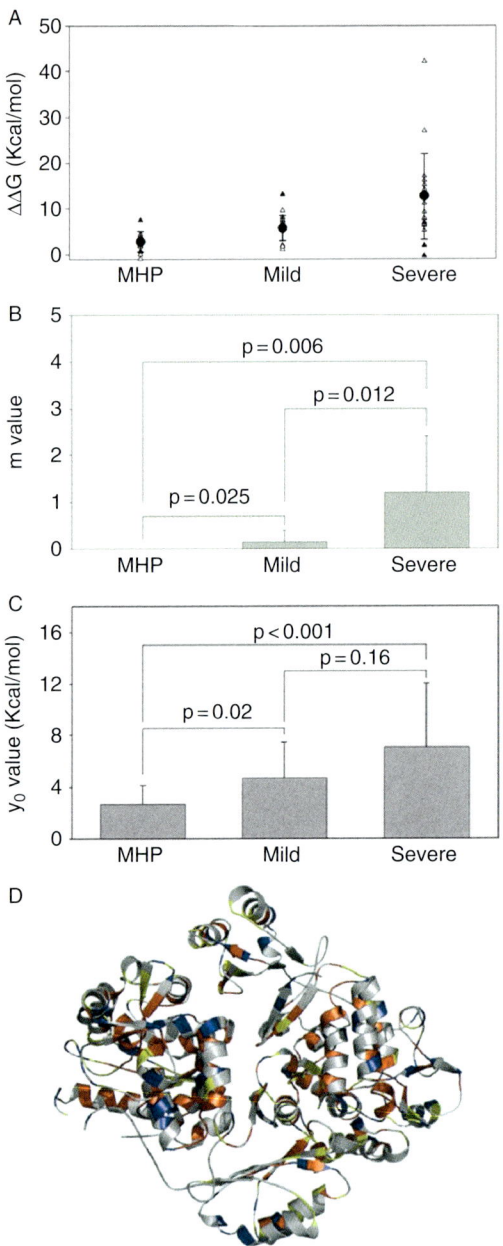

Martinez et al., Fig. 5. Mutation-dependent destabilization and *in vivo* patient phenotype. (A) Calculated effect on $\Delta\Delta G$ (in kcal/mol dimer) for 46 PKU mutants classified by phenotypic groups at a 5-kcal/mol penalty. The mean $\pm$ SD $\Delta\Delta G$ values (*blackened circles*) for the three phenotypic groups, calculated using 41 mutations (*unblackened triangles*), were 2.8 $\pm$ 2.2, 5.7 $\pm$ 2.7, and 13.0 $\pm$ 9.5 kcal/mol for MHP (group 1), mild (group 2), and severe (group 3) phenotypes, respectively. Five outliers (*blackened triangles*) were removed for the calculation of the mean values. (B) and (C) Means $\pm$ SDs of m and y_0 values for the different phenotypic groups, calculated using individual fits for each mutation. *P* values are obtained from one-way ANOVA; $p <$ 0.05 is considered statistically significant. (D) The predicted phenotype represented on the dimeric structure (PDB 2PHM). All misense PKU mutations (http://PAHdb, http://www.pahdb.mcgill.ca/) have been represented according to the predicted phenotype by FoldX analyses: MHP (blue), mild (yellow), and severe (*red*). Reproduced from Ref (37) with permission from Elsevier Limited.

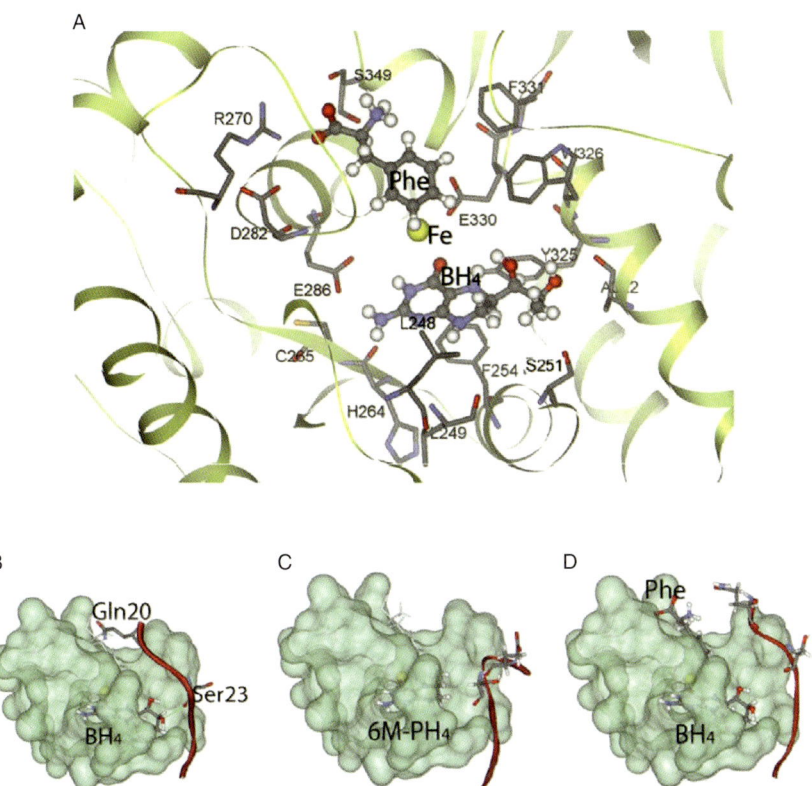

MARTINEZ ET AL., FIG. 6. Structural analyses. (A) Detailed structure of the active site region of hPAH with bound L-Phe and BH_4 based on NMR and docking analyses (149). The iron is shown as a yellow sphere. (B) PAH·BH_4 and (C) PAH·6M-PH_4 complexes obtained by MD simulations (150). The autoregulatory sequence is displayed as a red ribbon. (D) The PAH·L-Phe·BH_4 complex simulated at the same conditions as in (150). In (B) but not in (C) and (D) the carbonyl O in Ser23 interacts with the dihydroxypropyl side chain of BH_4 and Gln20 occupies the L-Phe-binding site.

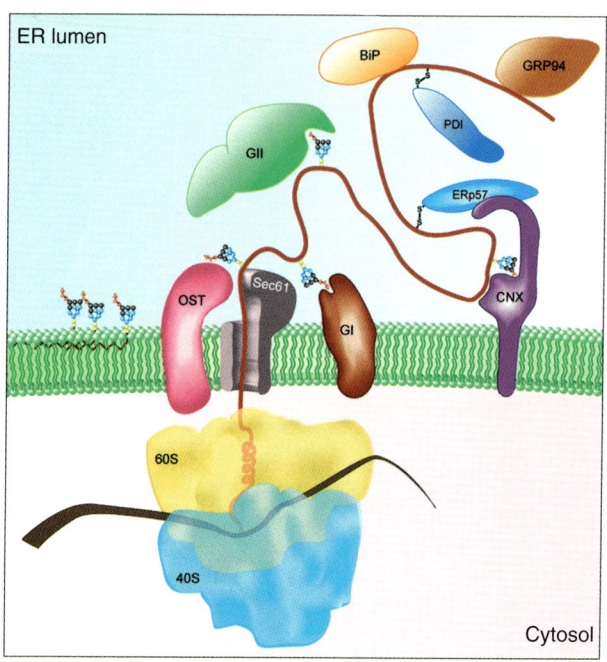

CALÌ ET AL., FIG. 1. Cotranslational protein translocation into the ER lumen. Nascent chains are cotranslationally injected into the ER lumen through the Sec61 translocon. The large and small ribosomal subunits at the cytosolic face of the ER membrane are labeled with 60S and 40S, respectively. Asparagine residues in appropriate sequons are covalently modified with the addition of preassembled oligosaccharides (the 2 N-acetylglucosamine residues are in yellow, the 4 α1,2-bonded mannose residues in gray, and the three terminal glucose residues in red. The shape and color code of the saccharide units in the protein-bound glycan are the same in Figs. 1–5). Nascent chains associate with a variety of ER-resident molecular chaperones and folding enzymes. OST is oligosaccharyl transferase; GI and GII are glucosidase I and II, respectively; Cnx is calnexin, PDI is protein disulfide isomerase.

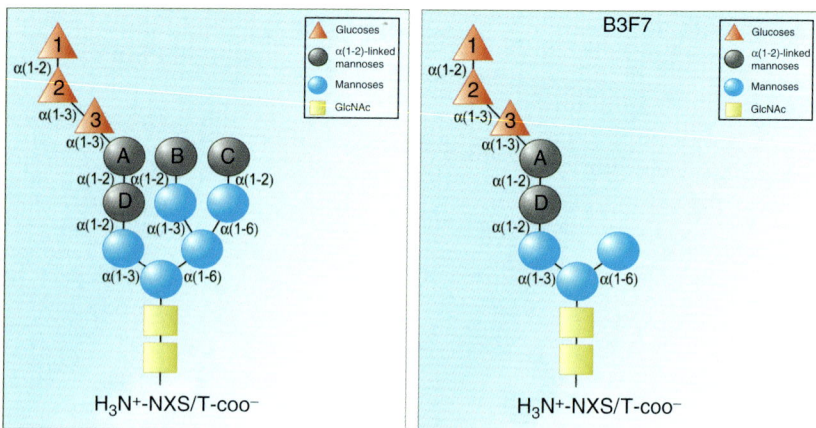

CALÌ ET AL., FIG. 2. Structure of core oligosaccharides. The panel on the left shows the three-antennary oligosaccharide covalently attached to the side chain of an asparagine in the N-X-S/T consensus sequence for N-linked glycosylation. Branches A, B, and C are those that display the terminal mannose residues A, B, and C, respectively. The panel on the right shows the aberrant oligosaccharide used in cell lines with defective synthesis of mannosylphosphoryldolichol (e.g., B3F7).

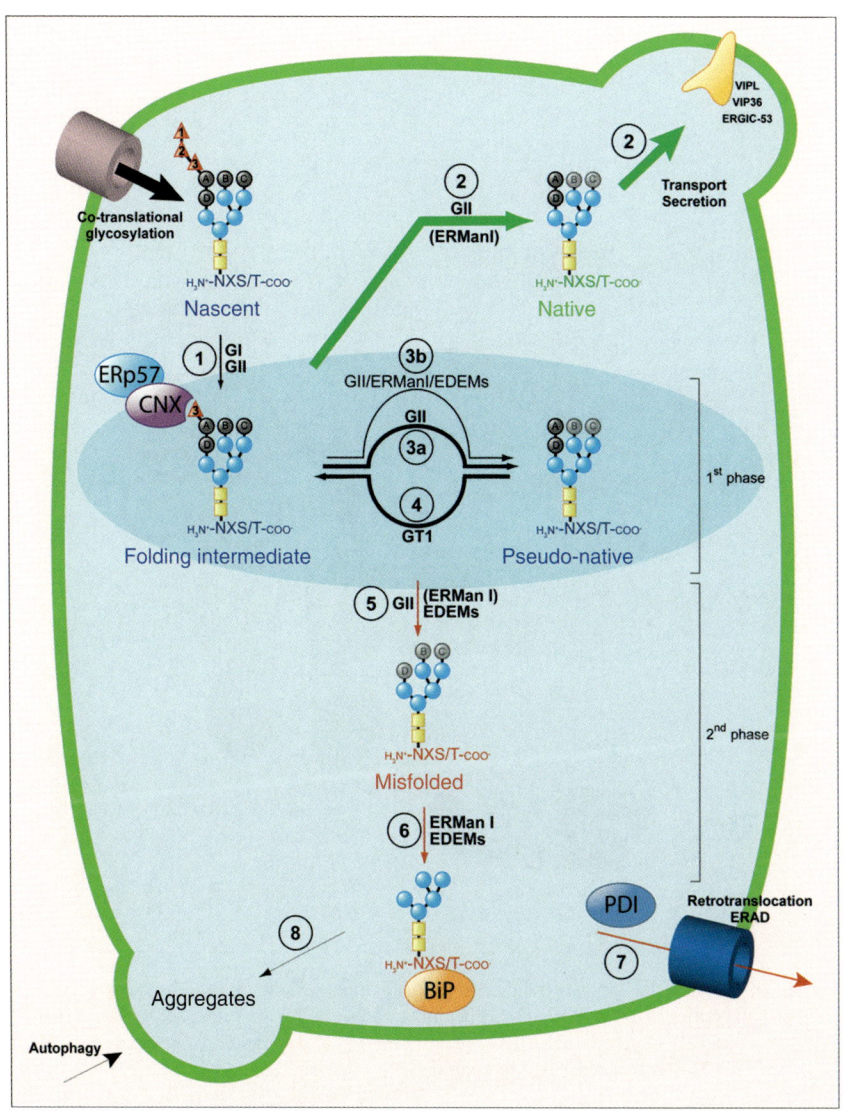

CALÌ ET AL., FIG. 3. The fate of folding-competent and folding-defective glycoproteins. The model shows the fate of newly synthesized, folding competent glycopolypeptides that are eventually transported at their site of activity through the secretory pathway (green arrows). Folding-defective polypeptides are trapped in a first phase of retention-based ER quality control (the calnexin cycle) and, eventually in a second phase of retention-based ER quality control (the BiP/PDI system). Terminally misfolded polypeptides are subjected to extensive de-mannosylation in the mammalian ER. ERManI is ER α1,2-mannosidase I; EDEMs stays for EDEM1, EDEM2, and EDEM3.

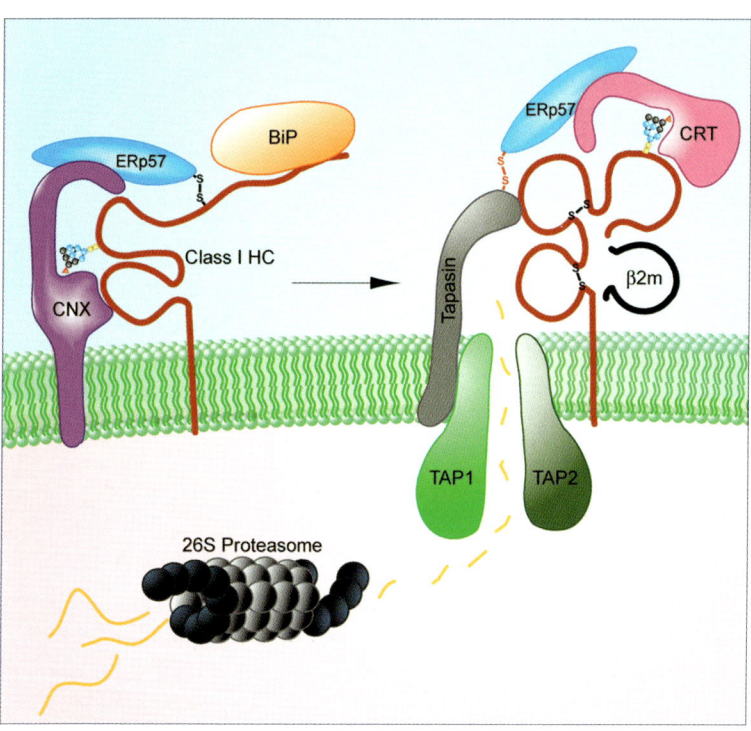

CALÌ ET AL., FIG. 4. The antigen-loading complex. The model shows biogenesis and loading with immunogenic peptides of the MHC class I complex. Antigenic peptides are generated by cytosolic proteasomes and are imported in the ER lumen through the TAP complex. Note that in the loading complex ERp57 forms a stable, covalent complex with tapasin.

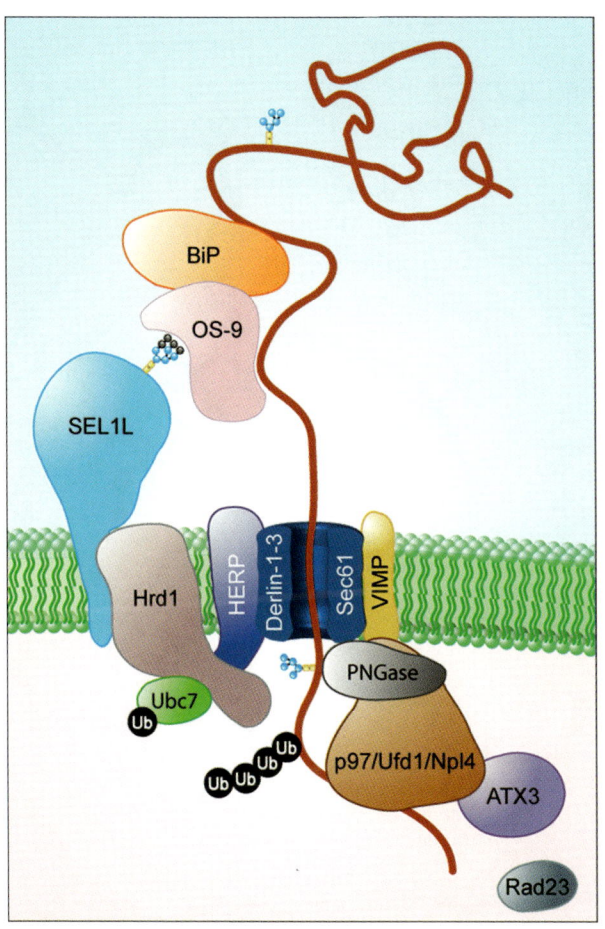

CALÌ ET AL., FIG. 5. A putative model of a complex regulating translocation of terminally misfolded polypeptides from the ER lumen into the cytosol.

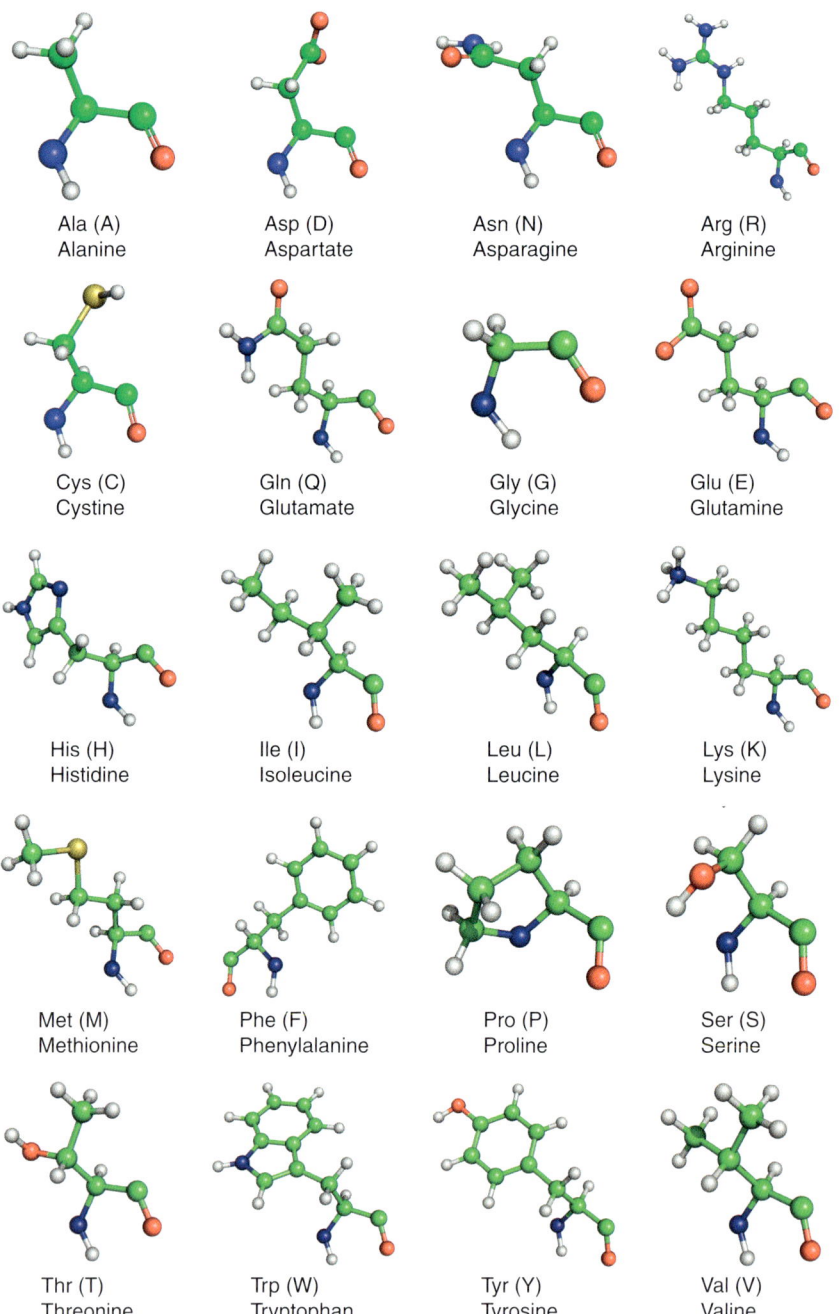

VERMA ET AL., FIG. 1. Twenty naturally occurring amino acids: their structure, name, three-letter code and one-letter code. The structures are color coded with carbon (green), nitrogen (blue), oxygen (red), hydrogen (white), and sulfur (orange).

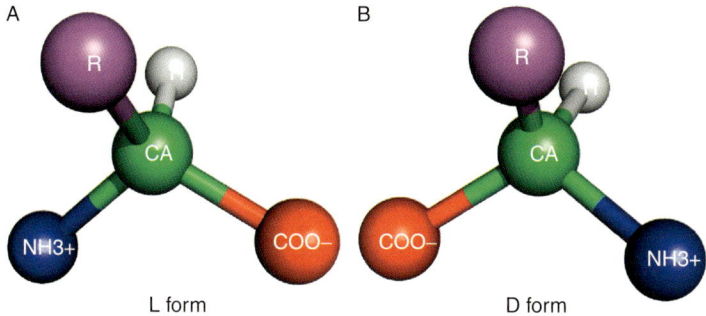

VERMA ET AL., FIG. 3. The L- and D-chiral forms of amino acids. R (magenta) represents the side chain.

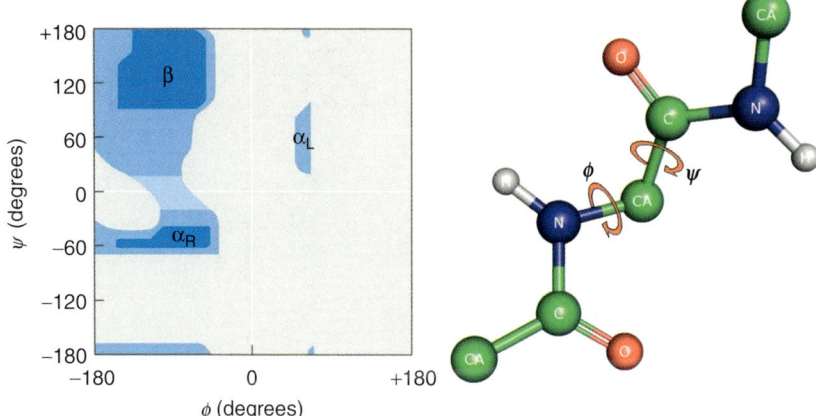

VERMA ET AL., FIG. 5. Sasisekharan–Ramakrishnan–Ramachandran plot (the figure for Ramachandran plot was taken from the internet from http://www.bifi.unizar.es/).

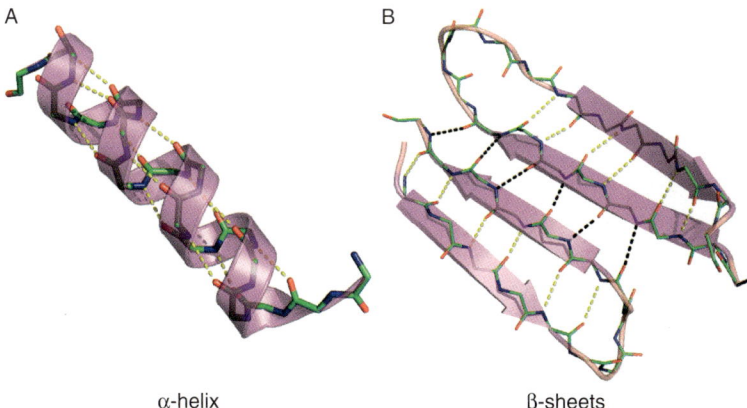

VERMA ET AL., FIG. 6. Secondary structural elements. Dashed lines indicate the presence of hydrogen bonds. In (b), yellow/black bonds are shown for antiparallel/parallel β-sheets, respectively.

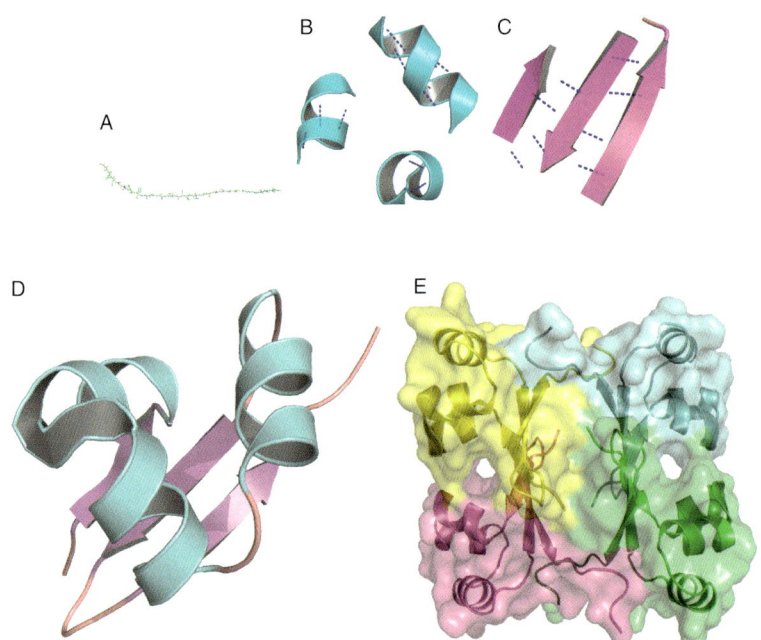

VERMA ET AL., FIG. 7. Protein structure classifications: (A) the primary structure, (B) and (C) the secondary structural elements, helices and sheets, respectively, (D) the tertiary structure, and (E) the quaternary structure of a protein.

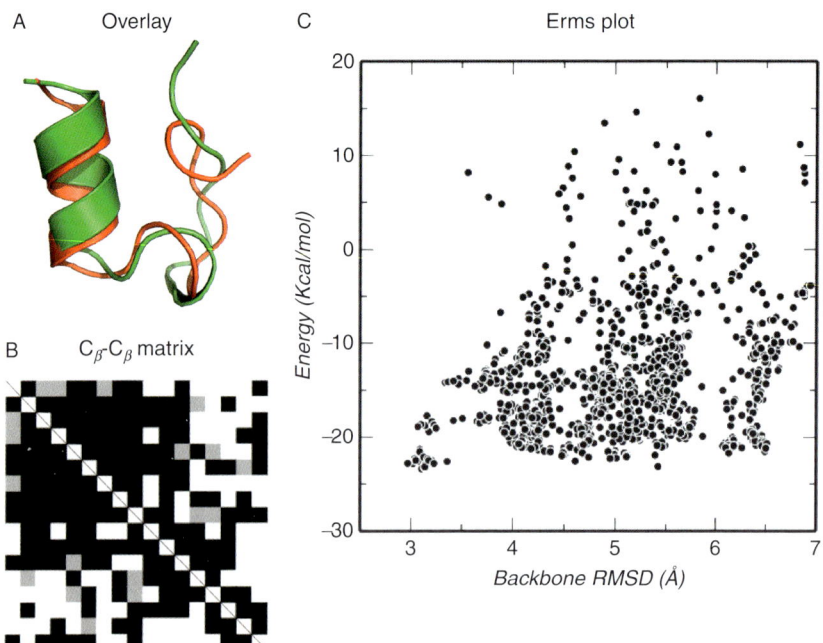

VERMA ET AL., FIG. 12. 1L2Y: overlay of predicted (red) structure to experimental (green) structure. The overlay of the C_β–C_β distance matrix and energy versus RMSD plot.

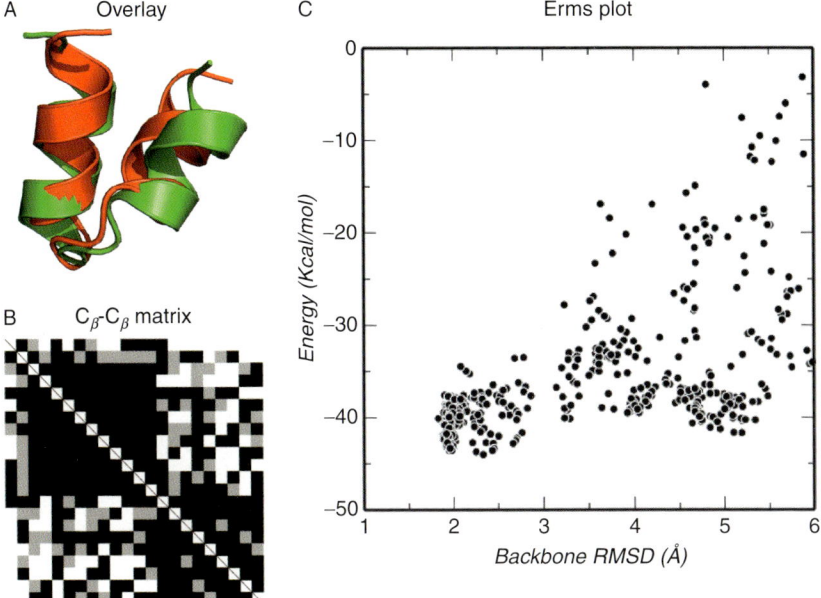

VERMA ET AL., FIG. 13. 1WQE: overlay of predicted (red) structure to experimental (green) structure. The overlay of the C_β–C_β distance matrix and energy versus RMSD plot.

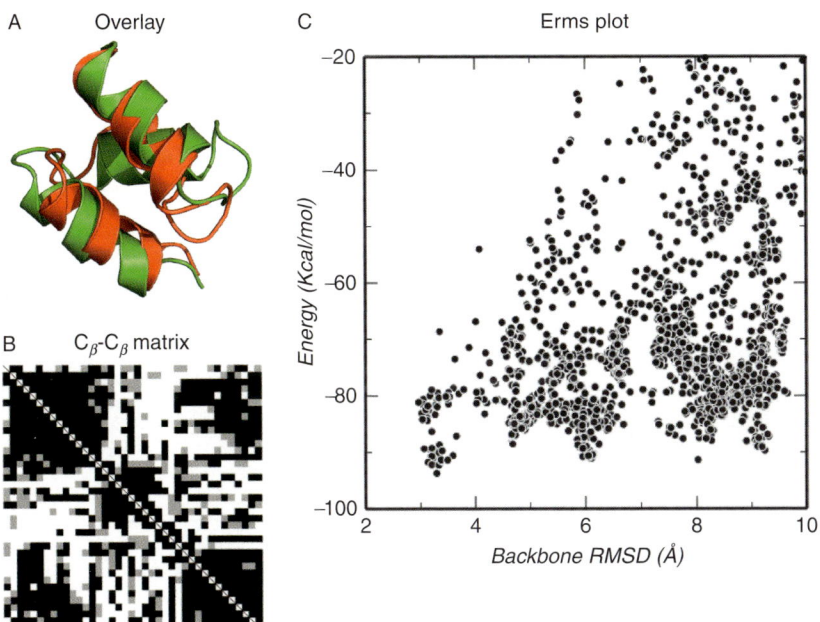

VERMA ET AL., FIG. 14. 1F4I: overlay of predicted (red) structure to experimental (green) structure. The overlay of the C_β–C_β distance matrix and energy versus RMSD plot.

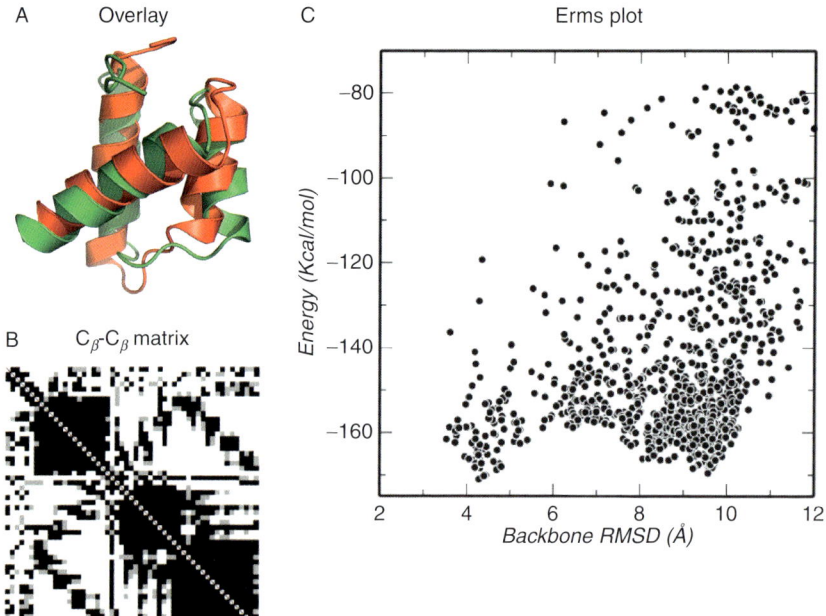

VERMA ET AL., FIG. 16. 1ENH: overlay of predicted (red) structure to experimental (green) structure. The overlay of the C_β–C_β distance matrix and energy versus RMSD plot.

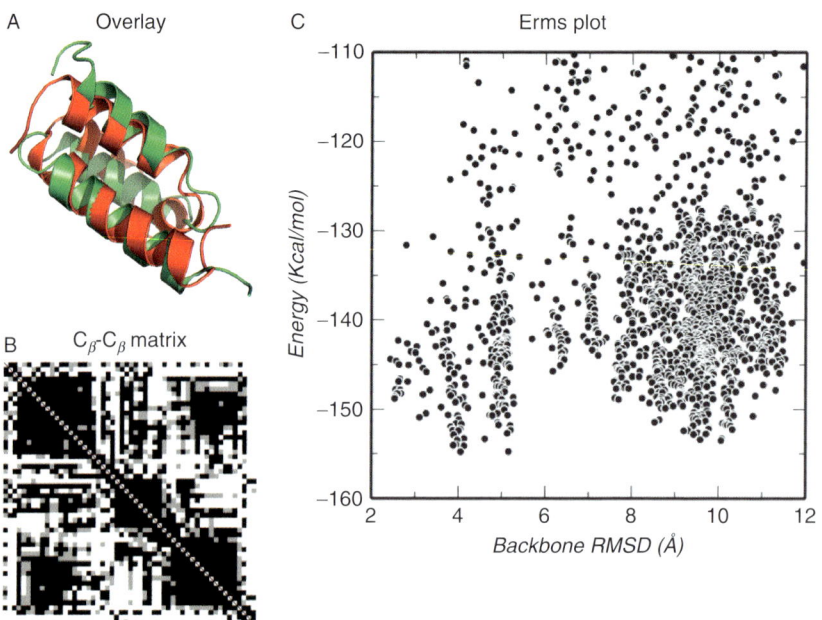

VERMA ET AL., FIG. 18. 1EDK: overlay of predicted (red) structure to experimental (green) structure. The overlay of the C_β–C_β distance matrix and energy versus RMSD plot.

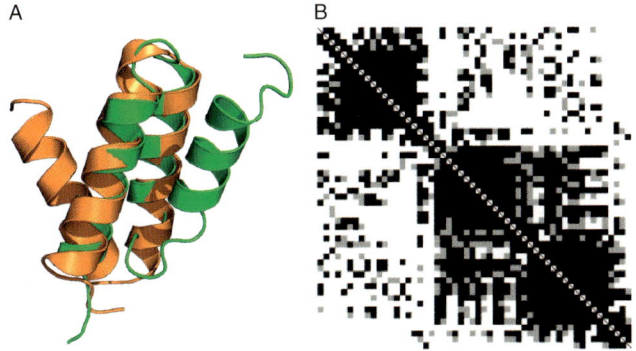

VERMA ET AL., FIG. 19. 1EDK: overlay of misfolded (orange) structure to experimental (green) structure and the overlay of the C_β–C_β distance matrix.

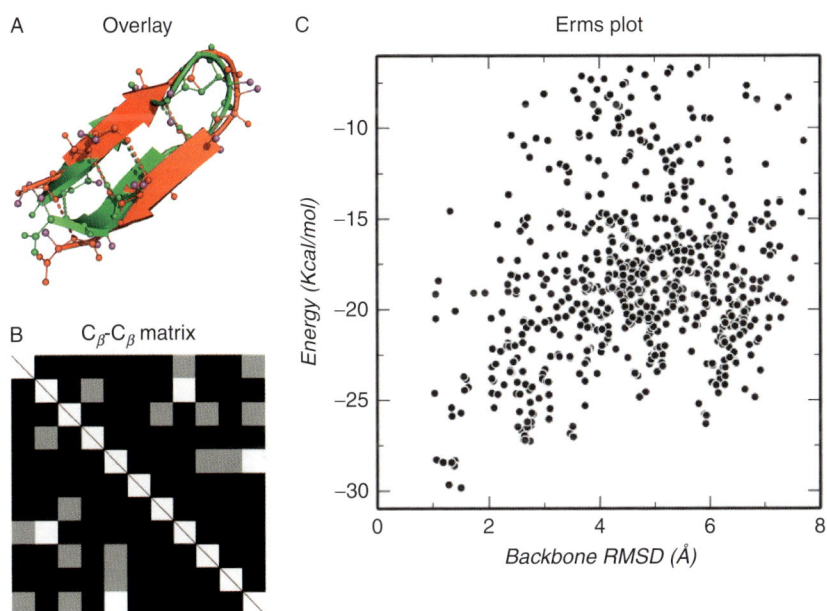

VERMA ET AL., FIG. 20. 1LE0: overlay of predicted (red) structure to experimental (green) structure. The overlay of the C_β–C_β distance matrix and energy versus RMSD plot.

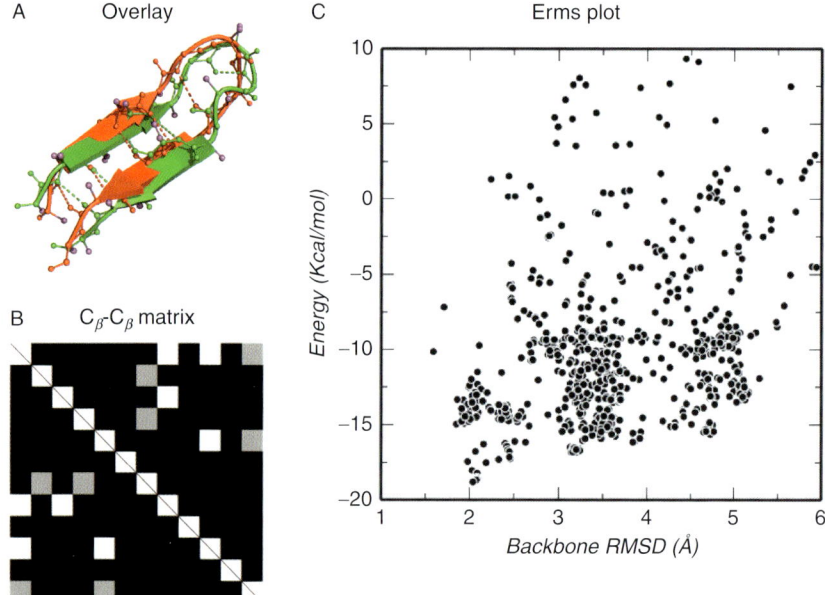

VERMA ET AL., FIG. 21. 1NIZ: overlay of predicted (red) structure to experimental (green) structure. The overlay of the C_β–C_β distance matrix and energy versus RMSD plot.

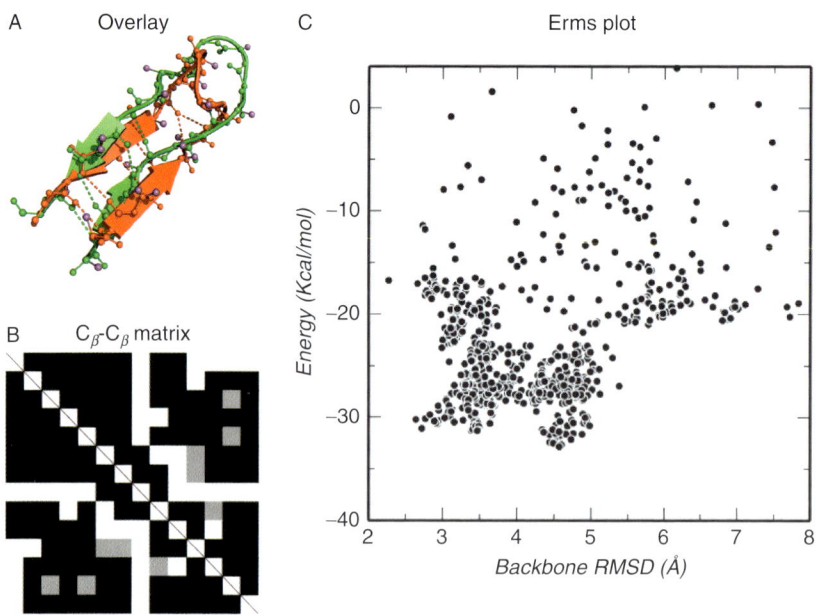

VERMA ET AL., FIG. 22. 1U6U: overlay of predicted (red) structure to experimental (green) structure. The overlay of the C_β–C_β distance matrix and energy versus RMSD plot.

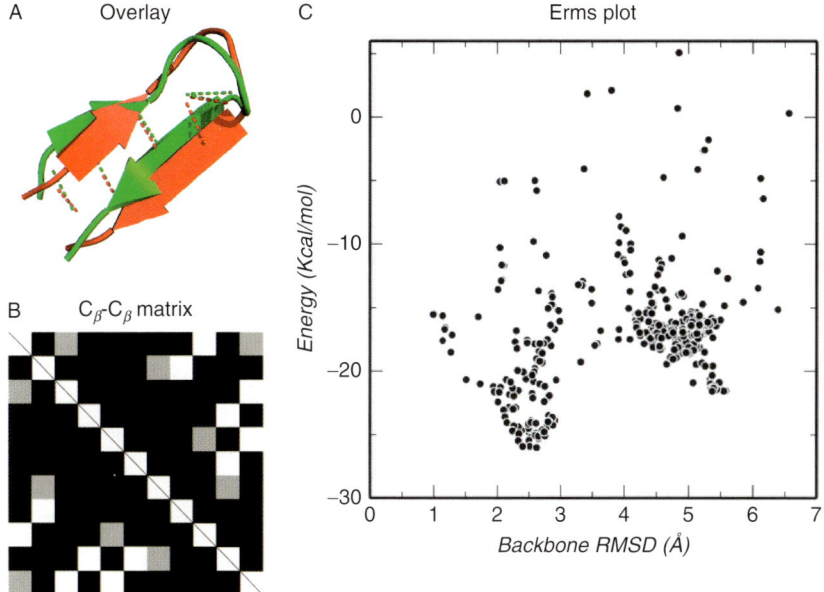

VERMA ET AL., FIG. 23. 2EVQ: overlay of predicted (red) structure to experimental (green) structure. The overlay of the C_β–C_β distance matrix and energy versus RMSD plot.

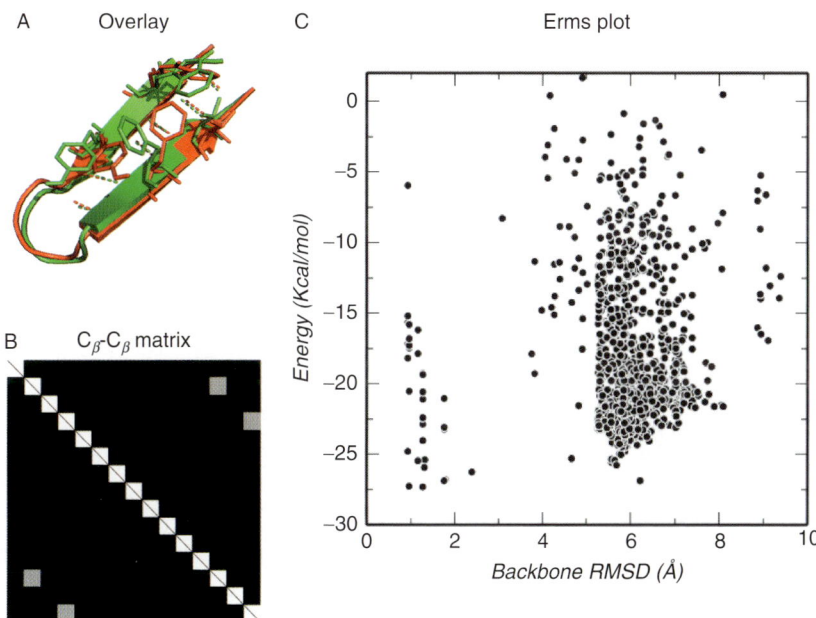

VERMA ET AL., FIG. 24. C-terminal hairpin of protein G: overlay of predicted (red) structure to experimental (green) structure. The overlay of the C_β–C_β distance matrix and energy versus RMSD plot.

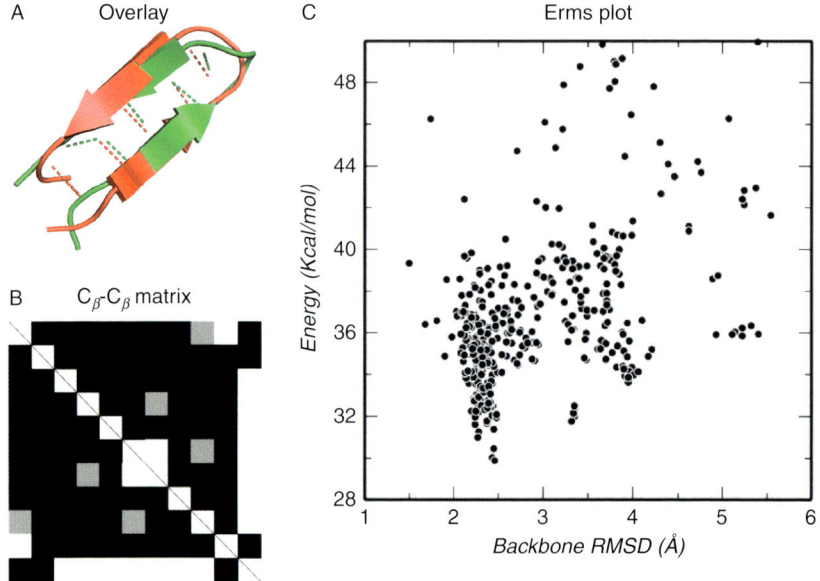

VERMA ET AL., FIG. 26. 1J4M: overlay of predicted (red) structure to experimental (green) structure. The overlay of the C_β–C_β distance matrix and energy versus RMSD plot.

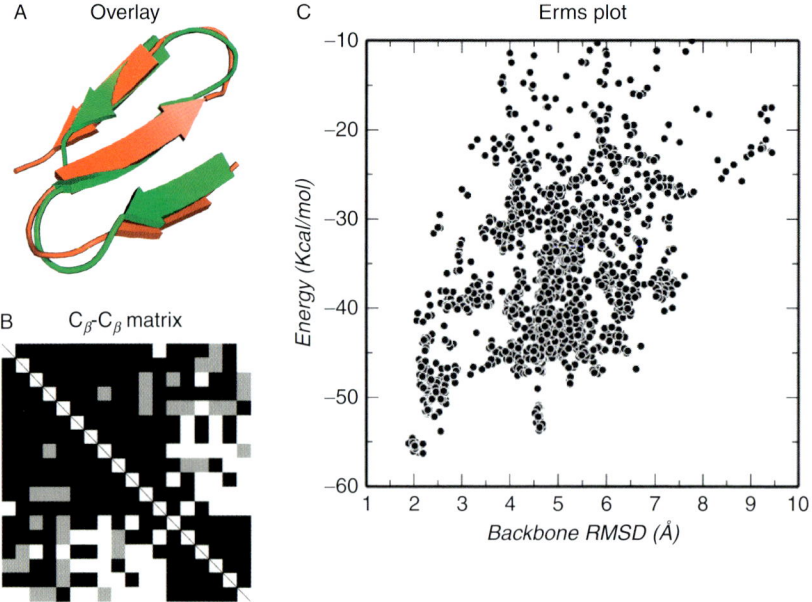

VERMA ET AL., FIG. 27. GSGS peptide: overlay of predicted (red) structure to experimental (green) structure. The overlay of the C_β–C_β distance matrix and energy versus RMSD plot.

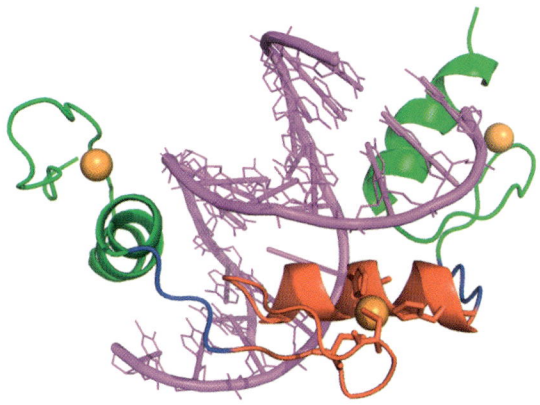

VERMA ET AL., FIG. 28. A classical Cys$_2$His$_2$ zinc finger motif with Zn-ion (orange) and DNA (magenta).

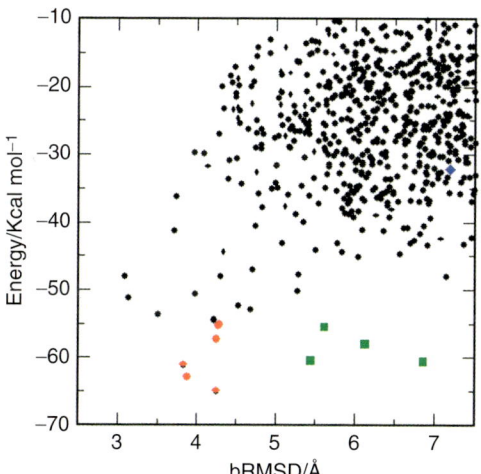

VERMA ET AL., FIG. 29. Free energy versus bRMSD of all accepted conformations in the simulation. The best 10 structures are highlighted as red circles (native-like), green squares (nonnative). The folding intermediate is denoted by blue diamond.

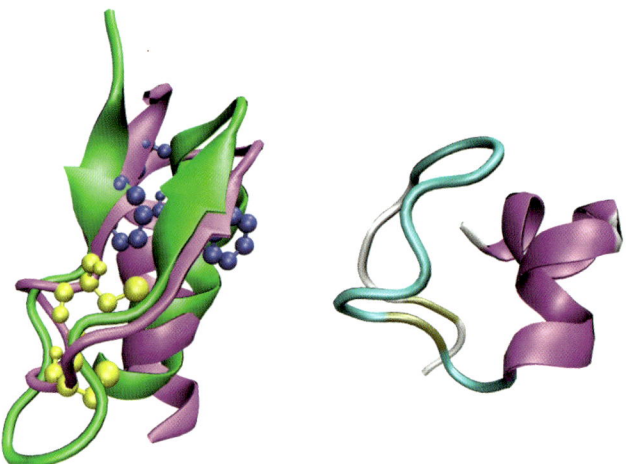

VERMA ET AL., FIG. 30. Left: overlay of the native (green) and folded (magenta) conformations. The conserved hydrophobic residues are shown in blue and Zn-binding cysteines are shown in yellow. Right: the intermediate conformation with partially formed helix and β-sheet.

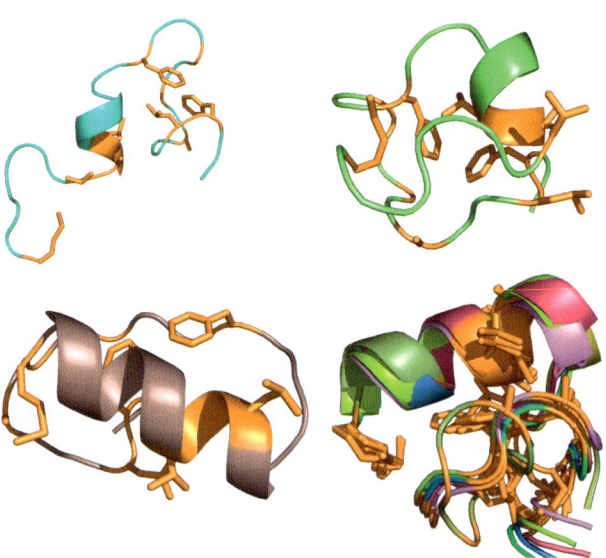

VERMA ET AL., FIG. 32. Key events in the folding: helix nucleation (top left), collapsed globular conformation (top right), fully formed helix (bottom left), partially formed β-sheets using the helix as a template (bottom right).